Sustainable Smart Manufacturing Processes in Industry 4.0

Edited by

Ramesh Kumar
Arbind Prasad
Ashwani Kumar

CRC Press
Taylor & Francis Group
Boca Raton London New York

CRC Press is an imprint of the
Taylor & Francis Group, an **informa** business

First edition published 2023
by CRC Press
2385 NW Executive Center Drive, Suite 320, Boca Raton, FL 33431

and by CRC Press
4 Park Square, Milton Park, Abingdon, Oxon, OX14 4RN

CRC Press is an imprint of Taylor & Francis Group, LLC

Library of Congress Cataloging-in-Publication Data
Names: Nayak, Ramesh Kumar, editor. | Prasad, Arbind, 1956- editor. | Kumar, Ashwani, 1989-editor.
Title: Sustainable smart manufacturing processes in industry 4.0 /
edited by Ramesh Kumar, Arbind Prasad and Ashwani Kumar.
Description: First edition. | Boca Raton : CRC Press, 2024. |
Series: Advances in manufacturing, design and computational intelligence techniques |
Includes bibliographical references and index.
Identifiers: LCCN 2023015717 (print) |
LCCN 2023015718 (ebook) | ISBN 9781032392790 (hbk) |
ISBN 9781032565576 (pbk) | ISBN 9781003436072 (ebk)
Subjects: LCSH: Industry 4.0. | Sustainable engineering. | Manufacturing processes.
Classification: LCC T59.6 .S869 2024 (print) |
LCC T59.6 (ebook) | DDC 670.286--dc23/eng/20230522
LC record available at https://lccn.loc.gov/2023015717
LC ebook record available at https://lccn.loc.gov/202301571

ISBN: 978-1-032-39279-0 (hbk)
ISBN: 978-1-032-56557-6 (pbk)
ISBN: 978-1-003-43607-2 (ebk)

DOI: 10.1201/9781003436072

Typeset in Sabon
by MPS Limited, Dehradun

Contents

Sustainable Smart Manufacturing Processes in Industry 4.0

The text discusses both theoretical and technological aspects of the Industry 4.0–based manufacturing processes. It covers important topics such as additive manufacturing, laser-based manufacturing processes, electromagnetic welding and joining processes, green manufacturing processes, and friction welding processes.

- Illustrates sustainable manufacturing aspects in robotics and aerospace industries.
- Showcases additive manufacturing processes with a focus on innovation and automation.
- Covers environment-friendly manufacturing processes resulting in zero waste and conserves natural resources.
- Synergizes exploration related to the various properties and functionalities through extensive theoretical and experimental modeling.
- Discusses impact welding for joining of dissimilar materials.

The text discusses the recent manufacturing techniques and methodologies such as impact welding for joining of dissimilar materials. It further covers techniques such as additive manufacturing and electromagnetic manufacturing, resulting in minimum or negligible waste. The text elaborates important topics such as friction stir welding energy consumption analysis, and industry waste recycling for sustainable development. It will serve as an ideal reference text for senior undergraduate, graduate students, and researchers in the fields including mechanical engineering, aerospace engineering, manufacturing engineering, and production engineering.

Advances in Manufacturing, Design and Computational Intelligence Techniques

Series Editor: Ashwani Kumar

Senior Lecturer, Mechanical Engineering, at Technical Education Department, Uttar Pradesh, Kanpur, India

The book series editor is inviting edited, reference, and textbook proposal submissions in the book series. The main objective of this book series is to provide researchers a platform to present state-of-the-art innovations, research related to advanced materials applications, cutting-edge manufacturing techniques, innovative design, and computational intelligence methods used for solving nonlinear problems of engineering. The series includes a comprehensive range of topics and its application in engineering areas such as additive manufacturing, nanomanufacturing, micromachining, biodegradable composites, material synthesis and processing, energy materials, polymers and soft matter, nonlinear dynamics, dynamics of complex systems, MEMS, green and sustainable technologies, vibration control, AI in power station, analog-digital hybrid modulation, advancement in inverter technology, adaptive piezoelectric energy harvesting circuit, contactless energy transfer system, energy-efficient motors, bioinformatics, computer-aided inspection planning, hybrid electrical vehicle, autonomous vehicle, object identification, machine intelligence, deep learning, control-robotics-automation, knowledge-based simulation, biomedical imaging, image processing, and visualization. This book series compiled all aspects of manufacturing, design, and computational intelligence techniques from fundamental principles to current advanced concepts.

Additive Manufacturing in Industry 4.0: Methods,Techniques, Modeling, and Nano Aspects
edited by Vipin Kumar Sharma, Ashwani Kumar, Manoj Gupta, Vinod Kumar, Dinesh Kumar Sharma, Subodh Kumar Sharma

Advanced Materials for Biomedical Applications
edited by Ashwani Kumar, Yatika Gori, Avinash Kumar, Chandan Swaroop Meena, Nitesh Dutt

Sustainable Smart Manufacturing Processes in Industry 4.0
edited by Ramesh Kumar, Arbind Prasad and Ashwani Kumar

For more information about this series, please visit: https://www.routledge.com/Advances-in-Manufacturing-Design-and-Computational-Intelligence-Techniques/book-series/CRCAIMDCIT

Aim and Scope

The sustainable development in manufacturing processes and related modeling processes upgraded the researchers to provide the industrial solution in a very lucid way in terms of cost, energy saving, and enhanced productivity without compromising with the quality. The cost-effective digital manufacturing solutions are needed to keep factories and supply chains running smoothly while producing high-quality products. Industry 4.0 has been defined as "a name for the current trends of automation and data exchange in manufacturing technologies, including cyber physical systems, the internet of things, cloud computing and cognitive computing and creating the smart factory". Many companies around the globe are looking for the solution to digitize and automate their factories and achieve zero defective parts per million. Our changing world calls for the industries that can quickly and fluidly adapt, and that means efficient, resilient manufacturing that could modernize operations and environmental impacts. Industry 4.0 provides the means for significant improvement in all aspects of manufacturing processes from material acquisition, usage and supply chain management to product design, engineering and delivery, predictive maintenance of equipment and other assets, and real-time monitoring of all systems.

"**Smart Manufacturing Processes for Sustainable Development in Industry 4.0**" provides in-depth knowledge to readers for the latest manufacturing technologies, enabling them with technologies such as electromagnetic welding, friction stir welding, gas tungsten arc welding, and joining technologies for dissimilar metal welds. The book consists of 12 chapters dedicated to manufacturing processes and modelling for fulfillment in achieving the concepts of Industry 4.0. The book recommends the advantages of technological intervention in engineering, machining, manufacturing, and pro duction technologies. The aim of the book is to inventory the latest achievements in the development and production of modern modelling and manufacturing processes that are used in industries and particularly fulfilling the relation with Industry 4.0. The authors covered all the aspects starting from introduction, technologies, processes, modelling approaches, and sustainability manufacturing for the prospective application and future scope and challenges in manufacturing process in industries. The titles covered in the book have the potential for further research.

Editors
Ramesh Kumar
Arbind Prasad
Ashwani Kumar

Preface

A paradigm shift can be achieved from mass production to customized production by incorporating the Industry 4.0 manufacturing processes. The modern advanced sustainable and green manufacturing processes enable the production of products on the concept of just-in-time. Sustainable financial and environmental benefits can be achieved by incorporating the industry 4.0 manufacturing processes. "**Smart Manufacturing Processes for Sustainable Development in Industry 4.0**"contains 12 chapters that focus on the present status and research prospects related to the manufacturing processes of the fourth industrial revolution. These manufacturing processes are the real foundation of Industry 4.0 in terms of the intelligentization and digitalization of the manufacturing processes. In this book, both theoretical and technological aspects of the Industry 4.0–based manufacturing processes have been discussed. The main topics discussed in the proposed book are additive manufacturing, laser-based manufacturing processes, electromagnetic welding and joining processes, green manufacturing processes, and friction welding processes. The modeling aspects of the advanced sustainable and green manufacturing processes were also discussed in this book. Sustainable manufacturing processes are economically sound processes that minimize negative environmental impacts while producing the objects. The main aim of this proposed book is to provide a communication channel to disseminate the knowledge of recent advancements in the advanced sustainable and green manufacturing processes which have great potential to be included in Industry 4.0.

Chapter 1 deals with the introduction to the sustainable manufacturing for Industry 4.0. In this regard, exciting and vibrant manufacturing ideas and innovations are getting attention as a result of this more academia-industry joint research and collaboration projects are coming out. Industry 4.0 is leading the revolution in manufacturing digitalization and exponential growth. In continuation, **Chapter 2** focuses on various welding techniques and their effect on microstructures of the dissimilar metal welds. Other joining techniques such as mechanical joints, adhesive joints, etc. are not discussed in this chapter. **Chapter 3** deals with the latest laser-based manufacturing technologies. The chapter discusses the various application of laser in machining operation. The ongoing evolution of laser technology and the development of other sophisticated assisting tools established laser material processing technology as one of the trusted allied manufacturing operations.

In **Chapter 4**, electromagnetic joining process is highlighted. In this chapter, the aluminium tube was crimped over the aluminium, copper, and brass core by using the pulse discharge energy of the capacitor bank. A unique type of double solenoidal coil is used to carry out the research work on electromagnetic crimping. The compression-shear test was used to test the strength of the joints. The increase in the discharge energy also increases the joining strength. **Chapter 5** highlights the advance composite materials and their manufacturing techniques suitable for biomedical applications, which is an emerging area of precise manufacturing. **Chapter 6** deals with the manufacturing processes for motor cores in electric vehicles.

Chapter 7 deals with an upcoming joining process in which the heat energy is produced with the help of microwave hybrid heating. The benefit of using microwave hybrid heating as a joining process is the ability to concentrate the heat energy such that the joint can be achieved by utilizing comparative lower power. The mechanism of heating of the metals powder has been described in this chapter, followed by various research on the joining of similar and dissimilar metals and alloys by microwave heating. **Chapter 8** highlights the use of powder metallurgy (PM) in manufacturing of different industrial components. **Chapter 9** simulates electromagnetic impact welding on Ls-Dyna using an EM module. The field shaper and flyer tube material is copper, whereas tube and end-plug material are steel. This study performed many simulations with different joining parameters.

Chapter 10 briefly discussed the effect of varying different process parameters such as tool traverse/welding speed (TS/WS), tool rotational speed (TRS), plunge depth, and tool tilt angle on the output responses such as temperature distribution, heat generation, and axial force variation, etc. These variations in the input responses ultimately change the welded joint's microstructure, hardness, and weld strength, applications in several industrial sectors, and the ability to weld various similar and dissimilar materials. A rigorous literature survey on tool design, heat transfer, material flow behavior, thermo-mechanical effect due to the rotational heat generation effect of pin and shoulder, understanding defect formation due to varying process parameters, etc., was done. **Chapter 11** highlights a cobalt-based superalloy called Stellite 6 for use in a variety of industries, particularly for hardfacing or weld overlay, principally for wear applications involving unlubricated systems or at high temperatures. In this chapter, Stellite 6 alloy covers the main features of the Stellite alloy family will be overlaid on commonly used mild steel substrate AISI 1020 using GTA cladding method and is studied for microstructure, hardness, and wear resistance at the room as well as high temperature.

The variations of these performances with temperature would be the emphasis of the research.

Chapter 12 concentrates on an analytical study-cum-evaluation of machinability behaviour of gravity die-casted aluminium matrix composites (GDC-AMCs) during the milling process in which chill cast aluminium alloy LM25 is reinforced with boron carbide ceramic particles. Material removal rate (MRR) and resultant cutting forces are calculated during the process. Finally, the best specimen with optimal material properties is selected from a series of composite combinations using a specialized subsect in response surface methodology (RSM)–based optimization called historical data design (HDD).

The book will be helpful for undergraduate and postgraduate students of mechanical engineering and production engineering. Different dissimilar joining techniques in this book can be an excellent case study for learning the green and advanced sustainable welding or joining processes. We hope that this book can be of interest to the academicians or industrialists with an interest in the advanced joining and welding techniques. At last, due to its focused treatment and concise size, it will be very useful for practicing engineers as well.

Editors
Dr. Ramesh Kumar
Dr. Arbind Prasad
Dr. Ashwani Kumar

About the Editors

Dr. Ramesh Kumar is an assistant professor in the Department of Mechanical Engineering at Saharsa College of Engineering, Saharsa, Bihar (under Department of Science and Technology, Government of Bihar), India. He received his PhD degree from the Indian Institute of Technology Guwahati, India. He received his MTech degree from the Indian Institute of Technology (Indian School of Mines) Dhanbad, Jharkhand, India. He has published several papers in peer-reviewed journals and international conferences in electromagnetics manufacturing. His main research interests include electromagnetics manufacturing, advanced welding technology, and bio-mechanics.

Dr. Arbind Prasad has received his PhD (mechanical engineering) from the Indian Institute of Technology Guwahati, Assam, India. He has earned a gold medal in his M Tech. He is currently working as an assistant professor and head (mechanical engineering) in the Department of Science and Technology, Government of Bihar, posted at Katihar Engineering College, Katihar, Bihar, India. He has one granted patent and filled three patents out of his research work. He has 15 international Journal papers, edited six books (Wiley Scrivener Publishing and CRC Press (Taylor & Francis) USA), 17 book chapters, and 15 reputed international conference papers to his credit. Dr. Arbind Prasad has obtained various prestigious awards, such as the Sponsored Research Industrial Consultancy (SRIC) award from IIT Kanpur, best oral presentation from American Chemical Society, and best paper award from IIT Guwahati during Research Conclave. He has completed numerous projects funded by the state government. He is an associate editor for *International Journal of Materials, Manufacturing and Sustainable Technologies (IJMMST)* and an early career editor for I*nternational Journal of Mathematical, Engineering and Management Sciences (IJMEMS) and i*ndexed in ESCI/Scopus and DOAJ. He is an editorial board member of various international journals and acts as an active review board member of 10 prestigious (Indexed in SCI/SCIE/Scopus) international journals with a high-impact factor i.e., *Bioengineering, Journal of design, Materials, Journal* *of functional materials, Polymer, Materials, Molecules,* etc. He has been invited to deliver talks at various organizations of repute. He has coordinated various faculty development programmes, short-term courses, symposium, national seminars, and completed research projects sponsored under various government schemes in India. His main areas of research include resorbable polymer, processing and characterizations, scaffolds, hydrogels, manufacturing, composites, biomaterials, nano materials processing, internal fixation devices, bioimplants, etc. He is also a lifetime member of the Society of Polymer Science India, Material Research Society of India, Society of Biomaterials and Artificial Organs of India, Asian Polymer Association, and Indian Society of Technical Education.

Dr. Ashwani Kumar received a PhD (mechanical engineering) in the area of mechanical vibration and design. He is currently working as senior lecturer, mechanical engineering (Gazetted Officer Group B) at the Technical Education Department, Uttar Pradesh (under Government of Uttar Pradesh) Kanpur, India since December 2013. He has worked as an assistant professor in the Department of Mechanical Engineering, Graphic Era University Dehradun India from July 2010 to November 2013. He has more than 13 years of research and academic experience. He is a series editor of four book series: *Advances in Manufacturing, Design and Computational Intelligence Techniques, Renewable and Sustainable Energy Developments, Smart Innovations and Technological Advancements in Mechanical and Materials Engineering, Computational Intelligence and Biomedical Engineering* published by CRC Press (Taylor & Francis USA) and Wiley Scrivener Publishing USA. He is editor-in-chief for *International Journal of Materials, Manufacturing and Sustainable Technologies (IJMMST)*. He is guest editor and editorial board member of eight international journals and acts as review board member of 20 prestigious (Indexed in SCI/SCIE/Scopus) international journals with a high-impact factor i.e. Applied Acoustics, Measurement, JESTEC, AJSE, SV-JME, and LAJSS. In addition, he has published 100+ research articles in journals, book chapters, and conferences. He has authored/co-authored cum edited 20+ books of mechanical, materials, and energy engineering. He has published two patents. He is associated with international conferences as an invited speaker/advisory board/review board member/program committee member. He has delivered many invited talks in webinar, FDP, and workshops. He has been awarded as a best teacher for excellence in academic and research. He has successfully guided 12 B.Tech., M.Tech., and PhD theses. In administration, he is working as a coordinator for AICTE, E.O.A., Nodal officer for PMKVY-TI Scheme (Government of India) and internal coordinator for CDTP scheme (Government of Uttar Pradesh). He is currently involved in the research area of AI and ML in mechanical engineering, smart materials and manufacturing techniques, thermal energy storage, building efficiency, renewable energy harvesting, sustainable transportation, and heavy vehicle dynamics.

Contributors

T. Aizawa, Surface Engineering Design Laboratory, Shibaura Institute of Technology, Japan

S. Aravindan, Indian Institute of Technology Delhi, Hauz Khas, New Delhi, India

Archana, Jindal Global Business School Sonipat, O. P. Jindal Global University Sonipat, Haryana, India

N. Baskar, Saranathan College of Engineering, Tiruchirapalli, India

Rituraj Bhattacharjee, Department of Mechanical Engineering, Indian Institute of Technology Guwahati, Assam, India

Pankaj Biswas, Indian Institute of Technology Guwahati, Assam, India

Gourhari Chakraborty, National Institute of Technology, Andhra Pradesh, India

Susmita Datta, Indian Institute of Technology Guwahati, Assam, India

Sriparna De, Brainware University, Kolkata, India

S. Janakiraman, Department of Mechanical Engineering, Indian Institute of Technology Indore, Madhya Pradesh, India

Kamal Kishore, National Institute of Technology Hamirpur, H.P., India

Sachin D. Kore, Indian Institute of Technology Goa, Goa, India

Ashwani Kumar, Technical Education Department, Kanpur, Uttar Pradesh, India

Ramesh Kumar, Department of Mechanical Engineering, Saharsa College of Engineering (under Department of Science & Technology, Govt. of Bihar) Saharsa, Bihar, India

Tanmoy Medhi, Indian Institute of Technology Guwahati, Assam, India

Vivek Pandey, School of Engineering & Technology, Mody University, Lakshmangarh, Sikar Rajasthan, India

P. Jothi Palavesam, Saranathan College of Engineering, Tiruchirapalli, India

Arbind Prasad, Katihar Engineering College (under Department of Science & Technology, Govt. of Bihar), Katihar, Bihar, India

Adarsh Prakash, School of Mechanical Engineering, Indian Institute of Technology GOA, India

Ashish Kumar Rajak, Department of Mechanical Engineering, Indian Institute of Technology Indore, Madhya Pradesh, India

P. V. Rajesh, Saranathan College of Engineering, Tiruchirapalli, India

R. Rekha, Saranathan College of Engineering, Tiruchirapalli, India

T. Shiratori, Faculty of Engineering, University of Toyama, Japan

Amir Shaikh, Graphic Era Deemed to be University, Dehradun, Uttarakhand, India

Pankaj Sharma, National Institute of Technology, Hamirpur, H.P., India

Amandeep Singh Shahi, Department of Mechanical Engineering, Sant Longowal Institute of Engineering and Technology Punjab, India

Alok Singh, Indian Institute of Technology Guwahati, Assam, India

Manoj Kumar Sinha, National Institute of Technology Kurukshetra, Haryana, India

Sonika, Rajiv Gandhi University, Rono Hills, Doimukh, Itanagar, India

Manoj Kumar Srivastava, D.A.V.P.G College, Gorakhpur, UP, India

Siddharth Tamang, Indian Institute of Technology Kharagpur, Kharagpur, West Bengal, India

Abhishek Tiwari, Department of Metallurgical and Materials Engineering, Indian Institute of Technology Ropar, Punjab, India

Tanuj Vishwakarma, Department of Mechanical Engineering, Indian Institute of Technology Indore, Madhya Pradesh, India

Chapter 1

Introduction to Sustainable Manufacturing for Industries 4.0

Gourhari Chakraborty[1], *Vivek Pandey*[2], *Arbind Prasad*[3], *and Ashwani Kumar*[4]

[1]National Institute of Technology, Andhra Pradesh, India
[2]School of Engineering & Technology, Mody University, Lakshmangarh, Sikar Rajasthan, India
[3]Katihar Engineering College (under Department of Science & Technology, Govt. of Bihar), Katihar, Bihar, India
[4]Technical Education Department, Kanpur, Uttar Pradesh, India

CONTENTS

1.1 INTRODUCTION

'Industry 4.0' was a German initiative in 2011 and continuously gathered attention from the business giants for smart and automated product manufacturing. Machines evolved from the era of steam engine-powered machines to the emergence of mass production through electrically powered machinery and presently smart automated machines. Manufacturing

DOI: 10.1201/9781003436072-1

processes including machinery, reactors, and instruments that are operated by advanced electronics and information technologies with flexibility of production, improved quality products, and high production output in the case of the fourth industrial revolution – Industry 4.0 [1,2]. The Internet of Things (IoT) is constructing the cyber-physical systems (CPS) that control the latest manufacturing processes. IoT concepts and the digitization of industries are changing the business scenario holistically [3]. 'Industry 4.0' allows more synchronization between the supply chain, production houses, and product delivery, which leads to the integration of companies and collaborative production. Smart products are also providing data for the continuous upgradation of technologies and products. 'Industry 4.0' can be visualized as a concurrence of emerging innovations, ideas, and technologies including big data analytics, cloud computing, simulation/optimization, additive manufacturing, machine learning, smart sensors, system integration, virtual/augmented reality, robotics, artificial intelligence, and cybersecurity [1,4–6].

The newfangled concept of 'Industry 4.0' deploys cybernetic technologies with the advantages of resource efficiency, efficient energy utilization, improved productivity, and waste reduction. It also promotes environment-friendly processes such as 'green manufacturing', which consists of reducing, reusing, and recycling (3R). Sustainable manufacturing (SM) implies the upgradation of three interrelated parameters of industry's product, process, and system standing on 6R methodology, which means apart from 3Rs remanufacture, redesign, and recover are taken care of in sustainable upliftment of the technologies [7].

Figure 1.1 depicts the integrated structure of sustainable manufacturing and Industry 4.0, which is very helpful in present-day manufacturing techniques. This shows the proper and efficient use of technology and human resources for enhancing smart manufacturing with less impact on the environment.

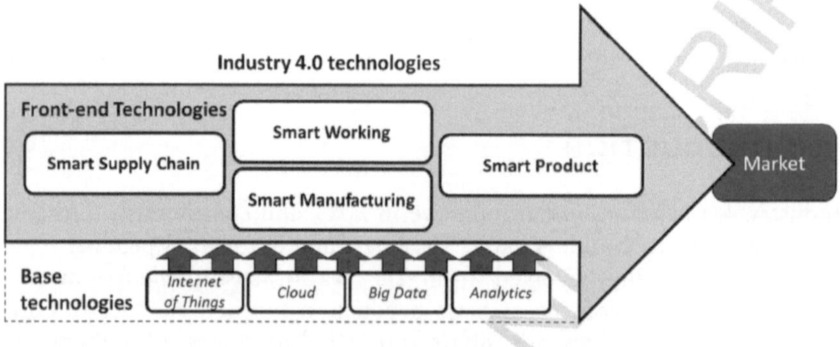

Figure 1.1 Integration of sustainable manufacturing and Industry 4.0.

This chapter is discussing the prime factors, opportunities, and prospects of sustainable manufacturing through the technology of 'Industry 4.0'. It also suitably depicts and clarifies major pillars behind this technology. Alongside attention to production systems, more focus is required on management, data handling, automation, the economy, the environment, and society in a holistic sense for the sustainability of the industry and humans. In addition to this, different theoretical and experimental outcomes of sustainable manufacturing metrics have been incorporated. Experimental studies reveal that the metrics most commonly explored by profitable businesses can be distilled into any number and are primarily concerned with the economic aspect, while the social and environmental metrics frequently appear to be negotiated with the sole intention of adhering to legal requirements. Scenarios, challenges, and future prospects of global sustainable manufacturing firms are also discussed and suitably incorporated in this chapter.

1.2 SUSTAINABLE MANUFACTURING

Smart and automated manufacturing technology is the centroid of 'Industry 4.0'; in association with customer information, this improves the manufacturing of products. Sustainable manufacturing is essential for industries for numerous external factors like stricter regulation, high consumer preferences, exaggerated energy costs, and environmental concerns for eco-friendly products [8]. Utilization of natural bio-based resources for various manufacturing processes needs to maintain regulation for raw materials as well as harmful emissions in the environment. The modification of materials and conversion into goods fulfillment of human needs is the definition of manufacturing [9]. Smart manufacturing can be subdivided into vertical integration, virtualization, automation, traceability, flexibility, and energy management. Apart from other parts, the business firm, including manufacturing, utilized the majority of the world's total delivered energy, which is almost half of the total energy produced [10]. All the leading countries, including United States, absorb more than 42% of the total energy consumption [11]. Similarly, the manufacturing sector in China utilizes 58% of the total energy consumption [12]. In addition to that, many strategies are implemented to minimize the environmental footprint of different production steps and numerous methods are followed to track and regulate variables like energy consumption [13], carbon emissions [14], and greenhouse emissions [8].

1.3 INDUSTRIAL 4.0

'Industry 4.0' (I4.0) has brought computerized, intelligent manufacturing with minimal harm to the environment and is noted to have the scope of production flexibility and significant efficiency [15]. It is popular among academia, government sector, and industry collaborators because of smart and automated connectivity between factories, distributors, and other chain members [16,17]. Santos et al. (2019) have regulated gadgets to minimize scrap and generate real-time and online records that lead to easy decision-making [18]. Along with this, the use of the IoT will address and resolve many global threats such as resource and energy efficiency for the growth of the industry [16] in order to achieve the target policies created by different countries based on their socioeconomic level. In the case of developed countries, the government sets strategies at the national level, whereas for developing countries, corporate houses holders the planning. Multinational companies those who are working globally want uniform policies. In addition to these different countries have their individual obstacles to implementing 'Industry 4.0' to a greater extent. In the market surveys of recent times it is reported that lack of logistics, intent of farms, or awareness about smart operation set the barrier towards the progress of sustainable industrialization. Interdependence of 'Industry 4.0' and the circular economy is also revealed in many current studies. The concept of 'Industry 4.0' has given great attention to the people associated with product development and continuous upgradation is taking place globally.

1.4 IMPACT OF INDUSTRY 4.0 TECHNOLOGIES ON SUSTAINABLE MANUFACTURING

'Industry 4.0' is an amalgam of various technological factors that can bring changes to industrial technology and upgrade it to better productivity. It leads to the collaborative approach of all different emerging technologies available in the market. In order to overcome difficulties and problems associated with TBL of sustainable manufacturing, industry professionals and researchers are opting for Industry 4.0 technologies in recent days. Universal topics such as climate change, depletion of non-renewable energy sources, and environmental protection will be managed to introduce Industry 4.0 technologies. However, for thriving implementation, there should be consistency and the confluence of Industry 4.0 technologies. Figure 1.2 implies the important technological enablers of the emerging Industrial Revolution; these factors are expected to give a pivotal role in the deployment of sustainable manufacturing in the near future.

In the advancement of Industry 4.0, various types of technology play an important role. But this can only be achieved through the employment of advanced technology used in sustainable manufacturing. The influence of the various latest technologies explored in Industry 4.0 is discussed in brief.

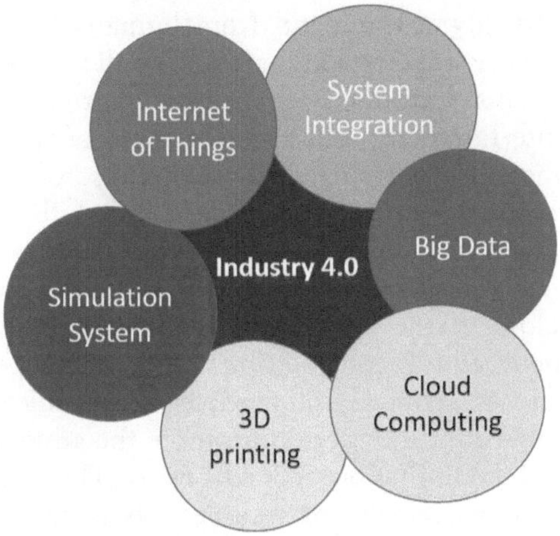

Figure 1.2 Various technologies associated with Industry 4.0.

1.4.1 Robots

A robot is a mechanical device that can accomplish a series of complex jobs, and operate by means of loaded computer programs. These are capable of working safely parallel to humans. Robots broadly can be fixed or mobile in nature (see Figure 1.3). Aquatic, terrestrial, and airborne are three different types of mobile robots. Fixed robots are mostly utilized in factories in jobs like soldering, painting, etc. that are repeating in nature. Robots also can be viewed through their respective application domain, majorly for industrial purposes and services. Mobile robots

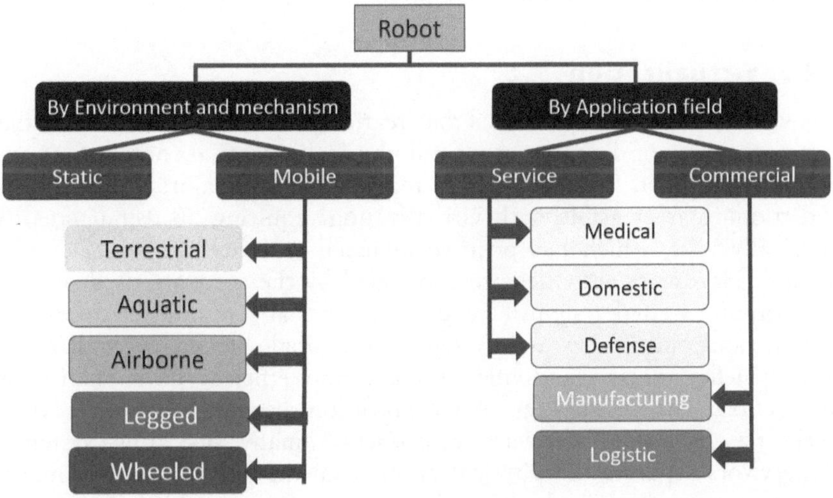

Figure 1.3 Classifications of robots.

can move automatically without aid from humans and are operable in unknown situations. Planetary exploration, medical operations, patrolling, rescue operations, industrial construction, automation of processes, transportation, entertainment, and care are some of the areas where the application of robots is increasing. Robots, because of their autonomous and collaborative nature integrated with human skills with technology, can improve the operation of processes or reduce human labor [5,19]. The employment of autonomous robots can aid in the creation of small, lucrative stores close to the customer, which will have positive effects on the environment and labor relations [20]. Advanced features like cognitive architecture, artificial intelligence, communication skills, and controllable human–robot interaction promote the adverse application of robots in industries, including big to small firms. The recent trends imply the technological advancement of mobile robots that are thought to spread out in all fields.

1.4.2 System Integration

Vertical and horizontal integration of manufacturing, management, and business systems are components of 'Industry 4.0'. It is observed that production visibility improved, manufacturing waste is reduced, and energy resources are utilized effectively because of the collaborative and integrated approach of the technology [20]. Corporate culture evolved with time through an integrated approach that promotes collaboration and communication between the stakeholders of the process of the companies to enhance production efficiency [21]. Systems integration empowers closed-loop production processes with better compatibility with customer demand in case of quantity and product qualities.

1.4.3 Virtualization

The virtualization of processes is the creation of an abstract version of the same consisting of digital content and physical industrial content that are mixed to establish an amalgamated reality. Remote maintenance, repair, and training are feasible with virtualization requiring no technicians to physically visit, which has positive financial and environmental effects [22,23]. Hardware activities are controlled by the software used for the creation of a virtual system. IT organizations that run multiple operating systems and multiple virtual systems use a single server. Virtualization adds benefits to the economic outcome and efficient functionality of production. This technology of virtualization also utilizes power effectively i.e., a single hardware can operate many operating systems. Hypervisors separate the physical event from the physical environment, resources are taken in the virtual environment and added as per the formulation, and finally, programs are run in comparison to the physical system to get a comparable virtual output. It can be subdivided based on the applications like network virtualization, storage virtualization, server virtualization, data virtualization, desktop virtualization, and application virtualization. Virtualization opens up the scope of single-minded servers, accelerated deployment and redeployment of the servers, efficient energy utilization, environment-friendly operation, cloud storage systems, and automated processing. In addition to this, the use of virtual reality alongside the training of workers at a low cost establishes a safe working environment in the industry [20]. Automation of a process takes high time to establish; the operation of some processes is not successful in automated conditions and the late solution of failure of one event connected with several parameters are some of the limitations that need to be addressed in the upcoming days.

1.4.4 3-D Printing

3-D printing is an additive manufacturing (AM) where layer-by-layer deposition of material creates one three-dimensional object from computer-aided design models (CAD models). A digital STL file of the object is used for printing any object and G-code language is used for programming (see Figure 1.4). This technology is designed specifically and is more beneficial in terms of applications. Inkjet printing, stereolithography, extrusion printing, laser-based sintering, bioplotting, fuse deposition modeling, etc. are some of the printing techniques that use solution, hydrogel, filament, powder, etc., for printing. 3D-printed items are increasing in all fields including household items, toys, electronics, and biomedical. The application of AM enhances

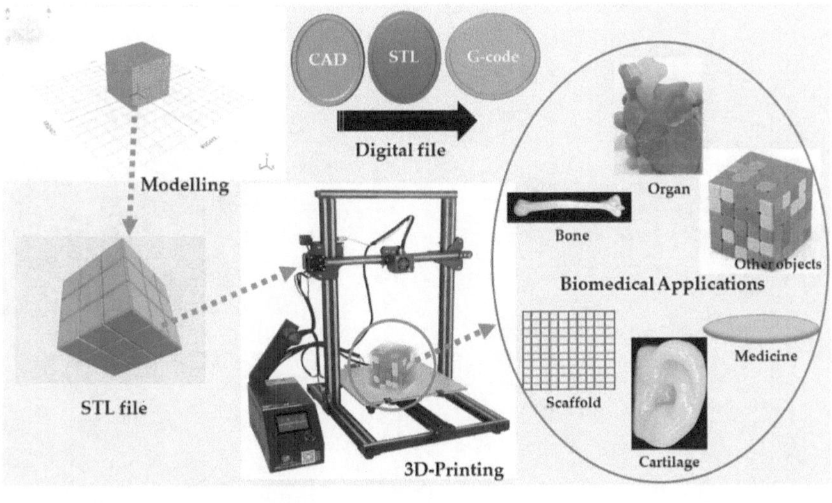

Figure 1.4 3D-printing techniques.

circular economy initiatives by minimizing the amount of ingredients used, the waste generated, and flexibility in production [20]. These days, many biomaterials, including metallic and polymeric implants, are fabricated using this process for orthopedic and biomedical applications [24–34]. 3D printing is cutting-edge technology with customized product development facilities and it allows the printing of prototypes more cheaply and rapidly, reducing the time to market.

1.4.5 Cloud Computing

Cloud computing is the availability of resources for computing and storage assistances on demand of the customer through the Internet anywhere and anytime. Cloud computing facilities are managed by servers (data centers) located remotely; thus, it accompanies profitability and energy savings (see Figure 1.5). Depending upon the amount of information companies use single or multiple data centers [20]. The advantage of using cloud computing for customers is payment for the space used for their purpose which helps in fast growth and scale-up of organizations. Cloud computing deployment can be done following three different models (i) public clouds, (ii) private clouds, and (iii) hybrid clouds. Cost-effectiveness, efficiency in service, flexibility, strategies, and information security are the benefits associated with cloud computing [35]. Cloud computing services like the service of infrastructure, service of platforms, and service of software help in the growth of industries; depending on the need, they can avail the services economically.

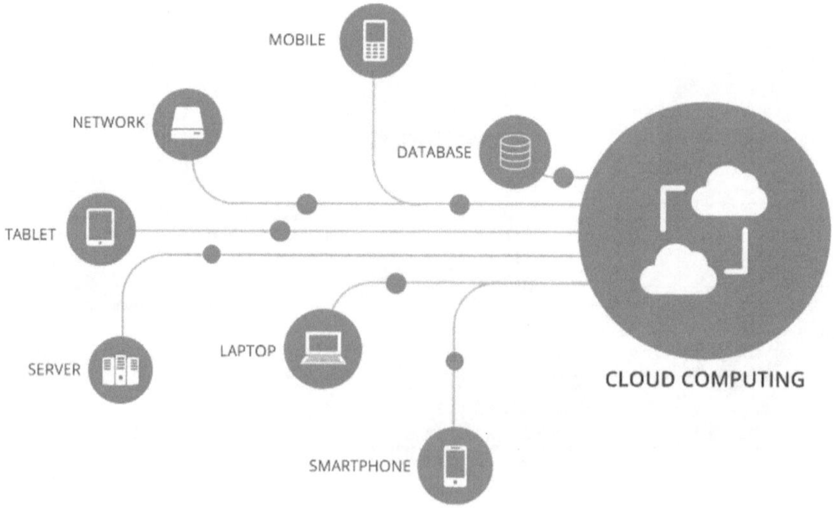

Figure 1.5 Cloud computing.

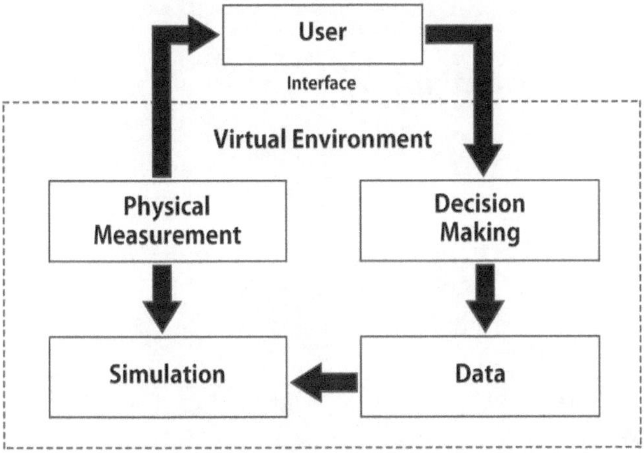

Figure 1.6 General block diagram of simulation.

1.4.6 Simulation System

Simulation systems imitate a process or service virtually and outputs of the process are incorporated suitably into real processes to enhance the delivery capacity of the manufacturing unit. Simulation applications can lead to a system with less energy consumption by eliminating unnecessary steps required during the foundation of a new process or the upgradation of a present unit [20]. Modeling is also used to optimize the operating conditions, comparison of different options for the same purpose, and validation of any new technology [36]. Model building is the primary requirement of simulation and these models are based on empirical information. Further, suitable mathematical computation with error correction can lead to a comparable theoretical model of any event or process (see Figure 1.6). In most of the cases, Monte Carlo simulation is conducted using a suitable programming language for getting the virtual information. The event scheduling facility was observed to improve the time management for the researchers and industry personnel. Thus proper modeling of a system followed by validation, verification, and calibration can lead to improved technologies that will continuously evolve by the researchers. A vast amount of software, such as ACSL, APROS, ARTIFEX SMPL, SimBank, SimPlusPlus, etc., is in use for the simulation of various systems and is rapidly increasing in number.

1.4.7 Industrial Internet of Things (IIoT)

IIoT is the extension and application of the Internet of Things (IoT) in the domain of industrial applications. It has applicability in robotics, medical, and software-related cases. It describes all interconnected sensors and equipment that communicate over the Internet with one another and with

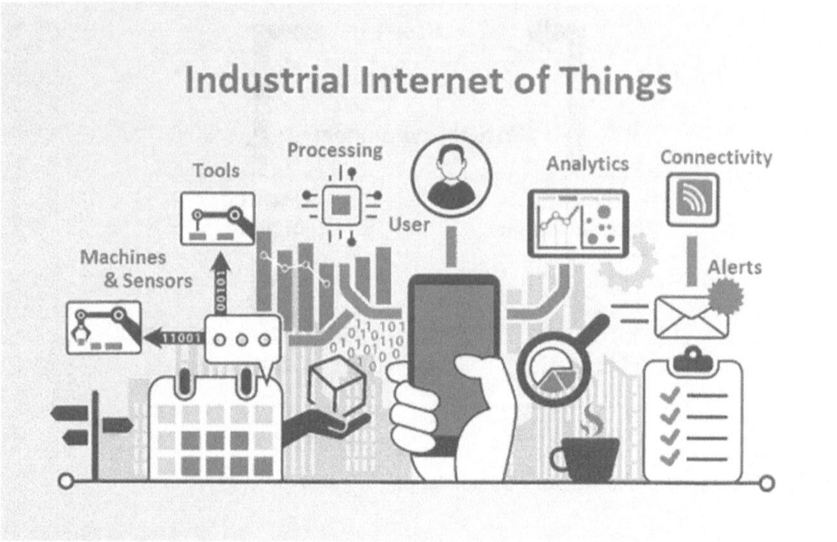

Figure 1.7 Industrial Internet of Things.

industrial applications. It is considered to have multiple architecture layers such as (i) content layer, (ii) service layer, (iii) network layer, and (iv) device layer. IIoT makes it simple to collect and analyse all forms of data from industrial operations, reducing waste and wasteful inventory and manufacturing procedures [20]. Smart device continuous surveillance using real-time problem solving raises awareness of energy consumption and its visibility. IIoT enhances equipment and operator safety by delivering real-time hazard warnings and better maintenance options [37]. Availability, scalability, and security are the prime factors that need to be focused on for the implementation of IIoT. Security is the key factor that holds the acceptability of the industrial operation. It has huge prospects in the oil and gas sector, automotive sector, and agricultural sector. Figure 1.7 shows a representative IIoT scheme composed of various components.

1.4.8 Big Data

Big data technology is employed for obtaining appropriate insightful and relevant information from large pools of structured and unstructured data that are available online. This assists in the decision-making of industries by gathering, examining, and arranging market data. Using the benefits of big data with data analytics, data mining, and other methods can improve data analysis and ultimately speed up production. Big data has the following features, including large volume, the velocity of data generation, variation, validity, authenticity, variability, and vagueness, which need to be taken care

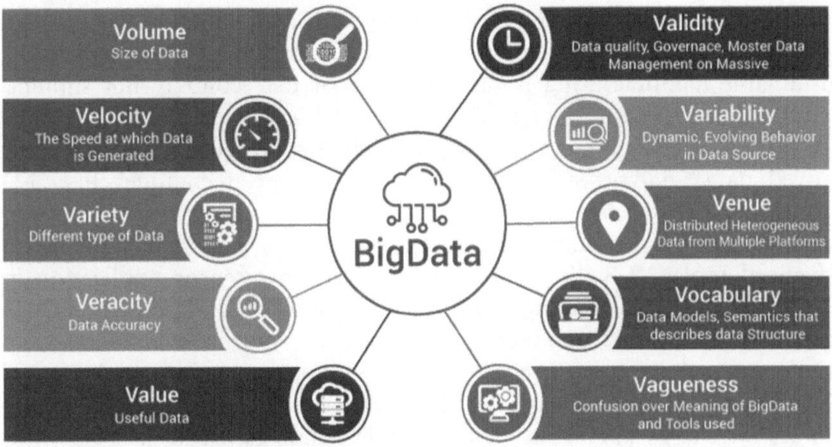

Figure 1.8 Features of big data.

during processing (see Figure 1.8). Data analytics can be (i) perspective analysis, (ii) diagnostic analysis, (iii) predictive analytics, and (iv) descriptive analytics depending upon the need. This helps in customer acquisition or retention, product development, risk analysis, price optimization, and decision-making for business firms. If these technologies were to properly collide, manufacturing organization would gain from process optimization, cost reduction, and operational processing improvements, which would immediately increase industry profitability [24–29,38–40]. Big data analytics will aid in growing reliability and effective prediction of critical situations and permitting real-time interventions [30–34].

1.5 INTEGRATIVE APPROACH OF SUSTAINABLE MANUFACTURING AND INDUSTRY 4.0

The economy of any country can be uplifted by the implementation of sustainable manufacturing for industries. This can only be possible through the adoption of advanced technology for manufacturing. Development of the industries lies on continuous upgradation of technology and implementation of automation techniques wherever possible [41–48]. Advanced technology including the above-mentioned factors can assist the integration of sustainable manufacturing and growth of 'Industry 4.0'. This integration will lead to a positive footprint on the environment, energy utilization, as well as on the economy of any country. One must have to recognize the important factor(s) for the development of industry 4.0 as per the standards. So, the integration of sustainable manufacturing and Industry 4.0 is a very important aspect of development and also for conservation of the environment [49–53]. In line with growing population pressure on the industry, especially the manufacturing sector for maintaining the demand and supply chain, is continuously increasing. In the era of globalization, sustainable manufacturing is the backbone of the economy but simultaneously industries will have to adopt new and advanced technologies like 'Industry 4.0'. This integrative approach, digitization, use of robotics, improved data analytics, and adaptation of modern techniques like 3-D printing will lead sustainable manufacturing to a new height in the upcoming years [54,55]. In the future, the study can be extended to the use of advanced tools ANN, fuzzy logic to analyze flow sensor application in buildings, advanced coating materials for photovoltaic applications and their thermal contact conductance analysis for rooftop solar plants and window paneling, nano techniques for energy conversion, and storage devices with special emphasis on high-performance piezoelectric nanogenerators for energy harvesting in sustainable buildings [56–67].

1.6 CONCLUSIONS

Sustainable manufacturing (SM) is an emerging and upcoming idea that is grabbing curiosity in the research domain and has reached from academia to industry because of adaptation in business firms, particularly in the manufacturing hubs. Despite this, there is still a wide range of explanations and notions connected to the idea of SM, leading many authors to claim that SM or its related sub-concepts are not sufficiently uniformly understood or accepted. 'Industry 4.0' is setting a new level for SM for the creation of newer products, economical product, or products with improved quality. Acceptance and adaptation of SM through the use of ICT infrastructure, IIoT, big data, new manufacturing technologies, and human-machine and machine-to-machine interaction will bring changes to the industry culture, production output, and to the society.

REFERENCES

[1] Rubmann, M.; Lorenz, M.; Gerbert, P.; Waldner, M.; Justus, J.; Engel, P.; and Harnisch, M. (2015). *Industry 4.0: The Future of Productivity and Growth in Manufacturing Industries*. Boston Consult, Group 9: Boston, MA, USA; pp. 54–89.

[2] Hermann, M.; Pentek, T.; and Otto, B. Design Principles for Industrie 4.0 Scenarios. In Proceedings of the 2016 49th Hawaii International Conference on System Sciences (HICSS), Koloa, HI, USA, 5–8 January 2016; pp. 3928–3937.

[3] Dalenogarea, L. S.; Beniteza, G. B.; Ayalab, N. F.; and Franka, A. G. (2018). The Expected Contribution of Industry 4.0 Technologies for Industrial Performance. *International Journal of Production Economics*, 204, 383–394.

[4] Wee, D.; Kelly, R.; Cattel, J.; and Breunig, M. (2015). *Industry 4.0: How to Navigate Digitization of the Manufacturing Sector*. McKinsey Company: Chicago, IL, USA.

[5] Machado, C. G.; Winroth, M. P.; and Da Silva, E. H. D. R. (2020). Sustainable Manufacturing in Industry 4.0: An Emerging Research Agenda. *Internation Journal of Production Research*, 58, 1462–1484.

[6] Sartal, A.; Bellas, R.; Mejías, A. M.; and García-Collado, A. (2020). The Sustainable Manufacturing Concept, Evolution and Opportunities within Industry 4.0: A Literature Review. *Advances in Mechanical Engineering*, 12.

[7] Gholami, H.; Abu, F.; Lee, J. K. Y.; Karganroudi, S. S.; and Sharif, S. (2021). Sustainable Manufacturing 4.0—Pathways and Practices. *Sustainability*, 13, 13956. 10.3390/su132413956

[8] Giret, A.; Trentesaux, D.; and Prabhu, V. (2015). Sustainability in Manufacturing Operations Scheduling: A State of the Art Review. *Journal of Manufacturing Systems*, 37, 126–140.

[9] Chryssolouris, G. (2006). *Manufacturing Systems: Theory and Practice*. Springer: New York, NY, USA.

[10] EIA (2018). International Energy Outlook 2018. Available online at: https://www.eia.gov/outlooks/ieo/ (last accessed, 10 April 2019).

[11] Alhourani, F.; and Saxena, U. (2009). Factors Affecting the Implementation Rates of Energy and Productivity Recommendations in Small and Medium Sized Companies. *Journal of Manufacturing Systems*, 28(1), 41–45.

[12] ESDNBS - Energy Statistics Division of National Bureau of Statistics (2013). China Energy Statistical Yearbook 2013.

[13] Doyle, F.; Duarte, M.; and Cosgrove, J. (2015). Design of an Embedded Sensor Network for Application in Energy Monitoring of Commercial and Industrial Facilities. *Energy Procedia*, 83, 504–514.

[14] Du, Y.; Yi, Q.; Li, C.; and Liao, L. (2015). Life Cycle Oriented Low-Carbon Operation Models of Machinery Manufacturing Industry, 91, 145–157.

[15] Zawadzki, P.; and Zywicki, K. (2016). Smart Product Design and Production Control for Effective Mass Customization in the Industry 4.0 Concept. *Management and Production Engineering Review*, 7(3), 105–112.

[16] Kagermann, H.; Helbig, J.; Hellinger, A.; and Wahlster, W. (2013). *Recommendations for Implementing the Strategic Initiative INDUSTRIE 4.0*. Forschungsunion: Berlin.

[17] Henao-Hernández, I.; Solano-Charris, E. L.; Muñoz-Villamizar, A.; Santos, J.; and Henríquez-Machado, R. (2019). Control and Monitoring for Sustainable Manufacturing in the Industry 4.0. *A literature Review: IFAC Papers OnLine*, 52-10(2019), 195–200.

[18] Santos, J.; Muñoz-Villamizar, A.; Ormazábal, M.; and Viles, E. (2019). Using Problem-Oriented Monitoring to Simultaneously Improve Productivity and Environmental Performance in Manufacturing Companies. *International Journal of Computer Integrated Manufacturing*, 32(2), 183–193.

[19] Romero Daz D.; Stahre J.; Wuest T.; et al. (eds.). (2016). Towards an Operator 4.0 Typology: A Human-Centric Perspective on the Fourth Industrial Revolution Technologies. In Proceedings of International Conference on Computers & Industrial Engineering CIE46. Tianjin, China: Curran Associates, Inc., pp. 608–618.

[20] Sartal, A.; Bellas, R.; Mejıas, A. M.; and Garcia-Collado, A. (2020) The Sustainable Manufacturing Concept, Evolution and Opportunities within Industry 4.0: A Literature Review. *Advances in Mechanical Engineering*, 12(5), 1–17.

[21] Kamble, S. S.; Gunasekaran, A.; and Gawankar, S. A. (2018). Sustainable Industry 4.0 Framework: A Systematic Literature Review Identifying the Current Trends and Future Perspectives. *Process Safety and Environmental Protection*, 117, 408–425.

[22] Keller M.; Rosenberg M.; Brettel M.; et al. (2014). How Virtualization, Decentralization and Network Building Change the Manufacturing Landscape: An Industry 4.0 Perspective. *International Journal of Mechanical Aerospace, Industrial, Mechatronic and Manufacturing Engineering*, 8(1), 37–44.

[23] Menon, S.; Shah, S.; and Coutroubis A. (2018). Impacts of I4.0 on Sustainable Manufacturing to Achieve Competitive Advantage. In Proceedings of the Eighth International Conference on Operations and Supply Chain Management (OSCM). Cranfield: Cranfield School of Management, pp. 379–387.

[24] Bhoi, S.; Prasad, A.; Kumar, A.; Sarkar, R. B.; Mahto, B.; Meena, C. S.; and Pandey, C. (2022). Experimental Study to Evaluate the Wear Performance of UHMWPE and XLPE Material for Orthopedics Application. *Bioengineering*, 9, 676. 10.3390/bioengineering9110676.

[25] Bhoi, S.; Kumar, A.; Prasad, A.; Meena, C. S.; Sarkar, R. B.; Mahto, B.; and Ghosh, A. (2022). Performance Evaluation of Different Coating Materials in Delamination for Micro-Milling Applications on High-Speed Steel Substrate. *Micromachines*, 13, 1277. 10.3390/mi13081277.

[26] Gupta, A. et al. (2017). Multifunctional Nanohydroxyapatite-Promoted Toughened High-Molecular-Weight Stereocomplex Poly(Lactic Acid)-Based Bionanocomposite for Both 3D-Printed Orthopedic Implants and High-Temperature Engineering Applications. *ACS Omega*, 2(7), 4039–4052.

[27] Prasad, A. (2021a). Bioabsorbable Polymeric Materials for Biofilms and Other Biomedical Applications: Recent and Future Trends. *Materials Today: Proceedings*, 44, 2447–2453. 10.1016/j.matpr.2020.12.489.

[28] Prasad, A. (2021b). State of Art Review on Bioabsorbable Polymeric Scaffolds for Bone Tissue Engineering. *Materials Today: Proceedings*, 44, 1391–1400. 10.1016/j.matpr.2020.11.622.

[29] Prasad, A.; Bhasney, S.; Katiyar, V.; and Sankar, M. R. (2017). Biowastes Processed Hydroxyapatite Filled Poly (Lactic Acid) Bio-Composite for Open Reduction Internal Fixation of Small Bones. *Materials Today: Proceedings*, 4(9), 10153–10157. http://linkinghub.elsevier.com/retrieve/pii/S2214785317311410.

[30] Prasad, A.; Devendar, B.; Sankar, M. R.; and Robi, P. S. (2015). Micro-Scratch Based Tribological Characterization of Hydroxyapatite (HAp) Fabricated through Fish Scales. *Materials Today: Proceedings*, 2(4–5), 1216–1224. http://linkinghub.elsevier.com/retrieve/pii/S2214785315002795.

[31] Prasad, A.; Sankar, M. R.; and Katiyar, V. (2017). State of Art on Solvent Casting Particulate Leaching Method for Orthopedic Scaffolds Fabrication. *Materials Today: Proceedings*, 4(2), 898–907. 10.1016/j.matpr.2017.01.101.

[32] Prasad, A. (2022). Biomaterial-Based Nanofibers Scaffolds in Tissue Engineering Application. In *Functional Biomaterials*. Springer: Singapore; pp. 245–264.

[33] Gangwar, A. K. S.; Rao, P. S.; and Kumar, A. (2021). Bio-Mechanical Design and Analysis of Femur Bone. *Materials Today: Proceedings*, 44(Part 1), 2179–2187. ISSN 2214-7853 10.1016/j.matpr.2020.12.282.
[34] Gangwar, A. K. S.; Rao, P. S.; Kumar, A.; and Patil, P. P. (2019). Design and Analysis of Femur Bone: BioMechanical Aspects. *Journal of Critical Reviews*, 6(4), 133–139. ISSN-2394-5125.
[35] Garg, S. K.; and Buyya, R. (2012). Green Cloud Computing and Environmental Sustainability. In Murugesan, S., and Gangadharan, G. R. (eds.), *Harnessing Green IT: Principles and Practices*. 1st ed. Wiley & Sons: Chichester; pp. 315–340.
[36] Ferrera, E.; Rossini, R.; Baptista, A. J.; et al. (eds). (2017). Toward Industry 4.0: Efficient and Sustainable Manufacturing Leveraging MAESTRI Total Efficiency Framework. In *SDM: Smart Innovation, Systems and Technologies*. Springer: New York, NY, USA; pp. 624–633.
[37] Thoben, K. D.; Wiesner, S. A.; and Wuest, T. (2017). 'Industrie 4.0' and Smart Manufacturing – A Review of Research Issues and Application Examples. *International Journal of Automation Technology*, 11(1), 4–16.
[38] Lee, H. K.; and Yang, S. (2014). Service Innovation and Smart Analytics for Industry 4.0 and Big Data Environment. *Procedia CIRP*, 16, 3–8. Available: 10.1016/j.procir.2014. 02.001.
[39] Lu, Y. (2017). Industry 4.0: A Survey on Technologies, Applications and Open Research Issues. *Journal of Industrial Information Integration*, 6, 1–10. Available: 10.1016/j.jii.2017. 04.005
[40] Kumar, A.; and Nayyar, A. si3-Industry: A Sustainable, Intelligent, Innovative, Internet-of-Things Industry, Advances in Science, Technology & Innovation, 10.1007/978-3-030-14544-6_1
[41] Kumar, A.; Mamgain, D. P.; Jaiswal, H.; and Patil, P. (2015). Modal Analysis of Hand Arm Vibration (Humerus Bone) for Biodynamic Response Using Varying Boundary Conditions Based on FEA. *Advances in Intelligent Systems and Computing*, 308, 169–176. 10.1007/978-81-322-2012-1_18.
[42] Kumar, A.; Behmad, S. I.; and Patil, P. (2014). Vibration Characterization and Static Analysis of Cortical Bone Fracture Based on Finite Element Analysis. *Engineering and Automation Problems*, No. 3-2014, pp. 115–119. UDC-621.
[43] Kumar, A.; Jaiswal, H.; Garg, T.; and Patil, P. (2014). Free Vibration Modes Analysis of Femur Bone Fracture Using Varying Boundary Conditions Based on FEA. *Procedia Materials Science*, 6, 1593–1599. 10.1016/j.mspro.2014. 07.142.
[44] Kumar, A.; Gori, Y.; Rana, S.; Sharma, N. K.; and Yadav, B. (2022). FEA of Humerus Bone Fracture and Healing. In *Advanced Materials for Biomechanical Applications*. 1st ed. CRC Press. 10.1201/9781003286806.
[45] Kumar, A.; Datta, A.; and Kumar, A. (2022). Recent Advancements and Future Trends in Next-Generation Materials for Biomedical Applications. In Kumar, A., Gori, Y., Kumar, A., Meena, C. S., and Dutt, N. (eds.), *Advanced Materials for Biomedical Applications*. CRC Press: Boca Raton, FL, USA, Chapter 1; pp. 1–19. 10.1201/9781003344810-1
[46] Prasad, A.; Chakraborty, G.; and Kumar, A. (2022). Bio-Based Environmentally Benign Polymeric Resorbable Materials for Orthopedic Fixation Applications. In Kumar, A., Gori, Y., Kumar, A., Meena, C. S., and Dutt, N. (eds.), *Advanced*

Materials for Biomedical Applications. CRC Press: Boca Raton, FL, USA, Chapter 15; pp. 251–266. 10.1201/9781003344810-15.
[47] Datta, A.; Kumar, A.; Kumar, A.; Kumar, A.; and Singh, V. P. (2022). Advanced Materials in Biological Implants and Surgical Tools. In Kumar, A., Gori, Y., Kumar, A., Meena, C. S., and Dutt, N. (eds.), *Advanced Materials for Biomedical Applications*. CRC Press: Boca Raton, FL, USA, Chapter 2; pp. 21–43. 10.1201/9781003344810-2.
[48] Kumar, A.; Gangwar, A. K. S.; Kumar, A.; Meena, C. S.; Singh, V. P.; Dutt, N.; Prasad, A.; and Gori, Y. (2022). Biomedical Study of Femur Bone Fracture and Healing. In Kumar, A., Gori, Y., Kumar, A., Meena, C. S., and Dutt, N. (eds.), *Advanced Materials for Biomedical Applications*. CRC Press: Boca Raton, FL, USA, Chapter 14; pp. 235–250. 10.1201/9781003344810-14.
[49] Patil, P. P.; Gori, Y.; Kumar, A.; and Tyagi, M. (2021). Experimental Analysis of Tribological Properties of Polyisobutylene Thickened Oil in Lubricated Contacts. *Tribology International*, 159, 106983. 10.1016/j.triboint.2021.106983.
[50] Kumar, A.; Rana, S.; Gori, Y.; and Sharma, N. K. (2021). Thermal Contact Conductance Prediction Using FEM Based Computational Techniques. In *Advanced Computational Methods in Mechanical and Materials Engineering*. CRC Press: Boca Raton, FL, USA; pp. 183–220. ISBN 9781032052915. 10.1201/9781003202233-13.
[51] Kumar, R.; Kumar, A.; Kant, L.; Prasad, A.; Bhoi, S.; Meena, C. S.; Singh, V. P.; and Ghosh, A. (2023). Experimental and RSM-Based Process-Parameters Optimisation for Turning Operation of EN36B Steel. *Materials*, 16, 339. 10.3390/ma16010339.
[52] Prasad, A.; Chakraborty, G.; Kumar, A.; and Gajrani, K. K. (2022). Introduction to Biodegradable Polymers. In Prasad, A., Kumar, A., and Gajrani, K. K. (eds.), *Biodegradable Composites for Packaging Applications*. CRC Press: Boca Raton, FL, USA, Chapter 1; pp. 1–12. 10.1201/9781003227908-1.
[53] Chakraborty, G.; Prasad, A.; and Kumar, A. (2022). Processing of Biodegradable Composites. In Prasad, A., Kumar, A., and Gajrani, K. K. (eds.), *Biodegradable Composites for Packaging Applications*. CRC Press: Boca Raton, FL, USA, Chapter 3; pp. 33–48. 10.1201/9781003227908-3.
[54] Kumar, A.; Rana, S.; Gori, Y.; and Sharma, N. K. (2022). Thermal Contact Conductance Prediction Using FEM Based Computational Techniques. In Kumar, A., Gori, Y., Dutt, N., Singla, Y. K., and Maurya, A. (eds.), *Advanced Computational Methods in Mechanical and Materials Engineering*. CRC Press: Boca Raton, FL, USA, Chapter 11; pp. 183–220. 10.1201/9781003202233-13.
[55] Sharma, V. K.; Kumar, V.; Joshi, R. S.; and Kumar, A. (2022). Effect of REOs on Tribological Behavior of Aluminum Hybrid Composites Using ANN. In Sharma, V. K., Kumar, A., Gupta, M., Kumar, V., Sharma, D. K., and Sharma, S. K. (eds.), *Additive Manufacturing in Industry 4.0: Methods, Techniques, Modeling and Nano Aspects*. CRC Press: Boca Raton, FL, USA, Chapter 9; pp. 153–168. 10.1201/9781003360001-9.
[56] Bansal, A.; Bhardwaj, H. K.; Sharma, V. K.; and Kumar, A. (2022) Static and Dynamic Behavior Analysis of Al-6063 Alloy Using Modified Hopkinson Bar. In Sharma, V. K., Kumar, A., Gupta, M., Kumar, V., Sharma, D. K., and

Sharma, S. K. (eds.), *Additive Manufacturing in Industry 4.0: Methods, Techniques, Modeling and Nano Aspects*. CRC Press: Boca Raton, FL, USA, Chapter 6; pp. 107–124. 10.1201/9781003360001-6.

[57] Singh, S.; Kumar, A.; Behura, S. K.; and Verma, K. (2022). Challenges and Opportunities in Nanomanufacturing. In Singh, S., Behura, S. K., Kumar, A., and Verma, K. (eds.), *Nanomanufacturing and Nanomaterials Design: Principles and Applications*. CRC Press: Boca Raton, FL, USA, Chapter 2; pp. 17–30. 10.1201/9781003220602-2.

[58] Srivastava, S.; Verma, D.; Thusoo, S.; Kumar, A.; Singh, V. P.; and Kumar, R. (2022). Nanomanufacturing for Energy Conversion and Storage Devices. In Singh, S., Behura, S. K., Kumar, A., and Verma, K. (eds.), *Nanomanufacturing and Nanomaterials Design: Principles and Applications*. CRC Press: Boca Raton, FL, USA, Chapter 10; pp. 165–173. 10.1201/9781003220602-10.

[59] Singh, V. P.; Dwivedi, A.; Karn, A.; Kumar, A.; Singh, S.; Srivastava S.; and Srivastava, K. (2022). Nanomanufacturing and Design of High-Performance Piezoelectric Nanogenerator for Energy Harvesting. In Singh, S., Behura, S. K., Kumar, A., and Verma, K. (eds.), *Nanomanufacturing and Nanomaterials Design: Principles and Applications*. CRC Press: Boca Raton, FL, USA, Chapter 15; pp. 241–242. 10.1201/9781003220602-15.

[60] Gori, Y.; Verma, R. P.; Kumar, A.; and Patil, P. (2021). FEA Based Fatigue Crack Growth Analysis. Materials Today Proceedings, 46(20), 10575–10581. 10.1016/j.matpr.2021.01.319.

[61] Patil, P. P.; and Kumar, A. (2017). Design and FEA Simulation of Omega Type Coriolis Mass Flow Sensor. *International Journal of Control Theory and Applications*, 9(40), 383–387.

[62] Patil, P.; Sharma, S. C.; Paliwal, V.; and Kumar, A. (2014). ANN Modeling of Cu Type Omega Vibration Based Mass Flow Sensor. *Procedia Technology*, 14, 260–265. 10.1016/j.protcy.2014.08.034.

[63] Patil, P.; Sharma, S. C.; Jaiswal, H.; and Kumar, A. (2014). Modeling Influence of Tube Material on Vibration Based EMMFS Using ANFIS. *Procedia Material Science*, 6, 1097–1103. 10.1016/j.mspro.2014.07.181.

[64] Paul, A. K.; Prasad, A.; and Kumar, A. (2022). Review on Artificial Neural Network and Its Application in the Field of Engineering. *Journal of Mechanical Engineering: PRAKASH*, 1(1), 53–61. 10.56697/JMEP.2022.1107.

[65] Kumar, A.; and Patil, P. (2016). FEA Simulation Based Performance Study of Multi Speed Transmission Gearbox. *International Journal of Manufacturing Materials and Mechanical Engineering*, 6, 57–67. 10.4018/IJMMME.2016010103.

[66] Rana, S.; Kumar, A.; Gori, Y.; and Patil, P. Design and Analysis of Thermal Contact Conductance. *Journal of Critical Reviews*, 6(5), 363–370. ISSN-2394-5125.

[67] Rana, S.; and Kumar, A. (2019). FEA Based Design and Thermal Contact Conductance Analysis of Steel and Al Rough Surfaces. *International Journal of Applied Engineering Research*, 13(16), 12715–12724. ISSN 0973-4562.

Chapter 2

Recent Developments in Joining Technologies of Dissimilar Metal Welds

Abhishek Tiwari

Department of Metallurgical and Materials Engineering,
Indian Institute of Technology Ropar, Punjab, India

CONTENTS

2.1 INTRODUCTION

With the increasing demands in electric energy, advanced power plants both nuclear and conventional, are pushing the efficiency by operating at high temperature and pressure. From sub-critical to advanced ultra-super critical power plants, the operating temperature has increased from 500 to 700°C and pressure has increased from 180 bars to 350 bars. Pushing towards high temperatures and pressure surely increases the efficiency of the power plants, but it also demands high creep and oxidation resistance from the structural materials. Ni-based alloys, which are usually good in these properties, suffer from difficulty in manufacturing and high cost. A better alternative that has been developed in recent years is ferritic/martensitic steels, especially grades like P92, and boron-added MARBN, etc. [1] (Abe et al. 2021). These grades have very good creep resistance and oxidation resistance. However, there appears a need to join these grades with

DOI: 10.1201/9781003436072-2

austenitic stainless steel grades like SS304L or SS316L. High carbon containing SS grades suffer from the sensitization issues post welding and hence low carbon grades are preferred [2]. The SS grades are used on the low temperature side, such as heater tubes that are connected to a steam header made of P91 and similar grades [3].

2.2 DISSIMILAR METAL WELDS: DIFFERENCES

In a fusion reactor design, an equivalent grade with low activity known as reduced activation ferritic/martensitic grades are being studied for joints with SS316LN grades [4]. In such joints where on one side the material has high thermal conductivity and lower thermal conductivity on the other side, the heat-affected zone (HAZ) size is very thin on the lower thermal conductivity side.

Another problem arises due to the differences in the coefficient of thermal expansion; for instance the coefficient of thermal expansion of SS316 and similar grades are close to 18e-6 K-1, whereas the same for P92 and similar grades are close to 13.2e-6 K-1. This difference in thermal expansion results in thermal strains in the welded region and causes damage and cracks in service. To avoid this, a thin buttering layer of an alloy which has an intermediate value of coefficient of thermal expansion is applied before welding the two differently deforming materials. Alloy 625 has been used in many investigations on DMWs of P91 or P92 with SS316 grades.

The schematics of similar and dissimilar metal welds are shown in Figure 2.1(a) and (b).

It can be observed from Figure 2.1 that the HAZ is of different size when two materials with different thermal conductivity are welded due to the flow of heat on one side more than other. The different cooling rates achieved as a combination of heat flow i.e., temperatures and cooling methods, different zones of varying grain sizes are observed. These microstructural details are discussed in Section 2.3.

2.3 MICROSTRUCTURAL AND MECHANICAL PROPERTY GRADIENT IN DMWs

In Figure 2.1(b), the zones that form during the cooling are related with the phase diagram and only indicative to explain the inter-critical HAZ (ICHAZ) that forms due to the starting temperature in the region between AC3 and AC1 temperatures. The region next to the weld zone on the ferritic steel side that has higher thermal conductivity, starts from a temperature above A_{c3}. Due to this high temperature, the carbides are dissolved as the solubility of interstitials is more in an austenitic phase. This results in coarse austenitic grains.

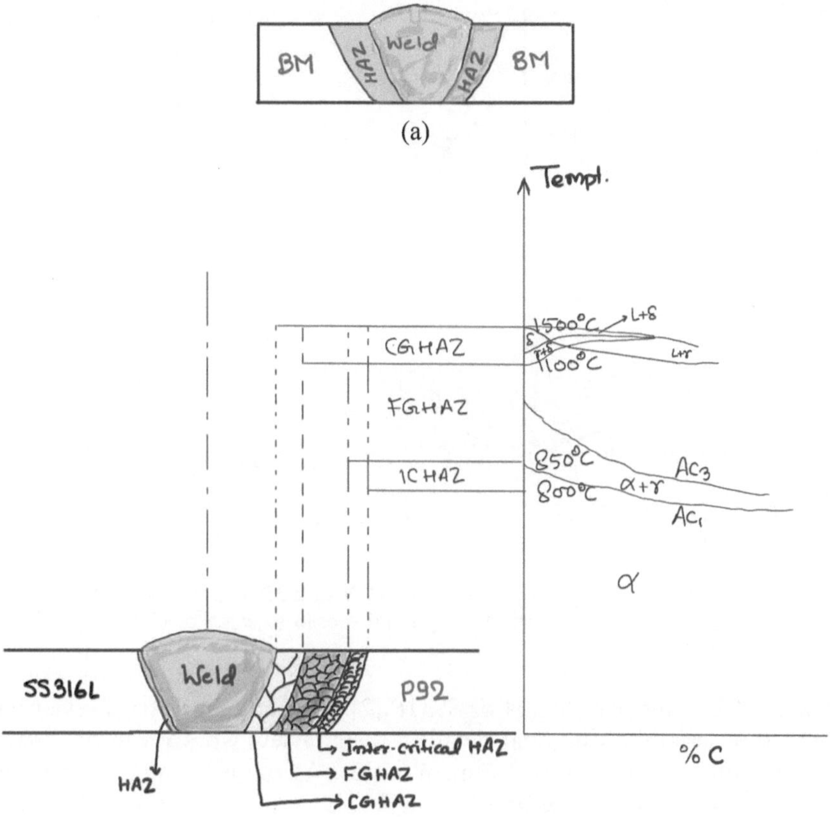

(b)

Figure 2.1 Schematic showing different zones in a (a) similar metal weld and (b) small and large HAZ in a DMW with ICHAZ, FGHAZ and CGHAZ along with the temperatures and corresponding regions in representative phase diagram (schematic based on [5]).

The adjacent region to CGHAZ is FGHAZ starts from a temperature slightly above A_{c3}. Because of this, it has only a few carbides are undissolved. This results in an un-tempered martensitic structure. The next zone, which is closest to the base metal on the ferritic steel side, starts in the inter-critical temperature region between A_{c3} and A_{c1}. This results in a partial transformation to austenitic phase with a few dissolutions of carbides and softer phase at room temperature, which causes the strengths to suffer and toughness to flourish in this region. This region also appears to be the weakest in terms of creep properties and Type IV cracking during operations.

In other examples of DMWs, such as 9Cr-1Mo welded with 2.25Cr-1Mo steel, a coupled region of carburized and decarburized zones was found by Karthik et al. [5]. Similar carbon diffusion or migration is noticed in

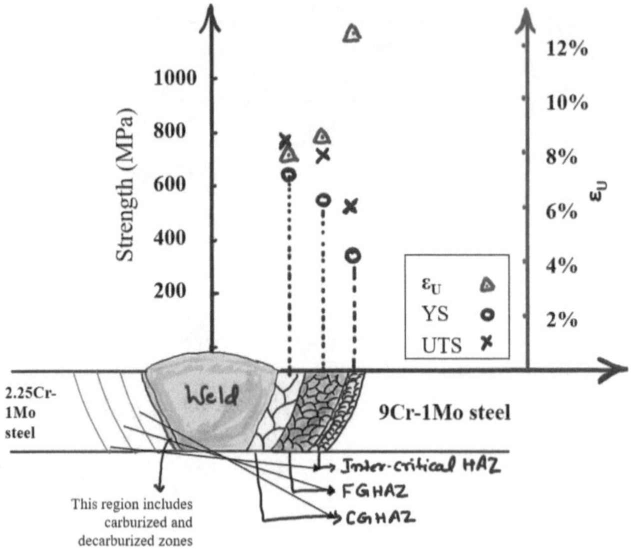

Figure 2.2 Schematic of microstructures in a DMW of 2.25Cr-1Mo steel and 9Cr-1Mo steel with ductility, yield strength and tensile strength of differently refined regions [5].

P92 or P91 welds with SS304 or SS316 [2]. Dak and Pandey [2] reviewed the carbon migration in great detail in their review work on ferritic/martensitic and austenitic steel dissimilar welds. As the percentage of carbon and Cr both have a huge gradient in P92/SS type of weld couples, and to add on to that the diffusion of carbon being an interstitial is well understood. This results in the carbon-depleted zone. The carbon diffuses to Cr-rich areas, which is also in the weld materials when ferritic/martensitic steel is used as filler, it forms $Cr_{23}C_6$ type of carbides on the SS side of the fusion boundary. This was also found by Karthik et al. [5] in a 9Cr-1Mo/2.25Cr-1Mo weld couple. The mechanical properties such as yield strength (YS), ultimate tensile strength (UTS) and uniform strain (ε_U) measured by Karthik et al. [5] are shown in Figure 2.2. Figure 2.2 is only indicative of the values of these properties with an aim to understand the gradient of strength and ductility on the conductive side of the weld couple. For the correct values, users are referred to the work by Karthik et al. [5].

2.4 VARIOUS WELDING METHODS AND APPLICATIONS

2.4.1 Laser Welding

Laser welding provides a very narrow heat-affected zone due to a concentrated heat source. This results in a small region of material property gradient, provides deeper penetration and no use of filler material [6,7].

Pa´ncikiewicz et al. [6] investigated dissimilar metal joints by laser welding between a drill rod of X12CrMoS17 steel also known as AISI 430 F and a tube of X5CrNi18-10, which is known as AISI 304. An AISI 430 F grade of steel is a stainless steel with MnS inclusions due to sulphur, which also causes reduced resistance to crevice and pitting corrosion. This grade is difficult to weld due to hydrogen-induced cold cracking, low weld ductility and intergranular corrosion. The welding process involved pre-heating at 150–200°C and post-weld annealing at 790–815°C. A low value of heat input was used to reduce the Cr depletion at grain boundaries. Apart from gas pores (max of size 0.16 mm), the two HAZ were of the same size. The ICHAZ, which was heated to δ ferrite + γ (austenite) + MnS region, revealed fine and medium grains with more carbides than the base metal. Accelerated cooling in the two phase fields in CGHAZ resulted in austenite transformed to martensite and δ ferrite. The resulting weld had increased ductile to brittle transition temperature and decreased tensile strength due to high δ ferrite and MnS inclusions.

Costa et al. [8], have investigated the dissimilar joint between hard metal K10 (WC 94% + CO 6%), K40 ((WC 88% + CO 12%), and steel of grade EN 10083 which has 0.254%C, 0.328%Cr, 0.168%Si, 1.231%Mn, 0.03–0.035% Al and Ti and P, V, B and S in lower amounts. The laser welding was performed using a CO_2 laser and (neodymium-doped yttrium aluminum garnet) Nd:YAG. The high melting point of hard metal resulted in a fusion zone by melting of steel only. The dendritic weld zone had an eutectic phase of austenite and complex carbides of Fe, W and Co. Again, in this case, similar to Cr diffusion and carbon diffusion in P92/SS weld couple, W and Co diffusion was observed from the hard metal.

2.4.2 Electron Beam Welding

The kinetic energy of electrons is used as a heating source in electron beam welding (EBW), which is generated by heating a cathode to its thermionic emission temperature. These electrons are accelerated through an electric field, under vacuum and strikes the material at 70% of speed of light. Approximately 95% of the kinetic energy is then converted into heat energy required for high precision welding of dissimilar metal welds. High depth to the width ratio can be obtained through EBW and it makes EBW suitable for deep penetration welding of thick plates. Materials with higher thermal conductivity can also be welded through EBW. For example, welding of Ti and Cu alloys can be done by focusing a beam on the Cu alloy to avoid melting the Ti and increase heat dissipation, as Cu has higher conductivity.

Because of the difference in melting temperature and physical properties, residual stresses can remain in other welding techniques. This can be avoided through EBW. The lower total heat input per unit length in EBW can substantially reduce the residual stresses as compared to arc welding.

Additionally, the small weld bead size minimizes the mixing of metals. The material combinations generally used to weld through EBW in steam turbines are 1/4Cr-1Mo-Mn steel and AISI422/Inconel 600/satellite alloy used in steam turbine blades.

Electron beam welding usually can be of high vacuum, medium and no vacuum environment. Reactive and refractory metals such as Ti or Ti-alloys are welded in high vacuum through EBW to avoid contamination [9]. The penetration capacity also decreases with the decreasing vacuum level. The quality of weld joints depends on accelerating current voltage, focusing lens current, welding speed and vacuum levels. Various dissimilar metal combinations are welded using the EBW technique. For example, 2.25Cr-1Mo steel with C-Mn steel for steam turbine diaphragms [10], tri-metallic joint of SS, Alloy800 and Cr-Mo steel as well as P91 and SS for fast breeder test reactor (Abburi Venkata et al. [11]), AISI 430 and AISI2205 [12] and Fe-Al alloy and plain C steel couple [13]. Abbduri Venkata et al. [11] investigated the residual strain after EBW between P91 and SS materials and found that the SS side was predominantly under tension and the P91 side was under compressive residual stress. The measurements were performed by lattice spacing in the stress free sample using neutron diffraction method.

2.4.3 Solid State Welding

Solid state welding processes have found their place since early 1990s and there has been a remarkable increase in research articles in this area. Additionally, friction welding has found its space in commercial application. The major advantage of this process is that a wide range of different materials can be joined together and as there is no melting, the inhomogeneities arising in microstructures are different. Friction welding of rotary and linear friction, friction stir welding, diffusion bonding, vapor foil actuator welding (VFAW), impact welding and explosive welding are some of the examples of solid state welding processes.

2.4.4 Friction Welding

Without going into the details of process parameters, for which the readers are referred to the review by Cai et al. [14], the different zones that form due to high strain rates and heat generated due to friction in friction welding are base metal or parent metal (BM or PM), HAZ, and thermos-mechanically affected zone (TMAZ) and weld zone. A major amount of studies in literature belongs to Ti-alloys and the texture related influences of linear friction welding (LFW) on similar joints of Ti-alloy and continuous drive friction welding (CDFW) of Al6082 and Cu [15,16]. In similar metal joint studies by [17,18], and [19] apart from [16], the welded part had comparable or more strength than the base metal. However, it is important to understand that for structural integrity of the welded joint even overmatching properties could be dangerous because it acts as a material inhomogeneity and can cause the crack to propagate faster in some cases [20]. [21], in his work on FSW of 1100 Al alloy and 6061-T6 discussed and showed using transmission electron microscopic images that the friction welding with rotation resulted in very fine dynamically recrystallized (DRX) grains in the weld zone. Benavides et al. [22,23] investigated the effect of temperature by performing FSW at –100°C and at room temperature (RT). The grain size at –100°C of the WZ was close to the steady state grain size of dynamic recrystallization, which was approximately 0.7 microns, whereas for the RT FSW it was 1.5 microns.

The microhardness measured from left to right when the pin rotated clockwise is shown schematically in Figure 2.3 for both –100°C and for RT. The hardness drops at the weld zone boundaries. The decline in hardness was attributed to the softening due to recrystallization. The stir affected zone (SAZ), also known as TMAZ, is also shown in Figure 2.3. The detailed profile can be found in the review by Murr [24]. A contrasting example of FSW where the welde zone shows more hardness is the 62Be-38Al composite, also known as AlBeMet AM16 [25]. In this material, the DRX results in a fine-grained heavily deformed structure that has a hardening higher than the PM by a factor of 3. The same is schematically shown in Figure 2.4. The details of the hardness values can be found from the review of Murr [24]. This hardening is because of the mechanical alloying of Be and Al, which have a melting point of 1080°C, unlike the Al matrix.

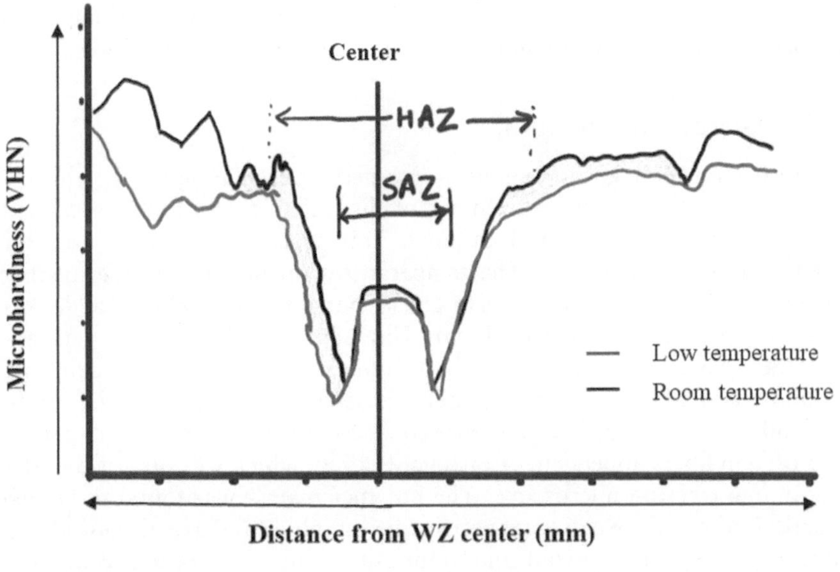

Figure 2.3 Schematic representation of hardness profile along an FSW plate of 1100 Al.

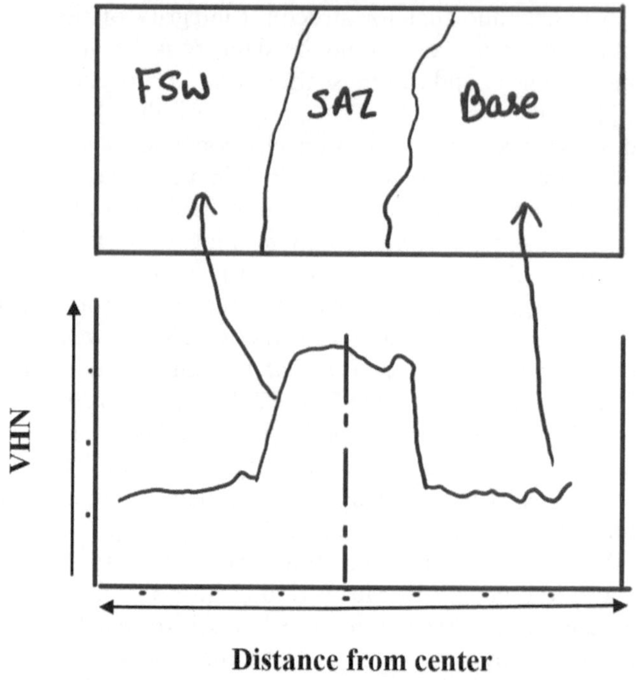

Figure 2.4 Schematic representation of hardness profile along the weld zone in an AlBeMet two-phase material.

Sometimes welding parameters such as rotational speed of pin and temperatures can introduce turbulence within the WZ, which results in a vortices structure also commonly known as "onion ring" structures.

2.4.5 Ultrasonic Welding

In ultrasonic welding, the joining is obtained by oscillating shear, which is generated by the relative movement of dissimilar pieces that are clamed together between a horn and an anvil. This process is generally used for plastics and soft materials. The temperatures do not exceed the melting temperature and so the problems of phase changes and metallurgical defects due to re-solidifications are absent. Ultrasonic welding, due to its advantages, has been used extensively in lithium ion battery manufacturing. Al-alloys and CFRPs have been joined and studied by Wagner et al. [26]. In his study, it was revealed that intense connection between metallic materials and carbon fibers appears in ultrasonic welding, which was observed under a scanning electron microscope. The polymer matrix was found to be displaced out of the welding zone. Krüger et al. [27] have discussed the ultrasonic welding of metal and composite joints. It was found in their study that the E-glass/PolyAmide composites could be welded successfully using ultrasonic welding to achieve a strength of 23 MPa without any cross-sectional damage.

2.4.6 Impact or Explosive Welding

This is a very special type of welding where two pieces of metals undergo high-velocity collision by impact. The surface contamination such as oxides and oils are scoured off and the fresh metal surface makes metallic bonds and join together. This was first observed when a piece of shrapnel was found to be welded to an armor plate of tank. The process was first patented in 1964. Explosive welding needs a large space and explosives to create such high pressure and velocity and its welds from one side to another; however, small scale impact welding can be performed by electromagnetic forces, vaporizing foils, gas guns and laser-induced shocks [28–30].

The microstructure of this type of welds are wavy in all types of impact techniques. The size of waves is easily visible with the naked eye in the case of explosion welding as the thickness of the plates are large. On the other hand, the waves that form in magnetic pulse welding and vaporizing foil actuator welding are 10 times and in the case of laser-induced shocks are 50 times smaller. This waviness helps in enhanced crack resistance of the material. Any intermetallic, if available, spread over this interface discontinuously which also helps in higher crack resistance in comparison to cases where a continuous layer or band of such phases forms.

Grain refinement due to dynamic recrystallization is observed due to adiabatic shear and due to sub-grain formations by dislocations. The time for impact welding is found to be smaller so that grains of 100 nm can only form. Such zones of nanoscale grain structures have been reported by Fan et al. [31]. Nano-indentation methods and other micro-mechanical tests have revealed a drop in hardness [32] at the weld interface and an absence of HAZ [33].

2.4.7 Diffusion Bonding

This type of joining is obtained by atomic level bonding of two materials due to high pressure, temperature and plastic deformation. Usually under an applied force, the interfacial voids decrease due to plastic deformation. This is followed by a rise in pressure and temperature, which causes large creep strains and grain boundary migration. The contact area also increases due to expansion and materials diffuse across the interface and a welding zone forms where two different materials join at atomic level by eliminating pores which are formed due to high temperature and expansion.

Li et al. [34] have investigated the effect of creating a large amount of grain boundaries on the surface by high energy shot peening (HESP) to create surface nano-crystallization (SN) and have compared their properties by welding SN Ti-alloy TC17 with TC4 and coarse-grained TC17 with TC4. He and Liu [35] investigated the intermetallics which form at the interfaces in this type of welding and which can act as sites for crack initiation. They found by conducting experiments by welding Ti-alloy TC4 and Ni that the intermetallics that form are $TiNi_3$, TiNi, Ti_2Ni in sequence. They also experimented with TC4 and 00Cr18Ni9Ti alloy weld and found TiC, Ti_3Al+FeAl+$FeAl_2$ and TiAl to form in sequence. The mechanism proposed was the flux-energy principle, i.e., if there is a flux along an interface with multicomposition diffusion couple, the phase with maximum driving force forms first based on the diffusion flux ratio at the interface. Two or more intermetallics with similar flux-energy may nucleate and grow simultaneously like Ti_3Al +FeAl+$FeAl_2$.

2.5 STRUCTURAL INTEGRITY OF DMW COMPONENTS

In general, the mechanical properties of DMWs are measured by cutting the tensile and fracture samples along with creep samples if the operational temperatures are high. The yield strength, tensile strengths and creep parameters etc. are obtained by cutting miniature tensile and creep samples from WZ, HAZ and base metals. Lindqvist et al. [36] in their study on Ferritic 18MND5 steel welded with AISI 316 L steel with Ni-based Alloy 52 as weld metal without buttering layer, examined the mechanical properties by miniature tensile specimens of 1 mm thickness and 2 mm Width. HAZ cracks were found to be deviating to a region of lower strength and/or hardness by Wang et al. [37] and Sarikka et al. [38].

In the presence of a crack, the DMW joints have shown problems of cracks deflecting at the interfaces. Lindqvist et al. [36] found that the HAZ crack deviates to a fusion boundary with a vertical change in crack growth plane. This transition can be gradual or instantaneous. Cracks starting from the weld zone do not deviate to the fusion boundary, even if the distance between them is less than 40 µm.

An experimental and numerical study has been performed by Štefane et al. [39], where similar metals were welded with two different filler materials, resulting in an overmatched (OM) and an undermatched (UM) region apart from BM and HAZ regions. The overmatching and undermathcing is described by yield strength of the region being larger or smaller than the base metal, respectively.

Upadhyaya and Tiwari [40] have numerically modelled the OM and UM interfaces and calculated the crack driving forces using the configurational force concept and found that the J vector deviates from the expected crack growth direction and crack deflects.

2.6 SUMMARY AND CONCLUSIONS

With the rise in material property demands, it has become inevitable for a few applications to avoid joining of differently deforming materials. The dissimilar metal joints by welding the two materials or in some cases more than two materials together have been discussed in this chapter. With inert gas conventional welding methods and electron beam, laser welding as well as diffusion and friction welding, it becomes evident that the microstructure locally changes either in WZ, HAZ or in plastically deformed and dynamically recrystallization zones. In the case of impact welding, a wavy interface and in the case of diffusion welding, a layer of continuous or discontinuous intermetallics decorate the interface between two materials.

It is unavoidable to have a continuity between two differently deforming materials and therefore at least one interface exists that can influence the structural property of the component. Based on the numerical analysis using configurational force approaches as performed by Upadhyaya and Tiwari [40], it is possible to design the component in a way where the crack resistance is maximum.

REFERENCES

[1] Abe, F., Tabuchi, M. and Tsukamoto, S., 2021. Alloy design of MARBN for boiler and turbine applications at 650 °C. *Materials at High Temperatures*, 38(5), pp. 306–321.

[2] Dak, G. and Pandey, C., 2020. A critical review on dissimilar welds joint between martensitic and austenitic steel for power plant application. *Journal of Manufacturing Processes*, 58, pp. 377–406.

[3] Dak, G. and Pandey, C., 2021. Experimental investigation on microstructure, mechanical properties, and residual stresses of dissimilar welded joint of martensitic P92 and AISI 304L austenitic stainless steel. *International Journal of Pressure Vessels and Piping*, 194, p.104536.

[4] Patel, N. P., Badheka, V. J., Vora, J. J. and Upadhyay, G. H., 2019. Effect of oxide fluxes in activated TIG welding of stainless steel 316LN to low activation ferritic/martensitic steel (LAFM) dissimilar combination. *Transactions of the Indian Institute of Metals*, 72(10), pp. 2753–2761.

[5] Karthik, V., Laha, K., Kasiviswanathan, K. V. and Raj, B., 2002. Determination of mechanical property gradients in heat-affected zones of ferritic steel weldments by shear-punch tests. In *Small Specimen Test Techniques: Fourth Volume*. ASTM International.

[6] Pańcikiewicz, K., Świerczyńska, A., Hućko, P. and Tumidajewicz, M., 2020. Laser dissimilar welding of AISI 430F and AISI 304 stainless steels. *Materials*, 13(20), p.4540.

[7] Xu, W. F., Jun, M. A., Luo, Y. X. and Fang, Y. X., 2020. Microstructure and high-temperature mechanical properties of laser beam welded TC4/TA15 dissimilar titanium alloy joints. *Transactions of Nonferrous Metals Society of China*, 30(1), pp. 160–170.

[8] Costa, A. P., Quintino, L. and Greitmann, M., 2003. Laser beam welding hard metals to steel. *Journal of Materials Processing Technology*, 141(2), pp. 163–173.
[9] Sun, Z. and Karppi, R., 1996. The application of electron beam welding for the joining of dissimilar metals: An overview. *Journal of Materials Processing Technology*, 59(3), pp. 257–267.
[10] Akutsu, Y., 1980, May. Application of electron beam welding to steam turbine diaphragms, beam technoi, Lectures of the Int. In *Beam Technology Conference, Essen, Germany* (pp. 63–69).
[11] Venkata, K. A., Truman, C. E., Coules, H. E. and Warren, A. D., 2017. Applying electron backscattering diffraction to macroscopic residual stress characterisation in a dissimilar weld. *Journal of Materials Processing Technology*, 241, pp. 54–63.
[12] Madhusudan Reddy, G. and Srinivasa Rao, K., 2009. Microstructure and mechanical properties of similar and dissimilar stainless steel electron beam and friction welds. *The International Journal of Advanced Manufacturing Technology*, 45(9), pp. 875–888.
[13] Dinda, S. K., Sk, M. B., Roy, G. G. and Srirangam, P., 2016. Microstructure and mechanical properties of electron beam welded dissimilar steel to Fe–Al alloy joints. *Materials Science and Engineering: A*, 677, pp. 182–192.
[14] Cai, W., Daehn, G., Vivek, A., Li, J., Khan, H., Mishra, R. S. and Komarasamy, M., 2019. A state-of-the-art review on solid-state metal joining. *Journal of Manufacturing Science and Engineering*, 141(3), pp. 031012-1–031012-35.
[15] Karadge, M., Preuss, M., Lovell, C., Withers, P. J. and Bray, S., 2007. Texture development in Ti–6Al–4V linear friction welds. *Materials Science and Engineering: A*, 459(1-2), pp. 182–191.
[16] Uday, M. B., Ahmad Fauzi, M. N., Zuhailawati, H. and Ismail, A. B., 2010. Advances in friction welding process: A review. *Science and Technology of Welding and Joining*, 15(7), pp. 534–558.
[17] Damodaram, R., Raman, S. G. S. and Rao, K. P., 2013. Microstructure and mechanical properties of friction welded alloy 718. *Materials Science and Engineering: A*, 560, pp. 781–786.
[18] Preuss, M., Da Fonseca, J. Q., Steuwer, A., Wang, L., Withers, P. J. and Bray, S., 2004. Residual stresses in linear friction welded IMI550. *Journal of Neutron Research*, 12(1-3), pp. 165–173.
[19] Ma, T. J., Li, W. Y., Xu, Q. Z., Zhang, Y., Li, J. L., Yang, S. Q. and Liao, H. L., 2007. Microstructure evolution and mechanical properties of linear friction welded 45 steel joint. *Advanced Engineering Materials*, 9(8), pp. 703–707.
[20] Tiwari, A., Wiener, J., Arbeiter, F., Pinter, G. and Kolednik, O., 2020. Application of the material inhomogeneity effect for the improvement of fracture toughness of a brittle polymer. *Engineering Fracture Mechanics*, 224, p.106776.
[21] Murr, L. E., Liu, G. and McClure, J. C., 1998. A TEM study of precipitation and related microstructures in friction-stir-welded 6061 aluminium. *Journal of Materials Science*, 33(5), pp. 1243–1251.
[22] Benavides, S., Li, Y. and Murr, L. E., 2000. Ultrafine grain structure in the friction-stir welding of aluminium alloy 2024 at low temperatures. Proceedings of Ultrafined Grained Materials, TMS, pp. 155–168.

[23] Benavides, S., Li, Y., Murr, L., Brown, D. and McClure, J., 1999. Low-temperature friction-stir welding of 2024 aluminum. *Scripta Materialia*, 41(8), pp. 809–815.
[24] Murr, L. E., 2010. A review of FSW research on dissimilar metal and alloy systems. *Journal of Materials Engineering and Performance*, 19(8), pp. 1071–1089.
[25] Contreras, F., Trillo, E. A. and Murr, L. E., 2002. Friction-stir welding of a beryllium-aluminum powder metallurgy alloy. *Journal of Materials Science*, 37(1), pp. 89–99.
[26] Wagner, G., Balle, F. and Eifler, D., 2013. Ultrasonic welding of aluminum alloys to fiber reinforced polymers. *Advanced Engineering Materials*, 15(9), pp. 792–803.
[27] Krüger, S., Wagner, G. and Eifler, D., 2004. Ultrasonic welding of metal/composite joints. *Advanced Engineering Materials*, 6(3), pp. 157–159.
[28] Daehn, G. S. and Lippold, J. C., 2011 Ohio State University. Low-temperature laser spot impact welding driven without contact. U.S. Patent 8,084,710.
[29] Ngaile, G., Lohr, P., Lowrie, J. and Modlin, R., 2014. Development of chemically produced hydrogen energy-based impact bonding process for dissimilar metals. *Journal of Manufacturing Processes*, 16(4), pp. 518–526.
[30] Vivek, A., Hansen, S. R., Liu, B. C. and Daehn, G. S., 2013. Vaporizing foil actuator: A tool for collision welding. *Journal of Materials Processing Technology*, 213(12), pp. 2304–2311.
[31] Fan, Z., Yu, H. and Li, C., 2016. Interface and grain-boundary amorphization in the Al/Fe bimetallic system during pulsed-magnetic-driven impact. *Scripta Materialia*, 110, pp. 14–18.
[32] Hansen, S. R., Vivek, A. and Daehn, G. S., 2015. Impact welding of aluminum alloys 6061 and 5052 by vaporizing foil actuators: Heat-affected zone size and peel strength. *Journal of Manufacturing Science and Engineering*, 137(5), pp. 051013-1–051013-6.
[33] Vargo, A. and Prothe, C., 2008. Comparative tensile strength and shear strength of detaclad explosion clad products. *International Application of Magnetic Pulse Welding for Aluminum Alloys and SPCC Steel Sheet Joints*, pp. 119–124.
[34] Li, L., Sun, L. and Li, M., 2022. Diffusion bonding of dissimilar titanium alloys via surface nanocrystallization treatment. *Journal of Materials Research and Technology*, 17, pp. 1274–1288.
[35] He, P. and Liu, D., 2006. Mechanism of forming interfacial intermetallic compounds at interface for solid state diffusion bonding of dissimilar materials. *Materials Science and Engineering: A*, 437(2), pp. 430–435.
[36] Lindqvist, S., 2016. Tearing resistance of heterogeneous interface region of a dissimilar metal weld characterised with sub-sized single edge bend specimens. *Procedia Structural Integrity*, 2, pp. 1031–1038.
[37] Wang, G., Wang, H., Xuan, F., Tu, S. and Liu, C., 2013. Local fracture properties and dissimilar weld integrity in nuclear power plants. *Frontiers of Mechanical Engineering*, 8(3), pp. 283–290.
[38] Sarikka, T., Ahonen, M., Mouginot, R., Nevasmaa, P., Karjalainen-Roikonen, P., Ehrnstén, U. and Hänninen, H., 2016. Microstructural, mechanical, and fracture mechanical characterization of SA 508-alloy 182

dissimilar metal weld in view of mismatch state. *International Journal of Pressure Vessels and Piping*, 145, pp. 13–22.
[39] Štefane, P., Naib, S., Hertelé, S., De Waele, W. and Gubeljak, N., 2019. Crack tip constraint analysis in welded joints with pronounced strength and toughness heterogeneity. *Theoretical and Applied Fracture Mechanics*, 103, p.102293.
[40] Upadhyaya, R. and Tiwari, A., 2023. Effect of material inhomogeneity and crack driving force for the case of OM-OM and UM-OM interface. Materials at High Temperature Special Edition. Under Review.

Chapter 3

Laser-Based Manufacturing Processes

A Comprehensive Review

Kamal Kishore[1], *Pankaj Sharma*[1], *Archana*[2], *and Manoj Kumar Sinha*[3]

[1]National Institute of Technology Hamirpur, H.P., India
[2]Jindal Global Business School Sonipat, O. P. Jindal Global University Sonipat, Haryana, India
[3]National Institute of Technology Kurukshetra, Haryana, India

CONTENTS

3.1 INTRODUCTION

Light is always a fascinating subject to humankind from the beginning of the modern age. It fulfils the purpose of sighting and is used for various purposes such as in decorations, medical and many other research applications. Light is an electromagnetic radiation having both particle and wave nature. Some of the phenomena shown by the particle and the wave

DOI: 10.1201/9781003436072-3

Table 3.1 Difference Between Particle and Wave Nature of Light

Phenomenon	*Wave nature*	*Particle nature*
Reflection	Yes	Yes
Refraction	Yes	Yes
Interference	Yes	No
Diffraction	Yes	No
Polarisation	Yes	No
Photoelectric effect	No	Yes

nature of the light are listed in Table 3.1. The properties of these two natures of light are the core concept for various scientific applications. Laser is one of the most widely used technologies based on the properties and phenomenon of light. The word "laser" stands for light amplification by the stimulated emission of radiation. Essentially, it is a device that simulates an atom or molecule emitting light at a specific wavelength and magnifies that light in order to produce a narrow beam that falls in visible regions of the electromagnetic spectrum [1,2]. A brief discussion on the properties and operation mechanism of the laser are presented in the upcoming section.

3.1.1 A Brief Discussion on Properties and Operation Mechanisms of a Laser

The light coming out from a laser has unique properties, such as monochromaticity, coherence, collimation and sharp focus. These unique properties of lasers are due to their operating mechanism. The laser operates on the principle of population inversion, stimulated emission and amplification. The population inversion is the condition in which more electrons are found to be in the excited state or high energy level state than in the ground or stable state. It is a non-equilibrium state achieved by pumping methods, namely optical and electrical pumping. The population inversion is a necessary condition for stimulated emission. Stimulated emission is a process wherein photons are used to excite the atoms to fall at a lower energy level so that highly coherent and collimated beams can be obtained. As the phase and polarisation state of the simulated photons are identical, this phenomenon results in an increase in amplitude and the process is called amplification [2].

3.1.2 Effect of Laser Beam on the Material

Laser beams are always considered as supporting tools in the field of machining. It is merely considered a heat source that accelerates the machining operation. When a laser beam falls on the material surface, there are always three possibilities: (i) transmission, (ii) reflection and (iii) absorption. The interaction of the laser with the material alters the mechanical, physical and chemical characteristics of the materials. The material undergoes physio-chemical modifications as a result of laser light absorption. The absorption of laser beams induced thermal effects like heating, vaporisation, ablation and melting at the surface or inside the surface of the material. Among these, reflection and transmission don't have any effect on the properties of material, whereas reflection and transmission have an influence on a material's characteristics. The changes that occur due to the absorption of laser beams are as follows [3]:

I. *Mechanical effect*: When the intensity of the laser beam is moderate, and scanning velocity is high, the temperature generation at the interaction point is generally below the melting point of the material. This temperature generation leads to thermal softening and induces thermal stress in the material. As a result, local yielding, cracks, buckling and loss of stiffness may induce in the workpiece.
II. *Phase change effect*: When the power intensity of the laser is high enough, it is possible that at the point of interaction, the temperature may rise above the melting point of the workpiece, which results in phase change and evaporation of the material.
III. *Physio-chemical effect*: As a laser beam has high energy intensity associated with it, there is the possibility that physio-chemical interaction may take place between the workpiece and the assisting materials such as gases. These liquids are used in laser machining. This may result in alteration of chemical composition and physical properties of the workpiece due to alloying, doping, burning and sintering.

3.2 LASER BEAM MACHINING

In the 21st century, the application of lasers in machining grows rapidly with the development of advanced and sophisticated technologies. This requirement is more or less driven by the demand to improve the conventional machining efficiency and perform machining at a more accurate level. The use of a direct laser beam for machining is referred to as laser machining (LM), whereas laser-assisted machining (LAM) refers to the use of a laser beam in conjunction with traditional machining methods. Figure 3.1 indicates the different types of LM/LAM. In LM, laser beams are directly used to melt, burn and vaporise the material that needs to be removed. For material removal mechanisms, laser beams are focused at a point on the workpiece. During beam-material interaction, absorption of beam energy occurs in the vicinity of the workpiece and high heat is generated at the local level. The generation of high heat at localise

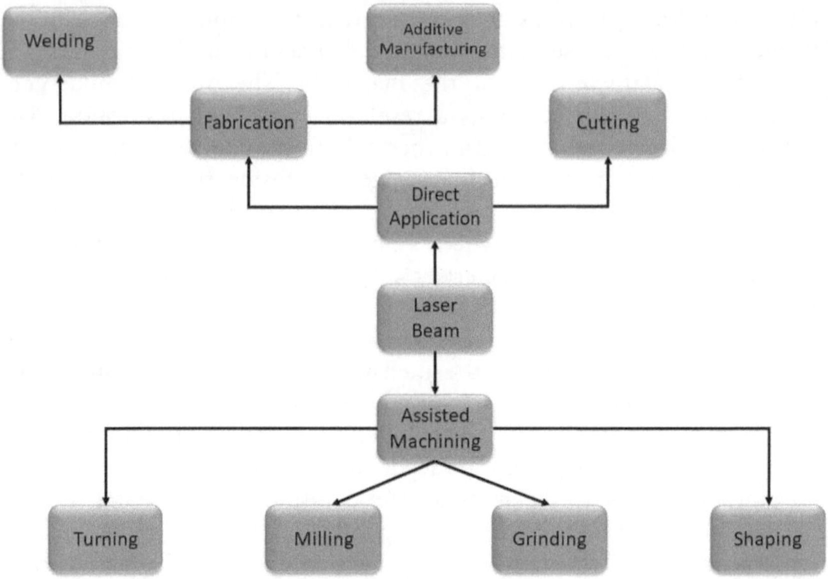

Figure 3.1 Laser beam application in different machining.

level results in softening, local yielding, melting, burning and vaporising of material [4]. The primary benefits of LM are the lack of tool chatter, tool damage, tool wear, machine deflection and any mechanically induced negative influence on the workpiece. Therefore, LM is widely used in almost every sector, starting from manufacturing, marine, defence and biomedical. Apart from that, laser beams are also used with conventional machining to increase the material removal rate. In LAM, laser beams are used to preheat the workpiece even before actual machining starts. Therefore LAM provides an alternative to conventional machining used to machine any materials such as ceramics, composites, glass and metals [5]. A thorough literature assessment was conducted by choosing reputed articles from databases such as Scopus, Web of Science and Science Direct to show the current scenario of laser machining. The literature from 2000 to 2022 has been chosen for this. From 2000 to 2022, the number of publications published on laser machining or LAM is shown in Figure 3.2.

The graph indicates a rising tendency from 2000 to 2022, with a decline in 2022 as more publications are added to the Scopus database over time. Further bibliographic research is carried out to identify terms that are commonly associated with laser machining; 1,448 recent publications titles and abstracts were extracted in format from the Scopus database for this purpose. The graph is plotted to find the status of various countries on the advancement of laser machining, as shown in Figure 3.3, and also a comma-separated value file was evaluated with

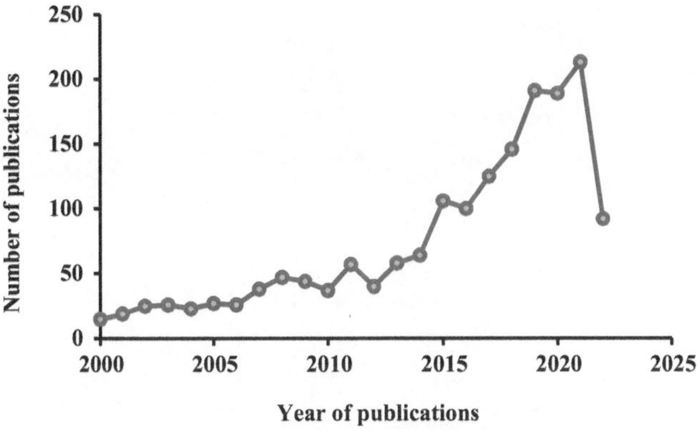

Figure 3.2 The graph represents the number of articles published from 2000 to 2022.

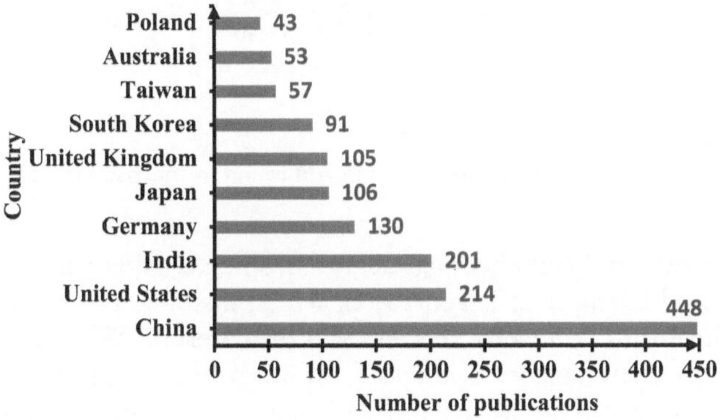

Figure 3.3 Countywide publication of an article on laser machining and related terms.

VOS view software to see whether there was any correlation between these terms. Figure 3.4 depicts the software-generated mapped picture. The size of the circles and the visibility of the text are related to the number of words found in the papers. The links between the terms indicate how often the terms appeared together.

3.2.1 Fundamentals of Laser Beam Machining

Laser beam machining (LBM) is a thermal process in which the thermal properties of the workpiece play a significant role in material removal effectiveness rather than the mechanical and chemical properties of the workpiece. Therefore, a material with high brittleness or hardness can be easily machined using the LBM technique if it has low thermal diffusivity and high thermal conductivity. Generally, the LBM process includes three stages by which material is removed from the workpiece: (i) melting,

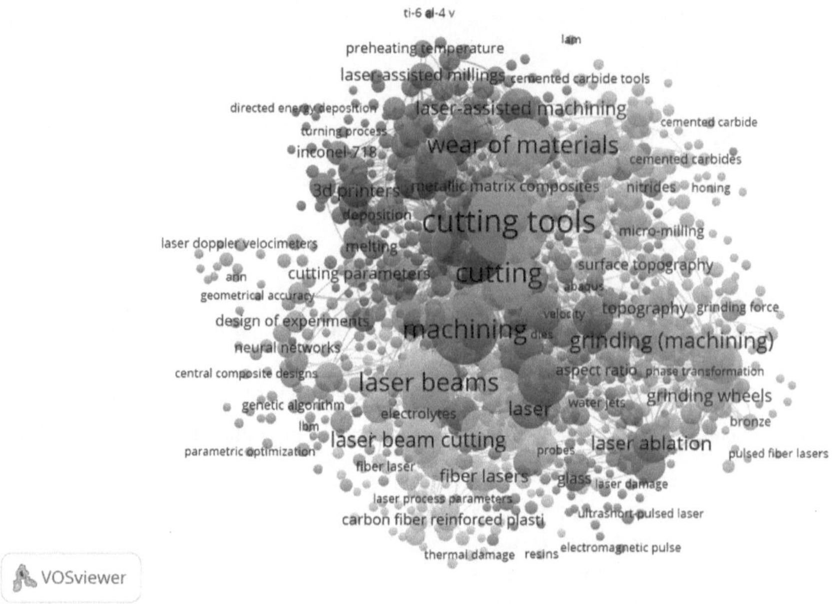

Figure 3.4 Different keywords related to LM/LAM found in the last five years.

(ii) vaporisation and (iii) chemical degradation or erosion (in this, due to a highly focused beam, heat is generated, which disintegrates chemical bond between the atoms and cause the removal of material). When a laser beam with a high energy density is focused on a workpiece, the workpiece absorbs the thermal energy associated with the heat; due to this, the material either melts, vaporises or chemically degrades from the surface and is taken away by the flowing gases [3,6].

3.3 TYPES OF LAM

LAM is a new technological advancement that helps to improve the machinability of various materials by integrating laser heating with conventional processes. In this method, work material is locally heated and softened by a laser beam in front of an advancing cutting tool. This helps to increase the temperature in the shearing zone and allows the ductile plastic deformation during machining by reducing the work hardening properties, flow stress and hardness of the work material [7]. The laser beam is focused in the machining zone in such a way that uniform temperature distribution is achieved without heating the cutting tool material. This integrated position of a laser beam depends on the machining process [8]. The different types of LAM are discussed next.

3.3.1 Laser-Assisted Grinding

Grinding is a machining operation in which abrasives are used to remove the material in the form of ultra-fine scratches. This inheritance nature of grinding is due to the cutting tool's high negative rake angle, i.e., grinding wheel. As a result, the rate of material removal during the grinding process is relatively low, and the specific grinding energy is excessively high compared to other traditional machining techniques [9]. To overcome this barrier, many researchers and scientists try to amalgamate the principle of other techniques such as vibration and lasers to increase the material removal rate of the grinding process [10]. A schematic diagram of laser-assisted grinding is shown in Figure 3.5.

Laser-assisted grinding (LAG) is one of those techniques that utilise the benefit of laser to increase the material removal rate in the grinding process. In the 1970s, the laser was first used in machining operations as a heat source [11]. LAG is a hybrid technique in which the workpiece is locally preheated before grinding occurs. This phenomenon makes the material softer, and a ductile regime can easily be achieved even in a brittle material. Generally, CO_2 laser, Excimer and Nd: YAG laser are widely used in LAG. Ma et al. [12] studied the effect of LAG on the surface integrity of Zirconia ceramic. Parameters like surface quality, morphology and subsurface damages were investigated. The obtained results were compared with the findings of

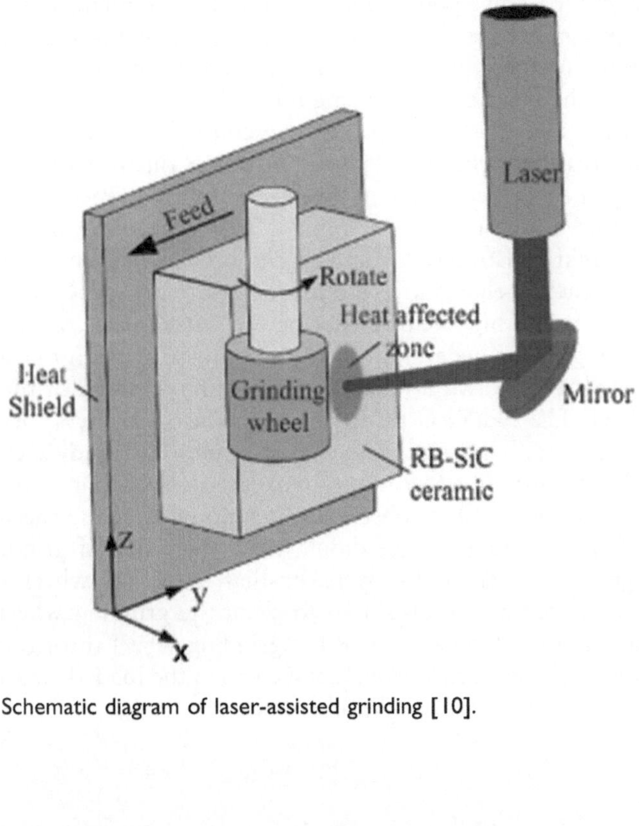

Figure 3.5 Schematic diagram of laser-assisted grinding [10].

conventional grinding. The outcome of the study suggested that ductile regime grinding was easily achieved in the case of LAG even in a high depth of cut, which leads to better surface integrity of the ground surfaces compared to conventional grinding. Likewise, Azarhoushang et al. [13] performed the LAG on silicon nitride (Si_3N_4) ceramic to overcome the poor surface integrity and limited material removal rate as obtained during the conventional grinding technique. To improve the material removal rate in the grinding process, the ultra-short pulsed laser was employed for ablation and to regulate thermal heating. These findings revealed a significant reduction in normal and tangential grinding forces. Moreover, significant improvement was seen in the surface integrity. Apart from that, lasers are also widely used in the case of micro grinding. Chang et al. [14] performed the micro-LAG on silicon nitride and aluminum oxide workpieces to make a groove on their surface. A diode laser with 800 nm wavelength was used to preheat the workpiece. The experimental analysis showed that better surface roughness and almost no subsurface damage occurred while making a deep groove using LAG compared to solo micro grinding.

Apart from the direct use of lasers for assisting the grinding process, lasers are also used to dress the grinding wheel. Dressing of a grinding wheel is a technique in which the surface topology of the worn-out wheel is modified by mechanical means so that new and sharp grits are protruded on the cutting surface of the wheel. The mechanical dressing is a process in which contact type single point, multi-point and diamond dressers are used. Although mechanical dressing is effective, it induces stress and deep cracks on the surface of the wheel. Also, the consistency of the dressing is not proper due to the wear of the dressing tool. Therefore, a non-contact type technique with a laser is developed to overcome the disadvantages of the mechanical dressing technique. In laser dressing, the surface topology of the grinding wheel is locally heated by a laser beam that melts the material, and after some time, this melt re-solidify. This rapid heating and re-solidification of material induced micro-cracks on the surface of the wheel. These micro-cracks help remove the loaded and work out grits when the wheel is subjected to high velocity during the initial phase of the grinding process, leading to the generation of fresh and new grits on the surface of the wheel [15]. Hosokawa et al. [16] studied the dressing of the bronze-bonded wheel. The Nd:YAG pulsed laser was used to dress the grinding wheel through rapid heating of the laser beam. The study's conclusion suggested that the grinding force was almost the same as that of an ordinary dressed grinding wheel. This observation explored that the thermal deterioration of the grinding surface didn't affect the value of grinding force even though high laser beams were irradiated on the wheel's surface. Moreover, laser beams are also used to clean the grinding wheel by specifically targeting the loaded area of the grinding wheel through the laser beam. Due to high heat generation, metal chips in the loaded area vaporise, which automatically cleans the wheel surface.

3.3.2 Laser Cutting

Laser cutting is a fabrication process in which focused, highly intense laser beams are used to cut the materials. The CNC laser cutting provides a wide range of cutting operations, from simpler ones to custom designs. The shape and design of the product obtained from the laser cutting can vary from simpler to highly complex. Laser cutting can be used for almost every material, such as wood, plastics, metals, stone, etc. The product's superiority, such as precession, accuracy and high quality with minimal material contamination and physical damage, is obtained compared to conventional mechanical cutting. The parameters such as surface roughness, dimensional accuracy, oxide generation, kerf width burr absence or presence etc., are used to evaluate the quality of the laser cutting. Figure 3.6 represents the schematic of laser cutting and the mechanism of material removal in laser cutting. Figure 3.7 shows the correlation between thermal damage and different performance parameters. Biffi et al. [17] reviewed the laser cutting performance parameter in terms of technological, function and metallurgical aspects for NiTi shape memory alloys. They concluded that the parameters like pulse duration and emission wave of laser beam significantly affect the heat conduction and quality of the cut edges as NiTi shape memory alloys are susceptible to temperature. Wang et al. [12] used Inconel 625 to conduct a laser cutting experiment to investigate the impact of laser parameters such as cutting speed, focal length and laser power on cut quality. The RSM (response surface method) approach was used to create a nonlinear regression model and perform an analysis of variance in the data. The result of the experiment suggested that better surface roughness was achieved when power was 1,100 W and the focal length was 4 mm. Based

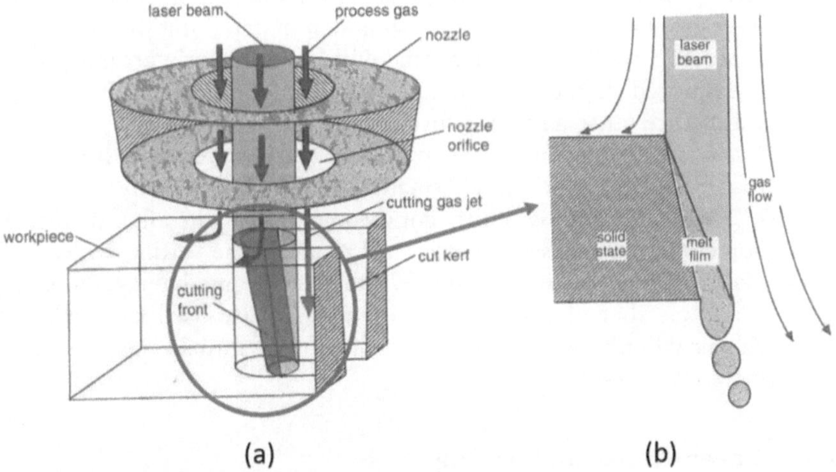

Figure 3.6 A pictorial representation of (a) laser cutting process and (b) material removal mechanism [17].

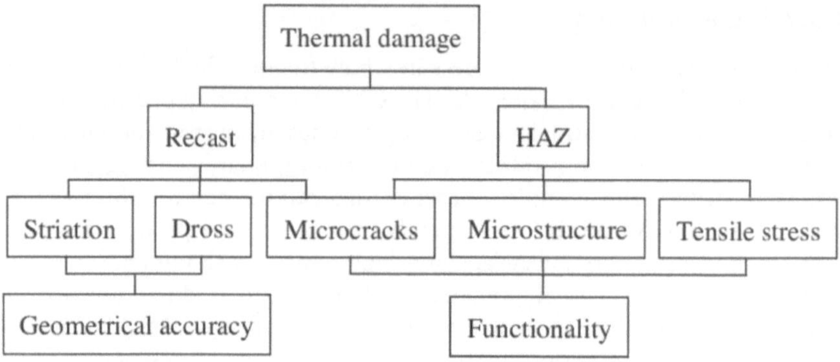

Figure 3.7 Relation between thermal damage and performance of a component in laser cutting.

on the requirement, specification and complexity, laser cutting can be classified into oxidation, fusion, vaporisation and cold cutting.

I. *Oxidation cutting*: As the name suggests, the cutting of material is performed in the presence of oxygen. In this cutting, when the focused and high-intensity laser beams fall on the workpiece surface, the material gets burned and vaporises as soon as the ignition temperature of the material is achieved. An exothermic reaction between the oxygen and high-temperature metals has occurred, further releasing the heat. This exothermic reaction supports the cutting process.
II. *Fusion Cutting*: Inert gases like nitrogen, argon, and helium are commonly utilised to aid the laser cutting process. In this cutting, the laser power is exclusively responsible for the cutting process. Only inert gases are employed to blow molten material out of the cutting zone. As a result, a laser machine with a significantly greater power is employed than oxidation cutting. Clean cutting or high-pressure cutting are other terms for fusion cutting.
III. *Vaporisation Cutting*: In this process, high-powered laser beams are focused on the workpiece's small spot, which evaporates the solid material directly without converting it to the liquid phase. Sometimes inert gases are also used to shield the cutting optics.
IV. *Cold Cutting*: In this technique, the laser beams having the same energy as that of bond energy of material are used to break the chemical bonds of the materials leading to the formation of powdery material as residue.

3.3.3 Laser-Assisted Turning

The method of LAM may be applied in a variety of different ways, one of which is laser-assisted turning. Since the turning tool remains in place during the process of laser-assisted turning, it is not difficult to include the laser beam in the standard turning procedure. The performance of laser-assisted turning is greatly impacted by a variety of factors, including laser locations, spot size, tool-beam distance, and incidence angle. The technique of turning with laser assistance makes use of both a single laser beam and a multi-laser beam at the same time. The single laser beam is included in the usual machining process in a variety of various arrangements, such as (a) normal to the workpiece and (b) normal to the chamfered surface [18]. The majority of work is reported when the laser beam is perpendicular to the work surface. This type of arrangement makes it easy for the laser-assisted turning operation as compared to the laser beam normal to the chamfered surface. The benefit of this setup is that it is easy to operate and does not allow heating of the work surface. However, laser-generated heat is not enough for a high depth of cut and less reduction in cutting forces in comparison to laser beam normal to the chamfered surface arrangement. This may be a disadvantage for cutting materials with high chemical reactivity with a tool. Alternatively, a laser beam incident normal to the chamfered surface can be used to produce homogeneous cutting force reduction in the X, Y and Z directions. It's a regulated beam position configuration that results in a higher rate of material removal and less influence on microstructure properties. A laser point is evenly distributed throughout the champer's surface, resulting in reduced tool wear [19]. Lei et al. [20] described work on the multi-laser unit to get a constant temperature distribution throughout the deep cut, hence extending the life of the tool. The placement of the laser beam in relation to a tool is crucial. Along with the cutting speed, the tool beam distance influences the period between laser heating and machining, and hence the temperature distribution of the cutting zone. By positioning the cutting tool closer to the laser spot during the machining of commercial titanium, hardened steel, and white cast iron, a considerable improvement is realised in the reduction of cutting forces. However, the tool should not be placed too near to the laser beam since overheating might harm it.

In laser-assisted turning, it is important to predict the required laser power and distance between the cutting tool and the laser spot for efficient machining. In this regard, many works have been reported on the development of 3-D numerical models. Arrizubieta et al. [21] developed a laser thermal model (LATHEM) to simulate the thermal field within the part and find out the correct position of the laser focus spot to determine the laser power and maximum cutting depth. The temperature field of the simulated part is shown in Figure 3.8. It is observed that the optimum distance between laser focus beam and cutting edge and required cutting depth can be predicted based on the desired temperature (450°C). Tian et al. [22] also developed a 3-D numerical model that predicts the transient temperature field with material removal rate. These models are experimentally validated by using different types of equipment, such as infrared cameras and a digital pyrometer system.

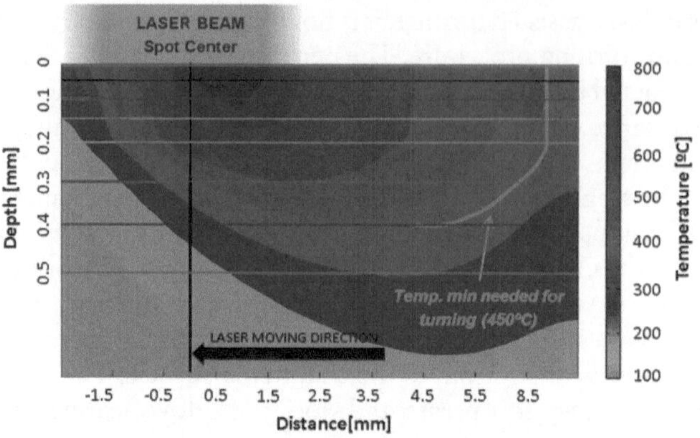

Figure 3.8 Thermal field prediction for determining laser beam position and maximum cutting depth [21].

3.3.4 Laser-Assisted Milling

Laser-assisted milling is based on the LAM principle. In laser-assisted milling, the workpiece material is preheated by a laser beam, followed by the advancement of a cutting tool towards the workpiece. It is one form of thermally assisted machining. Local heating of the workpiece using laser beam energy lowers its hardness and mechanical strength. LAM decreases cutting forces, tool wear, and machining time while enhancing surface properties. This method is ideal for the machining of difficult-to-cut materials such as superalloys, composites and ceramics. However, during laser-assisted milling, it is difficult for the laser source to continue to follow the cutting edge when the route is complicated., it is necessary to incorporate an extra control axis that accurately regulates the focus length and focus angle of the laser heat source [23]. Laser-assisted milling has mostly been used for one-direction milling of flat surfaces with a micro-end mill or face cutter. To meet industrial requirements, a 3D laser-assisted milling device has been developed to manufacture intricate parts.

Yang et al. [24] studied the operating temperature and edge chipping phenomenon during the laser-assisted milling of silicon nitrate ceramic material. A diode laser was integrated with the CNC milling operation. A laser beam was focused on a workpiece surface by optical fibre at a 60° angle. They observed that the laser source and cutter were not able to operate simultaneously for the required machining. A spindle tool system with a laser source was developed by Brecher et al. [23]. This device includes two powerful lenses with high energy with a five-axis machine tool. Wiedemann et al. [25] investigated the microstructure of titanium alloy during the laser-assisted milling machining (NCV750) integrated with the numerically controlled axis. A mirror deflected the laser heat source from the laser optics and into the cutting area. Bermingham et al. [26]

recommended two main approaches for laser-assisted milling development. The first approach required the use of a fixed-position laser and a revolving machine table with precise tool path selection. A rotary table moves to maintain a stationary laser at the front of the milling cutter in focus at all times. The second method incorporated a laser that rotated around the spindle, allowing for variations in machining direction. These techniques were used to adapt a laser-assisted milling system, especially for laser-assisted face milling [27].

3.3.5 Laser-Assisted Drilling

High-strength materials such as nickel-based superalloys, ceramics, titanium and austenitic stainless steel have been effectively employed in a broad range of applications, including aerospace, aviation and car parts. These materials, however, are difficult to machine due to brittle fracture and need the development of a novel machining method. Laser-assisted drilling is a non-contact, low-heat method for producing small holes with a smooth surface structure and a low taper. These difficult-to-cut materials have been successfully machined using a combination of laser energy and the drilling process [28]. Laser-assisted drilling is one of the new advanced technologies with different laser sources, such as Nd:YAG, CO_2, excimer and diode laser. In laser-assisted drilling, the workpiece can be directly heated in front of a cutting tool. After heating the workpiece with an intense laser beam to a softening temperature just below the melting point, the drilling operation is completed with a standard drill bit. This process was explored to reduce drilling time and energy loss and improve the quality of the finished products. The quality of the drilled hole is examined by the various parameters such as heat-affected zone, taper and internal form and its related geometric features. These output parameters are significantly affected by input parameters such as laser focus length, pulse length, laser energy and workpiece thickness. The correct selection of these parameters is critical for maximum laser drilling performance. Statistical techniques are fruitful in designing and optimising the relationship between input and output parameters [29].

Substantial studies have been conducted to investigate the laser-assisted drilling process. Laser heating and material erosion were the main focused areas during the modelling of the laser-assisted drilling process. It was also investigated that more than 90% of material is ejected in a liquid state at an intensity of 1,011 W/m^2. However, the assessment of material removal is more difficult during laser-assisted drilling and needs much more research efforts. Yilbas [30] studied the impact of laser focus setting on the resulting hole diameter. He concluded that the mean hole diameter increased by increasing laser energy and focal length above the workpiece surface. Takeno et al. [31] studied the use of laser beams to drill the electronic components. They concluded that the machining quality could be improved by using a high beam pulse of CO_2 laser. Following are the process parameters of the hole geometry [32–35]:

I. *Laser wavelength*: Shorter wavelengths of light allow the beam of laser light to be focused to a smaller spot size with the same beam quality. This allows drilling holes with a smaller diameter. However, short-wave lasers have other advantages as well. The energy binding to the workpiece is generally good, the plasma absorption is low, and the photon energy is high [18].

II. *Peak power:* The hole quality tends to enhance with increasing peak power and power density. This is primarily related to the transition from melt discharge to evaporation. This minimizes the recast layer and heat-affected zone and enhances the repeatability of the hole. However, vaporization-dominated drilling is generally slower than molten ejection.

III. *Pulse width*: Pulse duration or pulse width usually lies in the range of a fraction of a microsecond to one microsecond. It is proven that a shorter pulse gives better whole quality; however, it increases machining cost.

IV. *Focal length and position*: Focal length and position play an important role in the laser-assisted drilling process. A longer focal length, for example, will yield a greater depth of focus, but a larger spot size will yield a lower power density. Shorter focal lengths, on the other hand, produce larger power densities but are more difficult to manage. In terms of focal location, inappropriate positioning of the focal point above or below the surface will influence the hole's depth and taper.

V. *Process gases*: Process gases such as N_2, O_2, Ar and compressed air can help in many ways throughout the drilling process. The gas shields the lens from material discharged from the drilling hole and reduces debris build-up around the hole. The gas can also be required to retain a plasma or plume from forming over the hole (which absorbs and scatters a lot of the laser energy). Another benefit is that oxygen can help with the removal of material by generating an exothermic reaction in some materials.

VI. *Performance parameters*: Performance parameters of laser-assisted drilling are based on its quality, geometrical and metallurgical characteristics.

VII. *Hole taper*: The hole taper is greatly affected by the number of pulses, the pulse energy and the focus setting of the laser beam. When going towards the depth, the laser beam's converging-diverging characteristic causes non-uniform dimensioning of the drilled hole. Taper angle (θ) is calculated by $\tan \theta = (D_1\text{-}D_2)/2$, where D_1 is the entrance hole diameter, and D_2 is the exit hole diameter.

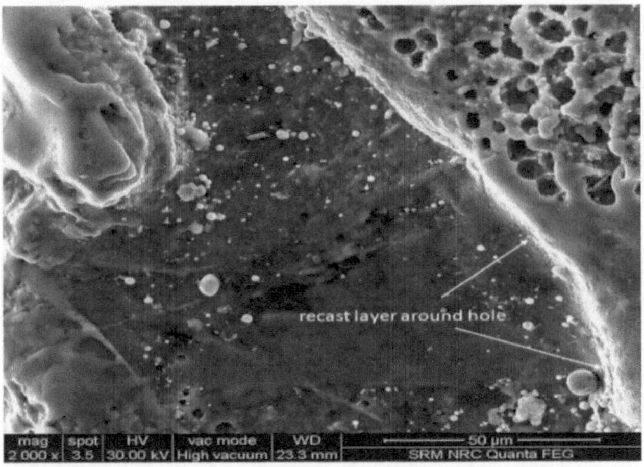

Figure 3.9 Recast layer formation in laser drilling at hole entry [36].

VIII. *Hole circularity*: In the laser-assisted drilling process, hole circularity is a critical metric. It typically indicates how round the machined hole appears. The hole circularity error is determined by the radius deviation around the circumference of the hole. The ratio of the lowest and the highest hole diameters is used to determine it.

IX. *Recast layer*: In a laser assisted-drilling operation, the recast layer is formed around the machined hole, as illustrated in Figure 3.9. The formation of the recast layer is not favourable for the laser-assisted drilling process. It's caused by re-solidified molten material that hasn't been eliminated. It results in micro-cracks and different mechanical properties to the parent material.

X. *Material removal rate*: In laser drilling, the metal removal rate (MRR) is less important as it is used for very accurate cutting and maximum material utilization. Usually, it is calculated by MRR= $(M_i\ M_f)/t$, where M_i and M_f are the initial and final weight of the workpiece before and after drilling and it is the machining time.

XI. *Micro cracks*: When drilling brittle or high-hardenability materials, micro-cracking is a typical phenomenon induced by fast cooling or temperature gradients.

3.4 LASER MACHINING AND LAM

In almost every LAM/LM, there are five stages involved in making the laser operation successful: beam generation, beam focusing, localised heating and melting, material removal and beam movement. Therefore,

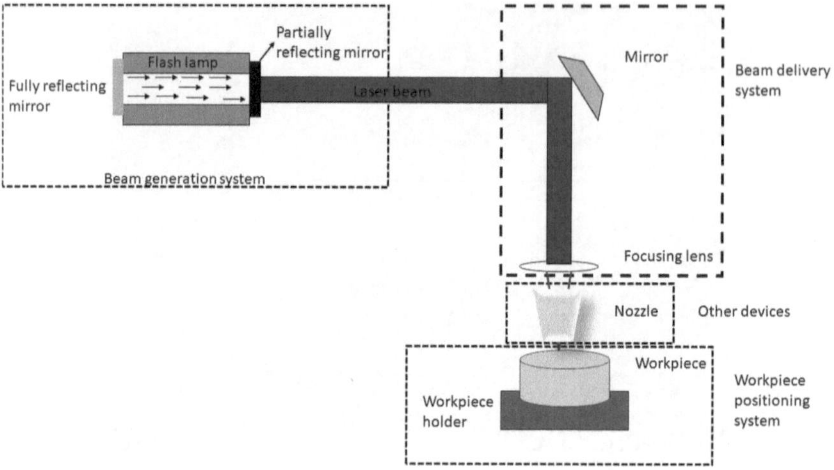

Figure 3.10 Necessary system for LM/LAM.

implementing any laser machining/assisted process requires a number of components that belong to different domains of optics, electronic and mechanical. The main subsystems usually used in LM/LAM are shown in Figure 3.10 [5,37]:

- Beam generation system
- Beam delivery system
- Workpiece positioning system
- Other devices

I. *Beam generation system*: In industries, CO_2, Nd:YAG, fibre and diode lasers are the widely used laser beam generation. The laser beams are basically generated by following three principles: population inversion, stimulated emission and amplification.

II. *Beam delivery system*: Optical components such as a mirror, beam polariser, splitter, focusing lens and some fibre optics comprise the majority of a beam delivery system. The details of these components are presented in Table 3.2.

III. *Workpiece position system:* The control of the motion of the workpiece or laser head is a crucial task for the better quality of the product. Generally, three techniques control motion between the laser head and workpiece.
   - For large workpieces, both translation and rotation motion is provided to the laser head while keeping the workpiece fixed. This system provides more flexibility while machining complex shapes. However, the laser head is supported by other expensive

Table 3.2 Component and functionality of beam delivery system

Components	*Functions*
Mirror	Generally, metallic mirrors are preferred in the beam delivery system to keep minimal energy loss by the laser beams. Sometimes dielectric material is also applied to the mirror to enhance its optical properties. To avoid any excessive distortion of the mirror, the water cooling technique is used to release excessive heat. Copper is the most common material preferred for mirrors.
Beam polariser	Generally, two types of polarising beams, i.e., linear polarised beams and circularly polarised beams, are used in the machining operation. Laser Polarizers are used in a range of laser applications to isolate certain polarizations of light or to convert unpolarised light to polarised light. Laser Polarisers transmit a single polarisation state across a variety of substrates, coatings, or a combination of the two.
Beam splitter	It is an optical device which has a flat surface on which some part of the incident beam is reflected, and the rest is transmitted. Splitters are generally used where multiple works have to be performed on different workstations through a single laser beam. Roof splitter and Inconel-coated ZnSe block are two types of splitter that are generally used in industries.
Focusing Lens	This is the convex lens that concentrates all of the laser beams into a single spot to obtain high energy density. In CO_2 and Nd:YAG lasers, the lens made up of KCl, NaCl, gallium arsenide and germanium are used.
Fiber Optics Coupling	Fibre optics coupling is used in the delivery system to overcome the disadvantages provided by the mirror delivery system. In the mirror delivery system, it is difficult to position the laser nozzle properly on the workpiece. This system requires the collection of subsystems to overcome these disadvantages. Therefore, optical fibres are used in the delivery system to provide multiple degrees of freedom which aim the laser properly at the workpiece due to their working principle of total internal reflection.

and sensitive optical systems, and the speed of laser speed is limited to a certain extent.

- For small workpieces, rotation and translation are provided to the worktable instead of a laser head. This method is commonly employed for laser cutting operations. The motion of the worktable is generally controlled by CNC programming, which provides a high degree of accuracy, smooth motion and high repeatability.
- Nowadays, with the development of technology, machines with multi-axis controls are the demand of the industry and the scientific community, which provide flexibility to a large range. Recently, various laser systems have been robotics manipulation beam

positioning systems that provide facilities to perform cutting, drilling and welding at any position. This multi-axis workstation is widely used in turbine, aircraft and automobile industries. The main advantages of using a multi-axis control system are the reduction of required floor space and high accuracy, repeatability, smooth motion, and beam power control. However, for complex geometry, the velocity of laser scanning and accuracy of the beam positioning is limited to a certain extent in a multi-axis system.

IV. *Other devices:* The other devices especially includes laser head subsystems which consist of various types of nozzle systems along with the focusing lenses. Generally, laser heads have coaxial nozzles that supply assisted gases to the machining zone and help keep the temperature of the system under control. The parallel and convergent flow passage nozzles are widely used by current laser systems. In laser machining, the assisted gas generally has two essential roles: to keep the debris away from the focusing lens and generate the pressure gradient at the machining zone for debris expulsion, resulting in an improvement in the surface quality. Nowadays, there are various types of nozzles available in the market, such as parallel nozzle, convergent nozzle, convergent-divergent nozzle and ring nozzle, among which convergent design is considered most effective for cutting purposes.

The laser used in industries and academics contains a high-energy beam and releases fumes when it falls on the workpiece, posing a threat to the operators or persons in the working area. Therefore, it is always recommended to use the laser safety equipment properly in the working zone. All the industrial lasers are kept in the Class IV category, which means it is always hazardous to view them directly and may also cause fire [38,39]. The following precautions should be taken during laser machining:

- Always use laser glasses while performing any LM/LAM operation as laser beams are harmful to the eyes, and any negligence may make you blind.
- Always work with a trained person who knows the entire system.
- A laser beam delivery system must be made with a thick tube that can sustain any leakage.
- The laser head, workpiece and workpiece positioning system must be enclosed or covered correctly to prevent the leakage of assisted gas and hot material or fumes.
- Entry and exit of any person should be controlled while the laser is in condition.
- A reflection absorber setup must be constructed around the workpiece to prevent any damage due to the reflection of the beam through the workpiece.

3.5 LASER IN ADDITIVE MANUFACTURING

Additive manufacturing (AM) is a layer-wise manufacturing process that tries to manufacture well at minimum wastage. It is the bright spot of the current manufacturing domain. The ability to produced well with greater customisation and feasibility makes this process more demanding in the current era [40]. The AM process such as fused deposition modelling, 3D printing, sheet lamination, direct energy deposition, etc. are continuously changing the world of manufacturing. The process of direct energy deposition uses laser power to fabricate product using metal powders. The laser-powered AM has basically subcategories into two types i.e., laser fusion powder based and laser-directed energy deposition techniques [41].

3.5.1 Laser Fusion Powder-Based Additive Manufacturing

The laser fusion powder-based additive manufacturing is one of the subtractive manufacturing technologies in which parts are fabricated by layer upon layer addition of metal. In the powder bed fusion method, a build plate is coated with small layers of powder, which are then fused at predetermined spots by a model of the required shape using an energy source (a laser or electron beam). As soon as one layer is completed, a fresh layer of powder is placed, and the process is carried on until a three-dimensional item is created [42]. The designs of metal laser powder bed fusion additive manufacturing systems are similar to those shown in Figure 3.11.

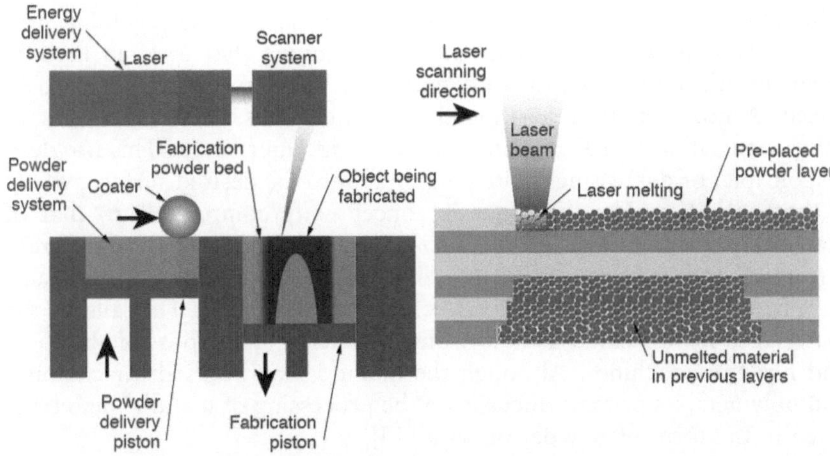

Figure 3.11 Schematic of the selective laser melting process [42].

They are made up of powder and energy delivery systems. A piston supplies powder, a coater creates the powder layer, and a piston keeps the created item in the powder delivery system. The energy delivery system consists of a laser (often a single-mode continuous-wave Ytterbium fibre laser operating at 1,075 nm wavelength) and a scanning system with optics that allow a focused spot to be delivered to all locations of the build platform. The laser implements a scanning or exposure technique during manufacturing. The length, direction, and separation (hatch spacing) of contiguous scan vectors determine the characteristics of the laser path methods [43].

Powder bed fusion is theoretically simple but the underlying physics is complicated and covers a wider range of length and time scales. Laser speeds are 1 m/s, while laser beam and powder layer thicknesses are 10 µm. The dimensions and accuracy of fabricated parts depend on many parameters such as laser speed and power, laser beam size and width and depth of the melt pool. The breadth, depth and length of the melt may have a significant impact on part density and via the cooling rate, the length can have an impact on the microstructure. Usually, it is important to control the geometry of the melt pool during fabrication. This technology is relatively inexpensive and best suited for prototypes and visual models. However, it is a slow-speed process with size limitations. The finishing of the final part is dependent on powder grain size. Any powder-based material may be utilized in the powder bed fusion process; however, the most commonly used materials are metal and polymer [44].

3.5.2 Laser Direct Energy Deposition

Laser direct energy deposition (LDED) is also known as directed light fabrication, laser-engineered net shaping and the 3D laser cladding process. It is a more complicated printing technique that is often used to repair or add more material to components that have already been produced. A nozzle that is positioned on a multi-axis arm is the primary component of an LDED machine, as shown in Figure 3.12. This nozzle is responsible for depositing molten material onto the desired surface, where it then solidifies. The technique is conceptually comparable to that of material extrusion; however, the nozzle is free to travel in several different directions and is not confined to a single axis. Lasers or electron beams may be used to melt the material before it is deposited. This allows the material to be deposited from any angle, which is made possible by four- and five-axis machines. Although the method may be used for ceramics and polymers, its primary focus is on the processing of metallic materials, often in the form of powder or wire [45].

In most circumstances, the arm moves and the item stays put, however, this may be flipped so the platform moves and the arm stays put. The choice depends on the application and printed item. Material cooling times are

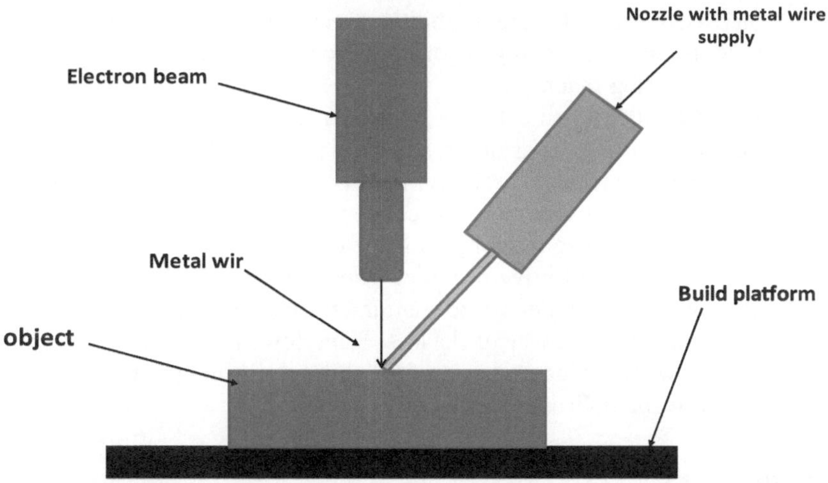

Figure 3.12 Schematic of laser direct energy deposition machine [45].

rapid, between 1000 and 5000°C. The cooling period affects the final grain structure of the deposited material, but overlapping material must also be addressed, as it may induce re-melting, resulting in a uniform yet alternating micro-structure [44].

3.6 CONCLUSION

The use of laser beam directly for machining or as an assisted technique with various other machining operations established the laser beam as a precise tool which makes the machining process easier and helps to readjust the characteristics of machine and material by improving their performances. Nowadays, several industries use CO_2 Laser, Nd: YAG laser, fibre laser and diode laser in their manufacturing units for fabrication and material removal processes. The ongoing evolution of laser technology and the development of other sophisticated assisting tools established laser material processing technology as one of the trusted allied manufacturing operations. Following are the key conclusions that can be drawn from the present exhaustive review work:

I. The laser beam can operate as a "supporting" or "assisting" tool to increase the efficiency of conventional machining processes or to assist in the machining of difficult-to-machine materials.
II. The laser system amalgamation with conventional technologies helps in the machining of new-generation material with flexibility to attend to complex shapes and accuracy. Laser-equipped stamping-cutting machines are an example of this type of combination to meet

increased needs for adaptability, precision or quality. Such technologies as laser caving or drilling would expand the functionality of the cutting machine.

III. It is difficult to grind brittle materials and composites using traditional techniques as grinding has to be performed in a ductile regime for these materials. Therefore, to solve this problem, LAG is introduced in which the workpiece is preheated with a laser before grinding to induce ductility to some extent.
IV. Use of laser in additive provides a sophisticated tool to explore the area of metal powder-based manufacturing domain.
V. Lasers are very powerful devices. Therefore, it is always necessary to follow the rules and protocol while performing LAM to avoid any unwanted circumstances.

REFERENCES

[1] R. E. Slusher, "Laser technology," *Rev. Mod. Phys.*, vol. 71, no. 2, p. S471, 1999.
[2] R. M. Herd, J. S. Dover, and K. A. Arndt, "Basic laser principles," *Dermatol. Clin.*, vol. 15, no. 3, pp. 355–372, 1997.
[3] G. Chryssolouris, N. Anifantis, and S. Karagiannis, "Laser assisted machining: An overview," *J. Manuf. Sci. Eng. Trans. ASME*, vol. 119, no. 4B, pp. 766–769, 1997.
[4] N. B. Dahotre and S. Harimkar, *Laser fabrication and machining of materials*. Springer Science & Business Media, 2008.
[5] G. Chryssolouris, *Laser machining: Theory and practice*. Springer Science & Business Media, 2013.
[6] A. K. Dubey and V. Yadava, "Laser beam machining – A review." *Int. J. Mach. Tools Manuf.*, vol. 48, no. 6, pp. 609–628, 2008.
[7] A. September, B. Rosi, A. Marinkovi, and A. Vencl, "4th International Conference" Research and Development in Mechanical Industry "RaDMI 2004," no. September, pp. 73–77, 2004.
[8] S. Sun, M. Brandt, and M. S. Dargusch, "Thermally enhanced machining of hard-to-machine materials-A review," *Int. J. Mach. Tools Manuf.*, vol. 50, no. 8, pp. 663–680, 2010.
[9] M. K. Sinha, "Experimental investigations in grinding of Inconel 718 using different environments and modelling of specific grinding energy," Doctoral dissertation, IIT Delhi, 2018.
[10] Z. Li, F. Zhang, X. Luo, W. Chang, Y. Cai, W. Zhong, and F. Ding, "Material removal mechanism of laser-assisted grinding of RB-SiC ceramics and process optimization," *J. Eur. Ceram. Soc.*, vol. 39, no. 4, pp. 705–717, 2019.
[11] I. Vilumsone-Nemes, *Industrial cutting of textile materials*. Woodhead Publishing, 2018.
[12] Z. Ma, Z. Wang, X. Wang, and T. Yu, "Effects of laser-assisted grinding on surface integrity of zirconia ceramic," *Ceram. Int.*, vol. 46, no. 1, pp. 921–929, 2020.

[13] B. Azarhoushang, B. Soltani, and A. Zahedi, "Laser-assisted grinding of silicon nitride by picosecond laser," *Int. J. Adv. Manuf. Technol.*, vol. 93, no. 5–8, pp. 2517–2529, 2017.

[14] W. Chang, X. Luo, Q. Zhao, J. Sun, and Y. Zhao, "Laser assisted micro grinding of high strength materials," *Key Eng. Mater.*, vol. 496, pp. 44–49, 2012.

[15] M. J. Jackson, A. Khangar, X. Chen, G. M. Robinson, V. C. Venkatesh, and N. B. Dahotre, "Laser cleaning and dressing of vitrified grinding wheels," *J. Mater. Process. Technol.*, vol. 185, no. 1–3, pp. 17–23, 2007.

[16] A. Hosokawa, T. Ueda, and T. Yunoki, "Laser dressing of metal bonded diamond wheel," *CIRP Ann. - Manuf. Technol.*, vol. 55, no. 1, pp. 329–332, 2006.

[17] C. A. Biffi, J. Fiocchi, and A. Tuissi, "Relevant aspects of laser cutting of NiTi shape memory alloys," *J. Mater. Res. Technol.*, vol. 19, pp. 472–506, 2022.

[18] E. Rahim, N. Warap, and Z. Mohid, "Thermal-assisted machining of nickel-based alloy," *Superalloys*, vol. 3, 2015.

[19] W. Chang and X. Luo, "Laser-assisted machining," in *Hybrid machining: Theory, methods, and case studies*. Academic Press, 2018, pp. 43–76.

[20] S. Lei, Y. C. Shin, and F. P. Incropera, "Experimental investigation of thermo-mechanical characteristics in laser-assisted machining of silicon nitride ceramics," *J. Manuf. Sci. Eng.*, vol. 123, no. 4, pp. 639–646, 2001.

[21] J. I. Arrizubieta, F. Klocke, S. Gräfe, K. Arntz, and A. Lamikiz, "Thermal simulation of laser-assisted turning," *Procedia Eng.*, vol. 132, pp. 639–646, 2015.

[22] Y. Tian and Y. C. Shin, "Laser-assisted machining of damage-free silicon nitride parts with complex geometric features via in-process control of laser power," *J. Am. Ceram. Soc.*, vol. 89, no. 11, pp. 3397–3405, 2006.

[23] C. Brecher, M. Emonts, C.-J. Rosen, and J.-P. Hermani, "Laser-assisted milling of advanced materials," *Phys. Procedia*, vol. 12, pp. 599–606, 2011.

[24] B. Yang, X. Shen, and S. Lei, "Mechanisms of edge chipping in laser-assisted milling of silicon nitride ceramics," *Int. J. Mach. Tools Manuf.*, vol. 49, no. 3–4, pp. 344–350, 2009.

[25] R. Wiedenmann and M. F. Zaeh, "Laser-assisted milling—Process modeling and experimental validation," *CIRP J. Manuf. Sci. Technol.*, vol. 8, pp. 70–77, 2015.

[26] M. J. Bermingham, W. M. Sim, D. Kent, S. Gardiner, and M. S. Dargusch, "Tool life and wear mechanisms in laser assisted milling Ti–6Al–4V," *Wear*, vol. 322, pp. 151–163, 2015.

[27] M. Anderson, R. Patwa, and Y. C. Shin, "Laser-assisted machining of Inconel 718 with an economic analysis," *Int. J. Mach. Tools Manuf.*, vol. 46, no. 14, pp. 1879–1891, 2006.

[28] T. Muthuramalingam, R. Akash, S. Krishnan, N. H. Phan, V. N. Pi, and A. H. Elsheikh, "Surface quality measures analysis and optimization on machining titanium alloy using CO2 based laser beam drilling process," *J. Manuf. Process.*, vol. 62, pp. 1–6, 2021.

[29] Y. Z. Liu, "Coaxial waterjet-assisted laser drilling of film cooling holes in turbine blades," *Int. J. Mach. Tools Manuf.*, vol. 150, p. 103510, 2020.
[30] B. S. Yilbas, "Parametric study to improve laser hole drilling process," *J. Mater. Process. Technol.*, vol. 70, no. 1–3, pp. 264–273, 1997.
[31] S. Takeno, M. Moriyasu, and S. Hiramoto, "Laser drilling by high peak pulsed CO_2 laser (LDHPCL)," in *International Congress on Applications of Lasers & Electro-Optics*, vol. 1992, no. 1, pp. 459–468, 1992.
[32] Y. Wang and W. Zhang, "Theoretical and experimental study on hybrid laser and shaped tube electrochemical machining (Laser-STEM) process," *Int. J. Adv. Manuf. Technol.*, vol. 112, no. 5, pp. 1601–1615, 2021.
[33] O. Kalantari, F. Jafarian, and M. M. Fallah, "Comparative investigation of surface integrity in laser assisted and conventional machining of Ti-6Al-4 V alloy," *J. Manuf. Process.*, vol. 62, pp. 90–98, 2021.
[34] A. K. Sahu, J. Malhotra, and S. Jha, "Laser-based hybrid micromachining processes: A review," *Opt. Laser Technol.*, vol. 146, p. 107554, 2022.
[35] G. Alsoruji, T. Muthuramalingam, E. B. Moustafa, and A. Elsheikh, "Investigation and TGRA based optimization of laser beam drilling process during machining of Nickel Inconel 718 alloy," *J. Mater. Res. Technol.*, vol. 18, pp. 720–730, 2022.
[36] G. D. Gautam and A. K. Pandey, "Pulsed Nd: YAG laser beam drilling: A review," *Opt. Laser Technol.*, vol. 100, pp. 183–215, 2018.
[37] J. Zuo and X. Lin, "High-power laser systems," *Laser Photon. Rev.*, vol. 16, no. 5, p. 2100741, 2022.
[38] R. Henderson and K. Schulmeister, *Laser safety*. CRC Press, 2003.
[39] D. H. Sliney, "Laser safety," *Lasers Surg. Med.*, vol. 16, no. 3, pp. 215–225, 1995.
[40] K. Kishore and M. K. Sinha, "A state-of-the-art review on fused deposition modelling process," *Advances in Manufacturing and Industrial Engineering*, pp. 855–864, 2021.
[41] C. Tan, F. Weng, S. Sui, Y. Chew, and G. Bi, "Progress and perspectives in laser additive manufacturing of key aeroengine materials," *International Journal of Machine Tools and Manufacture*, 170, p. 103804, 2021.
[42] W. E. King et al., "Laser powder bed fusion additive manufacturing of metals; physics, computational, and materials challenges applied physics reviews Laser powder bed fusion additive manufacturing of metals; physics, computational, and materials challenges," *Appl. Phys. Rev*, vol. 2, no. 4, p. 41304, 2015.
[43] T. DebRoy et al., "Additive manufacturing of metallic components–process, structure and properties," *Prog. Mater. Sci.*, vol. 92, pp. 112–224, 2018.
[44] S. Kolossov, E. Boillat, R. Glardon, P. Fischer, and M. Locher, "3D FE simulation for temperature evolution in the selective laser sintering process," *Int. J. Mach. Tools Manuf.*, vol. 44, no. 2–3, pp. 117–123, 2004.
[45] N. Ur Rahman et al., "Directed energy deposition and characterization of high-carbon high speed steels," *Addit. Manuf.*, vol. 30, p. 100838, 2019.

Chapter 4

Electromagnetic Joining of Aluminium Tube on Three Different Metal Cores Using Double Solenoid Coil

Ramesh Kumar[1], Ashish Kumar Rajak[2], and Sachin D Kore[3]

[1]Saharsa College of Engineering (under Department of Science & Technology, Govt. of Bihar) Saharsa Bihar, India
[2]Indian Institute of Technology Indore, Madhya Pradesh, India
[3]Indian Institute of Technology Goa, Goa, India

CONTENTS

4.1 INTRODUCTION

Electromagnetic crimping (EMC) or joining is one of the reliable methods that can be used for joining either similar or dissimilar metal parts. It provides an excellent tool to join electrically conductive metals such as aluminium, brass, or copper onto the metallic or non-metallic core. This technique comes under the category of high-velocity and strain rate joining process [1]. The weight of the automobile parts can be reduced by using

DOI: 10.1201/9781003436072-4

lightweight materials such as aluminium. But the joining of aluminium with other materials such as steel is difficult [2]. Hence, for the joining of dissimilar materials, this technique can be used. In this technique, the two parts which will be joined are known as flyer and base. The electrically conductive flyer tube was accelerated by using electromagnetic force. The solid-state joint was created due to the impact of the flyer on the core with high velocity [3].

In this technique, there are a few factors that will affect the quality of the joints. The factors that affect the quality of the joint is the material of the flyer tube, the thickness of the flyer tube, the gap between the tube and core, the gap between the tube and the coil and the magnitude of the discharged energy. The issue of the metallurgical compatibility between the dissimilar materials was greatly minimized by using this technique [4]. This technique can be used for joining either different or same thicknesses homogeneous or heterogeneous materials [5]. This technique was beneficial for joining in the lap configuration of cones, tubes, and rods [6]. This process has excellent repeatability and having great potential for automation. The strength of the joint produced by this process depends on the magnitude of impact velocity or the magnitude of electromagnetic discharge energy [7]. The advantage of the joint created by this technique is more than the strength of the weak parent materials [8]. To get the benefit of the dissimilar materials, joining this technique can be a substitute for the available conventional joining technique [9].

The manufacturing time for one joint is not more than 2 to 10 seconds; hence, the manufacturing rate of this process is very high. In this process, a large amount of energy is stored in capacitors, and the electromagnetic energy was discharged in a concise period below 100 μs [10]. The discharge energy allows passing through the double coil, which surrounds the outer tube, causing motion in the outer tube. The impact caused by the moving outer tube on the inner core generates electromagnetic joint by high-speed impact forming. The double coil used in this process does not come in contact with the tube, and it remains separated with the same. The separating distance between the coil and the outer tube is the coil gap. One similar gap is also there between the outer tube and the inner core, and this gap is known as the stand-off distance between the tube and the core. The magnitude of the velocity obtained in this process may be in the range of 100–500 m/s. The velocity of the base core in this process remains ideally zero.

In this work, numerical as well as experimental study was carried out to study the effect of the three types of core materials on the crimped joint strength. In the comparison of the three core materials, the same double solenoid coil was used. The same aluminium flyer tube was used in all three cases. LS-DYNA software was used to model and analyze the work with strong coupling between the coil and the flyer tube.

4.2 MATERIALS AND METHOD

4.2.1 Equipment and Sample

In the EMC process, a double solenoid coil made of copper was used. The schematic diagram of the EMC process with the double solenoid coil is shown in Figure 4.1. The numerical as well as experimental conditions were kept identical in the process. In the experiments, the samples used were the aluminium tube and the cores of aluminium, brass, and copper. The different dimensions of the outer tube and the inner core are shown in Table 4.1. The thickness of the outer tube used was 0.7 mm, and the materials used for inner core are copper, brass, and aluminium materials.

In this work, the gap between the coil and the outer tube was maintained at 0.5 mm. Prior to performing the experiments, the chemical composition of the materials was tested, and the same was compared and analyzed with the available standard composition of the materials. The energy-dispersive X-ray spectroscopy (EDX) equipment was used to analyze the chemical composition of the material. The chemical composition of the materials used in the experiments is shown in Table 4.2.

4.2.2 Numerical Model

For the purpose of numerical study, Ls-Dyna software was used. In the numerical study, Johnson's and Cook's material model was used. The main reason for using this model is that this model has the capacity

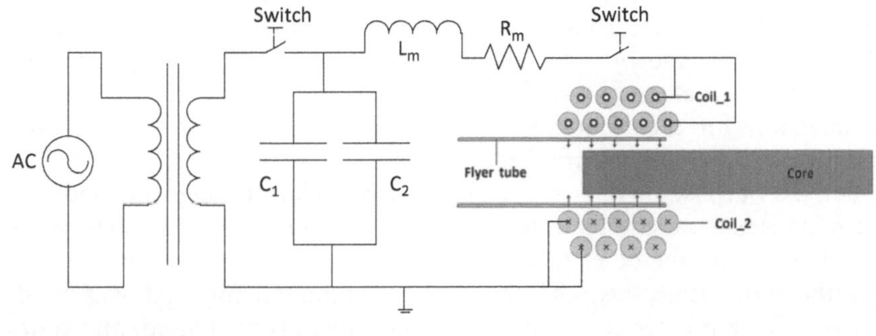

Figure 4.1 Experimental setup of magnetic pulse crimping with a double solenoid coil.

Table 4.1 Dimension of the Samples and Their Material

Specimen	*Outer diameter*	*Inner diameter*	*Length*	*Material*
Core	14 mm	–	50 mm	Al, Cu, Brass
Tube	16.4 mm	15 mm	100 mm	Aluminium

Table 4.2 Chemical Composition of Materials Used in the Experiment (in Weight %)

	Composition in (weight %)									
	Cu	Al	Zn	K	Fe	Si	S	Cl	Ca	Pb
Brass	87.1	6.9	2.3	1.1	0.8	0.7	0.3	0.3	0.3	0.2
	Zn	Fe	Cu	Pb	S	P	Sb	Sn	Si	
Copper	0.30	0.20	Balance	0.20	0.20	0.10	0.10	0.10	0.20	
	Al	Fe	Ti	Mn	Cu	Si	Zn			
Aluminium	Balance	1.10	0.10	0.10	0.20	0.60	0.20			

that it can accommodate the thermal softening of the material and the changes at the high strain rate together. It also has particular characteristics that it comprises the strain-rate hardening effect on the flow stress for high strain rates. Mathematically, the Johnson-Cook equation is given by equation (4.1).

$$S_y = (A + B\varepsilon^n)\,(1 + C\,ln\,\dot{\varepsilon}_p)\left(1 - \left(\frac{T - T_{room}}{T_m - T_{room}}\right)^m\right) \tag{4.1}$$

In equation (4.1), S_y is flow stress of the material, A, B are the yield strength parameters, $\dot{\varepsilon}$ is equivalent plastic strain, n is the strain hardening index, C is the strain-rate sensitivity, $\dot{\varepsilon}_p$ is plastic strain-rate, T is absolute temperature, T_{room} is the room temperature, T_m is the melting temperature of the material, and m is the thermal softening index. These parameters have different values for different materials and can be determined experimentally for each material. The Johnson-Cook parameters that were taken from the literature are listed in Table 4.3.

For the purpose of this study, both the boundary element method and the finite element methods were used. The three-dimensional models of the double coil with the outer tube and the inner core is shown in Figure 4.2. For the outer tube, i.e., the flyer tube, aluminium material was used, whereas for the inner core, three different materials were used, and those materials were brass, aluminium, and copper.

Table 4.3 Johnson-Cook Constant Values for Three Materials

Material	A *(MPa)*	B *(MPa)*	n	C	m	T_m *(K)*	ρ *(Kgm^{-3})*	E *(GPa)*	v
Al 1050 [11]	110	150	0.40	0.01	1	918	2705	69	0.33
ETP Copper (C11000) [12]	92	292	0.31	0.025	8.89	1338	8890	115	0.33
Brass [13]	206	505	0.42	0.01	1.68	1189	9059	115	0.31

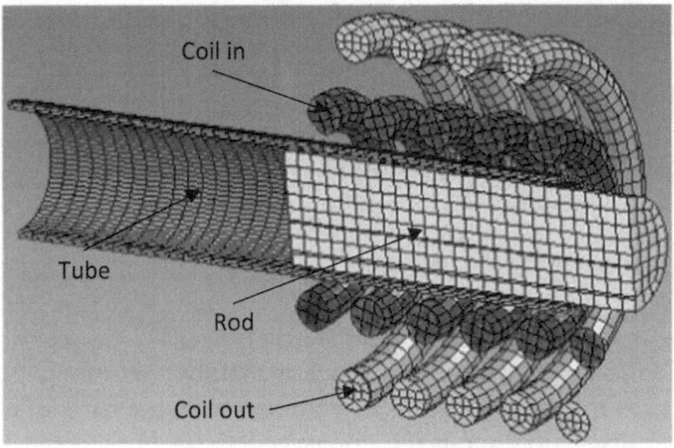

Figure 4.2 Three-dimensional model used for the numerical study.

4.2.3 Compression-Shear Test

To find the strength of the joints of the two combinations of the materials, i.e., the outer tube and the inner core, a particular type of test, i.e., compression-shear analysis, was used. In this typical test, a compressive load was being applied on the interface of the two materials, i.e., at the interface of the outer tube and the inner core. For this type of test, a universal testing machine was being used with some additional attachments to it.

When the samples were joined by using the electromagnetic joining technique, the successfully joined samples were used for testing purposes. Prior to testing the samples, compression-shear test samples were prepared by using the wire electric discharge machine. Many samples were prepared for the testing, and it was noted that at least three samples must be there for each energy and for each combination of materials. There was a minimum of 18 samples prepared, each one at 4.70 kJ, 5.40 kJ, and 6.20 kJ of the discharge energies. The three combinations of the materials, i.e., *Al+Al, Al +Brass, and Al+Cu* was used, and for each combination, three samples were prepared. For testing the samples on the universal testing machine, the speed of the movement of the jaw was maintained at a speed of 0.5 mm/min. The reason for selecting this speed is due to the height of the sample that was prepared, and it was 1.5 mm.

After performing the compression-shear tests, the strength of the joint can be obtained by using the formula given by equation (4.2). In this equation, the symbol τ was used for the joint strength, (P_{max}) represents the maximum load being applied by the jaw of the UTM, and A was used for the joint area, i.e., the circumferential area of the cut sample. The joint area A can be calculated by the product of the circumferential length to the height of the sample.

$$\tau = \frac{P_{max}}{A} = \frac{P_{max}}{\pi dh} \tag{4.2}$$

4.3 RESULTS AND DISCUSSION

To get the samples joined successfully, both the simulation as well as the experiments, were performed. All the parameters such as dimensions of the sample, the orientation of the coil, tube, core, and discharge current were kept identical in the simulations and the experiment. In the experiments and in the simulations, the parameter, which is mainly controlling the strength of the joints, is the discharge current obtained from the bank of capacitors. The detailed discussion about the discharge current in the simulations and in the experiments is discussed in the following subsection.

4.3.1 Discharge Current

The experiments were performed mainly on the three values of the discharge current, and these values of the energies are 4.7, 5.4, and 7.2 kJ. For the purpose of measuring the discharge current, which is flowing through the double solenoid coil, a Rogowski coil, and a digital oscilloscope were used in the experiments. The recorded discharge current curve from the experiments is shown in Figure 4.3. The nature of the measured discharge current is typically a damped sine curve. The approximated expression for the

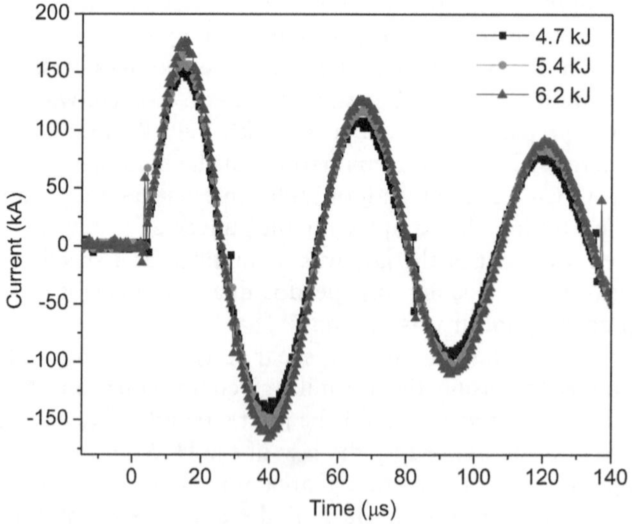

Figure 4.3 Discharge current measured in the experiments.

discharged current passing through the double coil can be approximated by equation (4.3).

$$I(t) = \frac{V_o}{\omega L} e^{-\beta\tau} \sin(\omega\tau) \tag{4.3}$$

In the expression, shown in equation (4.3), the discharge current is represented by *I(t)*, open-circuit voltage, i.e., initial discharge voltage is represented by V_o, the equivalent inductance of the circuit that includes the coil is given by L, the angular frequency of the current is given by ω, and the damping parameter of the current is given by β [14].

4.3.2 Compressive Strength of the Joint

The electromagnetic joining process was performed with an aluminium tube and three different materials core. The samples crimped by magnetic pulse crimping process with three different metal cores at different values of the discharge energies are shown in Figure 4.4.

From the crimped samples, compression-shear tests sample were produced. The thickness of the samples produced were 1.5 mm. The line drawing of the compression-shear test process with a punch die and a crimped sample is shown in Figure 4.5(a). The crimped samples with the three materials' core and die as well as punch used for the compression-shear test is shown in Figure 4.5(b).

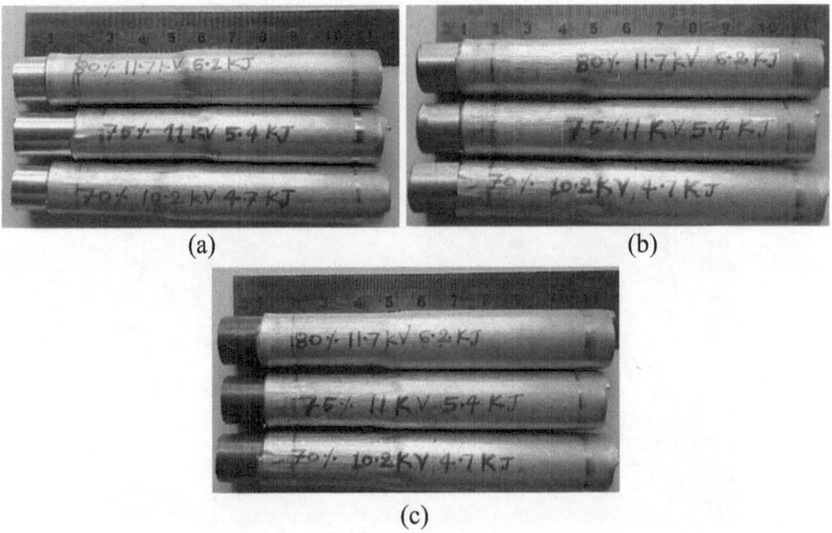

Figure 4.4 Al tube crimped on (a) Al core, (b) brass core, and (c) Cu core.

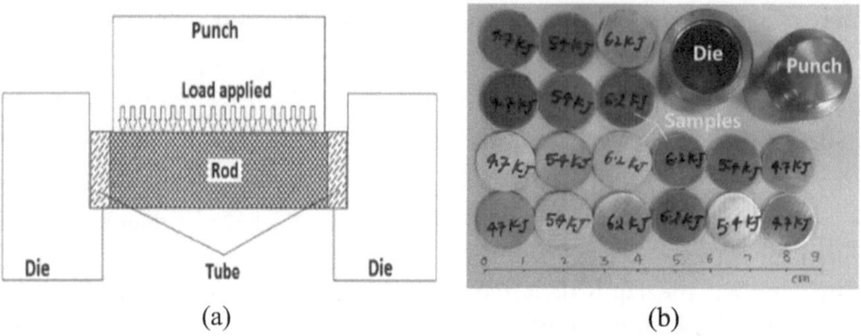

Figure 4.5 Compression-shear test: (a) line drawing and (b) samples with the fixture.

In the compression-shear tests, the samples were separated because of the compressive load on the crimped sample, and the inner core came out. The specimens that failed in the compression shear tests are shown in Figure 4.6. The variation in the collective strength with extension for three different amounts of discharge energy (a) 6.2 kJ, (b) 5.4 kJ, and (c) 4.6 kJ is shown in Figure 4.7. From the compression-shear tests at 6.2 kJ energy, it was found that the highest compressive strength was

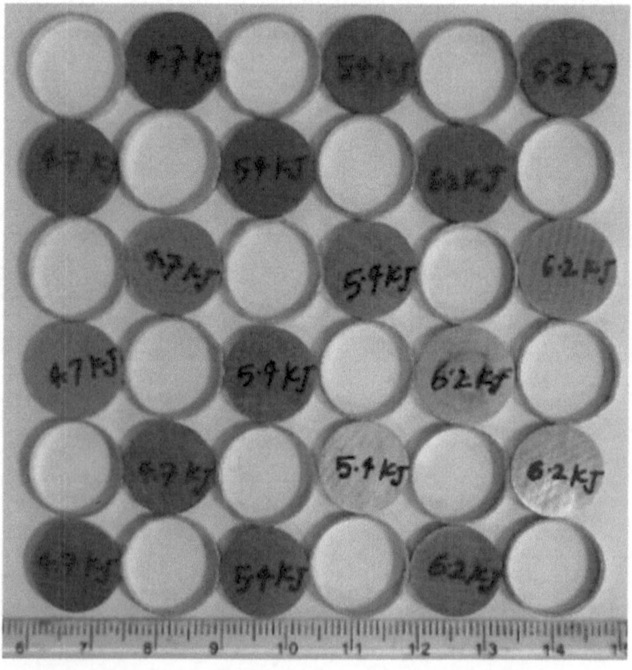

Figure 4.6 Magnetic pulse crimped samples failed after the compression-shear test at different discharge energies.

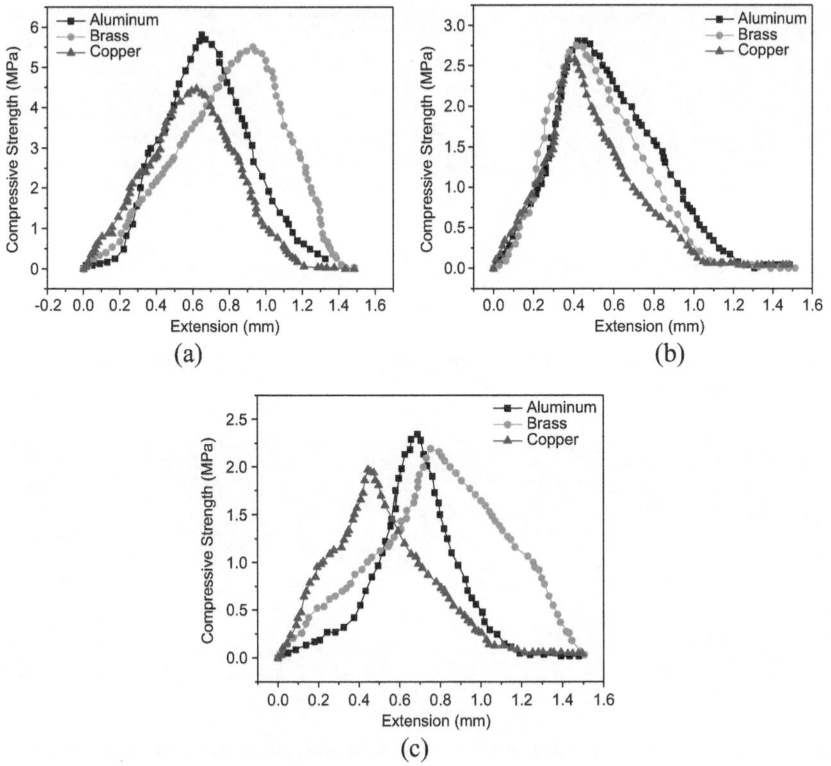

Figure 4.7 Compression-shear test results for three combinations of materials at (a) 6.2 kJ, (b) 5.4 kJ, and (c) 4.6 kJ.

5.82 MPa, 5.51 MPa, and 4.5 MPa with aluminium, brass, and copper as a core, respectively. It was also found that the increase in the magnitude of the energy causes an increase in the strength of the joint.

4.3.3 Analysis of Interface

A magnetic pulse crimped specimen, representative of the successful crimping experiments, was regarded on the microscale for Al-Al, Al-brass, and Al-copper. In the micro-section, it can be seen that there was a maximum gap of about 2–3 microns at some location in the case of Al-Cu, but the negligible gap was observed for the Al-Al and Al-Brass combination, where the flyer was crimped on the core with a plain profile at 6.2 kJ energy. The cross-section of the magnetic pulse crimped samples with three metal cores and at 6.2 kJ is shown in Figure 4.8. The interface of the crimped sample cross-section was analyzed, and it was found that the gap between the aluminium-aluminium materials combination is less than aluminium-brass and that of aluminium-copper.

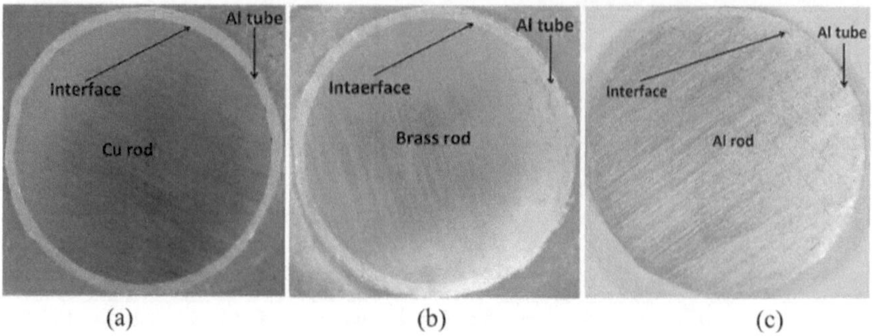

Figure 4.8 Cross-section of Al tube crimped on (a) copper core, (b) brass core, and (c) Al core.

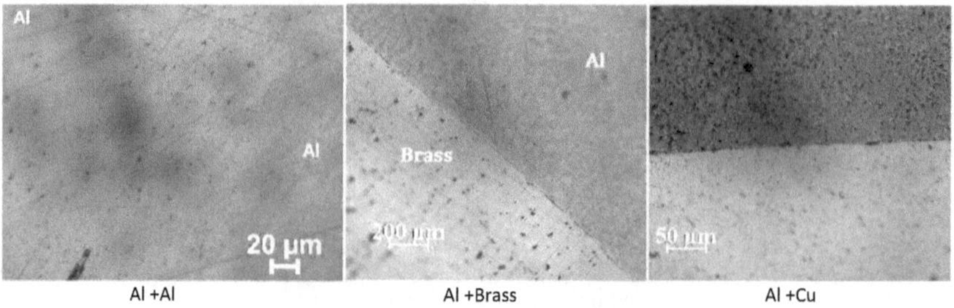

Figure 4.9 Microscopic view of the crimped sample with Al, brasss and copper core.

The microscopic view of the cross-section of the joined sample at 6.2 kJ energy is shown in Figure 4.9. In the microscopic analysis, it was found that the gap between the aluminium-aluminium is less compared to the aluminium-brass and that of aluminium-copper.

4.3.4 Hardness Test

Micro-hardness values obtained from Vicker's micro-hardness was analyzed. From this test, it was observed that near the interface of the joint, the magnitude of the hardness value is comparatively higher as compared to the other part of the sample. The increase in the hardness value near the interface from both sides may be due to the impact of the outer tube on the inner core with a high magnitude of speed. Vicker's micro-hardness tests were used to find hardness, and it was performed at four different points were selected. All the three different combinations, i.e., aluminium tubes with different core materials such as aluminium, brass, and copper, were tested for the hardness test. The average value of the four different values of hardness is shown in Figure 4.10.

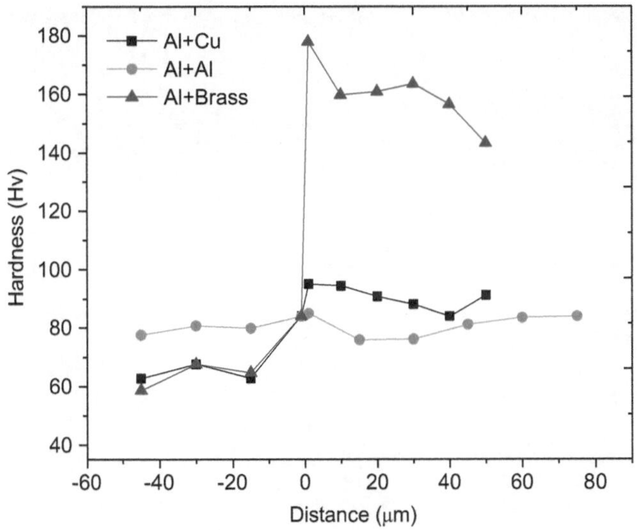

Figure 4.10 Vicker's micro-hardness for three combinations of joints.

4.3.5 Numerical Results

Numerical simulations were performed on the same parameters on which experiments were completed. The simulated model was validated on the basis of measured outer diameter in the experiments as well as in the simulations. There is a variation in the simulated and experimental values, and the variation was 2.3%, 1.5%, and 2.0% with aluminium, brass, and copper core, respectively. The comparison of the simulated and the crimped outer diameter of the samples is shown in Figure 4.11.

There is marginal variation in the two diameters, i.e., simulated diameter and in the experimentally measured diameter. This marginal variation may arise due to the size of the mesh and may be due to losses in the experiment. The size of mesh can be reduced, but due to this, the simulation time increases up to a large extent. The chosen size of the mesh is sufficient to fulfil the purpose of this study, and the obtained results can be considered for designing the experiments (Figure 4.12).

The fringe pattern of the deformation observed in the simulation at three different time steps is shown in Figure 4.13. Initially, the deformation in the tube was zero, and after the discharge of energy, the deformation was observed. In the simulation, the deformation of the coil and the core was zero as it was modelled as rigid.

In this study, the results of the electromagnetic field and the impact velocity obtained by the outer tube was also included. From the results of the electromagnetic field and the impact velocity, it was observed that due to the change in the inner core, these values also change. Figure 4.14 shows

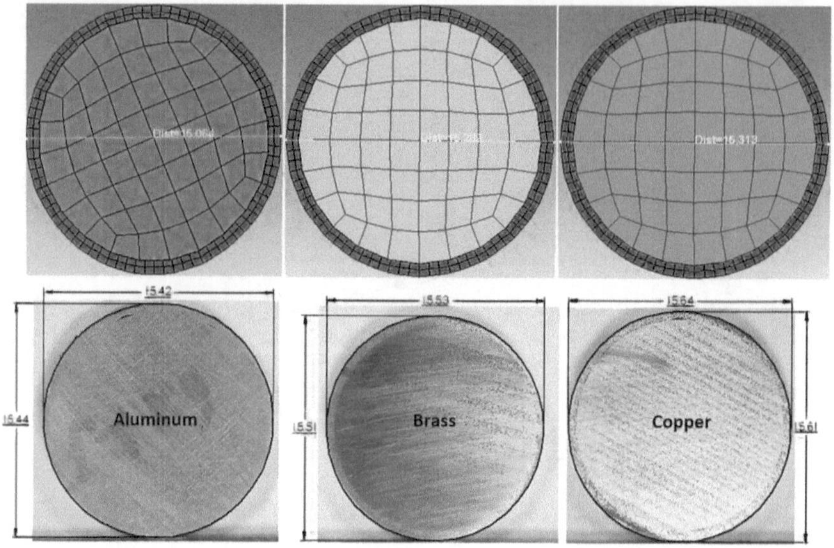

Figure 4.11 Cross-section of the outer diameter of joined samples in simulation and in the experiment.

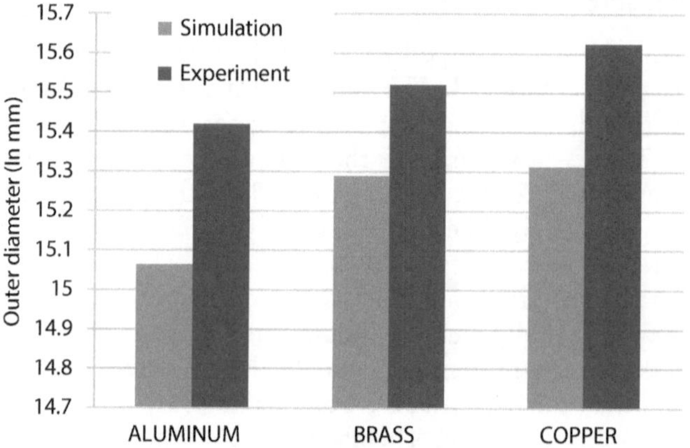

Figure 4.12 Comparison of simulated and experimental values.

the individual change in the velocity and magnetic field. But this change is marginally low. It can also be concluded that with the increase in the conductivity of the inner core, the impact velocity reduces. But this reduction in the results of post-process parameters is very low.

From the numerical analysis, the Lorentz force generated on the outer tube was also studied. The variation in the Lorentz force with three different types of cores is shown in Figure 4.15. From this study, it was observed that the magnitude of Lorentz force generated on the outer tube

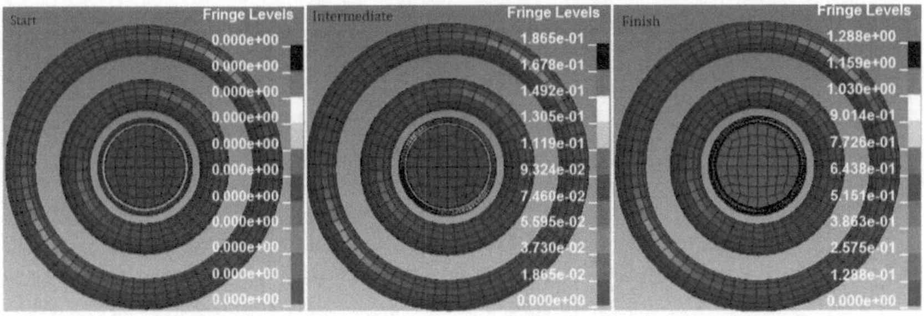

Figure 4.13 The movement of the flyer tube at three different time steps.

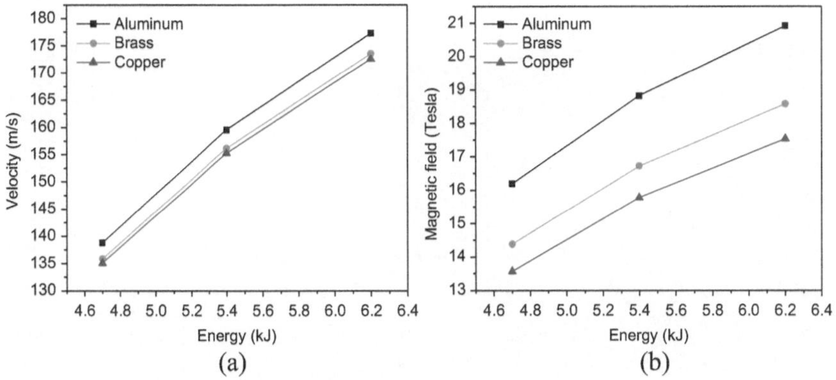

Figure 4.14 Post-process results: (a) impact velocity and (b) electromagnetic field generated in the flyer tube.

was relatively higher in the case of the aluminium rod as inner core as compared to the other two conditions. This change in the magnitude of the Lorentz force was due to the change in the conductivity of the inner core, i.e., due to the variation in the materials of the inner core.

4.4 CONCLUSIONS

In this study, the aluminium tube as the flyer was joined on the three different inner cores, such as the aluminium, brass, and copper core. For producing this dissimilar joint, a double solenoid coil was used. Both numerical and experimental works were carried out. Numerical values obtained were compared with the experimental values on the basis of the numerical and measured outer diameter of the crimped samples. There was a variation observed between the simulation results and that of the experimental results, the variation in the outer diameter was found to be 2.3%, 1.5%, and 2.0% with aluminium, brass, and copper core at 6.2 kJ

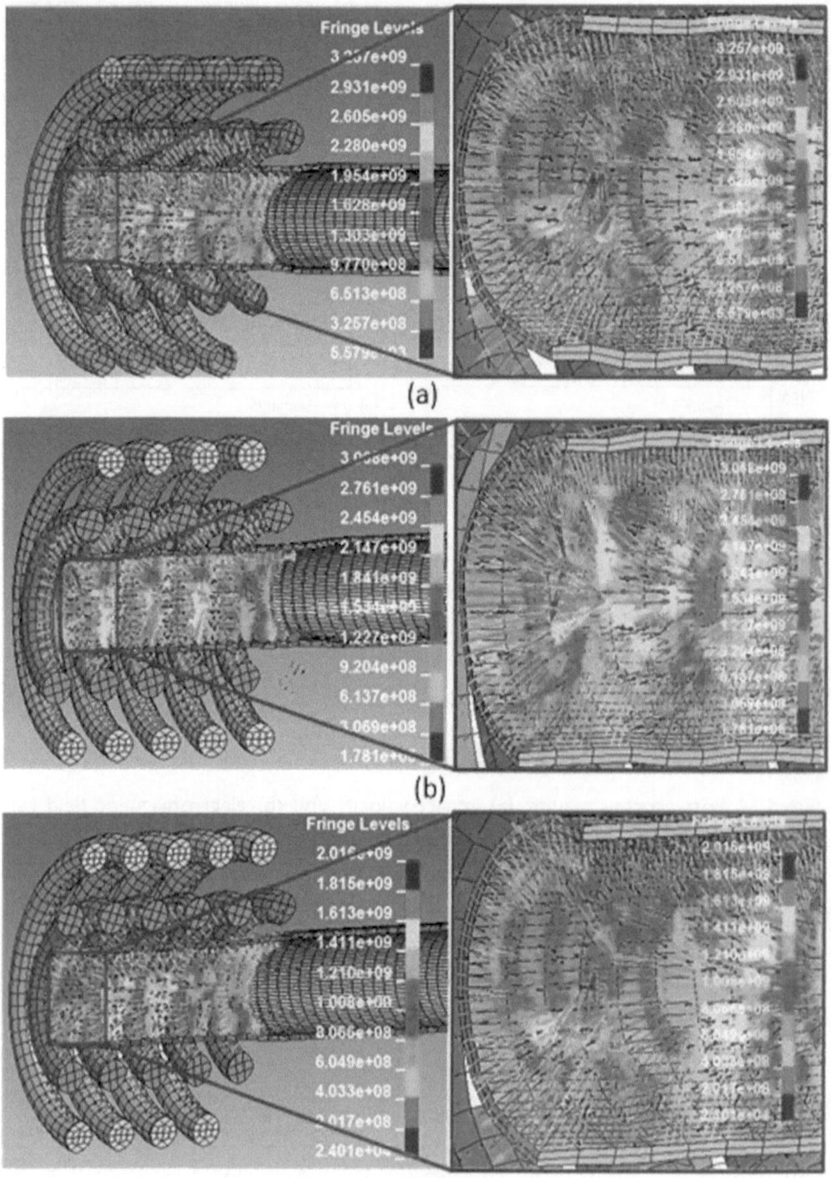

Figure 4.15 Variation in the Lorentz force generated with (a) Al, (b) brass, and (c) copper core.

energy. Vicker's micro-hardness test showed that there is an increase in the value of the hardness near the interface. Compression-shear tests revealed that the joining strength increase when the amount of energy discharged from the capacitor bank increases. The compressive strength was found to be 5.82 MPa, 5.51 MPa, and 4.5 MPa with aluminium, brass, and the copper core, respectively, at 6.2 kJ. In the simulation, the maximum velocity was found at 177.28 m/s, 173.585 m/s, and 172.595 m/s and the maximum magnetic field were found at 20.92 T, 18.59 T, and 17.546 Tesla with aluminium, brass, and copper core at 6.2 kJ.

ACKNOWLEDGEMENTS

The authors like to thank the Central Instrument Facility (CIF) Indian Institute of Technology Guwahati for providing the instruments to complete this work.

CONFLICT OF INTEREST

The authors declare that they have no conflict of interest.

REFERENCES

[1] V. Psyk, D. Risch, B. L. Kinsey, A. E. Tekkaya and M. Kleiner, "Electromagnetic Forming—A Review," *Journal of Materials Processing Technology*, vol. 211, no. 5, pp. 787–829, 2011.
[2] R. M. Miranda, B. Tomas, T. G. Santos and N. Fernandes, "Magnetic Pulse Welding on the Cutting Edge of Industrial Applications," *Soldagem & InspeçãoSão Paulo*, vol. 19, no. 1, pp. 69–81, 2014.
[3] J. G. Lee, J. J. Park, M. K. Lee, C. K. Rhee, T. K. Kim, A. Spirin, V. Krutikov and S. Paranin, "End Closure Joining of Ferritic-Martensitic and Oxide-Dispersion Strengthened Steel Cladding Tubes by Magnetic Pulse Welding," *Metallurgical and Materials Transactions A*, pp. 1–8, 2015.
[4] M. Marya, M. Rathod, S. Marya, M. Kutsuna and D. Priem, "Steel-to-Aluminum Joining by Control of Interface Microstructures— Laser-Roll Bonding & Magnetic Pulse Welding –," *Materials Science Forum*, vols. 539–543, pp. 4013–4018, 2007.
[5] S. D. Kore, P. P. Date and S. V. Kulkarni, "Effect of Process Parameters on Electromagnetic Impact Welding of Aluminum Sheets," *International Journal* of *Impact Engineering*, vol. 34, pp. 1327–1341, 2007.
[6] Y. Haiping, F. Zhisong and L. Chunfeng, "Magnetic Pulse Cladding of Aluminum Alloy on Mild Steel Tube," *Journal of Materials Processing Technology*, vol. 214, pp. 141–150, 2014.

[7] S. D. Kore, P. Dhanesh, S. V. Kulkarni and P. P. Date, "Numerical Modeling of Electromagnetic Welding," *International Journal for Applied Electromagnetics and Mechanics*, vol. 32, no. 1, pp. 1–19, 2010.
[8] R. Kumar and S. D. Kore, "Effects of Surface Profiles on the Joint Formation during Magnetic Pulse Crimping in Tube-to-Rod Configuration," *International Journal of Precision Engineering and Manufacturing*, vol. 18, no. 8, pp. 1181–1188, 2017.
[9] J. Lueg-Althoff, A. Lorenz, S. Gies, C. Weddeling, G. Goebel, A. E. Tekkaya and E. Beyer, "Magnetic Pulse Welding by Electromagnetic Compression: Determination of the Impact Velocity," *Advanced Materials Research*, vols. 966–967, pp. 489–499, 2014.
[10] B. T. Spitz and V. Shribman, "Magnetic Pulse Welding for Tubular Applications: Discovering New Technology for Welding Conductive Materials," *TPJ - The Tube & Pipe Journal*, pp. 1–3, 2000.
[11] H. O. S. Eide and E. A. Melby, "Blast Loaded Aluminium Plates Experiments and Numerical Simulations," *Master thesis Norwegian University of Science and Technology*, pp. 1–139, 2013.
[12] G. R. Johnson and W. H. Cook, "A Constitutive Model and Data for Metals Subjected to Large Strains, High Strain Rates and High Temperatures," *Proceedings of the 7th International Symposium on Ballistics*, vol. 21, no. 1, pp. 541–547, 1983.
[13] K. C. Jorgensen and V. Swan, "Modeling of Armour-Piercing Projectile Perforation of Thick Aluminium Plates," *In 13th Intern. LS-DYNA Users Conf*, vol. 8, pp. 1–15, 2014.
[14] Z. Fan, H. Yu and C. Li, "Plastic Deformation Behavior of Bi-Metal Tubes during Magnetic Pulse Cladding: FE Analysis and Experiments," *Journal of Materials Processing Technology*, vol. 229, pp. 230–243, 2015.

Chapter 5

Aspects of Bioabsorbable Polymeric Composites in Biomedical Applications

Arbind Prasad[1], Amir Shaikh[2], Sriparna De[3], Ramesh Kumar[4], Sonika[5], and Manoj Kumar Srivastava[6]

[1]Katihar Engineering College (under Department of Science & Technology, Govt. of Bihar), Katihar, Bihar, India

[2]Graphic Era Deemed to be University, Dehradun, Uttarakhand, India

[3]Brainware University, Kolkata, India

[4]Saharsa College of Engineering (under Department of Science & Technology, Govt. of Bihar) Saharsa Bihar, India

[5]Rajiv Gandhi University, Rono Hills, Doimukh, Itanagar, India

[6]D.A.V.P.G College, Gorakhpur, UP, India

CONTENTS

5.1 INTRODUCTION

The worldwide market for polymer-based clinical items and gadgets is strikingly different and evolving every day. While there is a reasonable and constant drive for the new following development, there stays a hearty interest for clinically validated, but economically driven choices. The rising patterns in composite materials are generally taken their surprising applications in different areas including packaging [1], food processing [2],

DOI: 10.1201/9781003436072-5

building, automobile, bioproducts, biosensors, commodities development, drug delivery, and biomedical applications [3,4]. In biomedical applications, there are few elements of bioabsorbable polymeric composite that set up a good foundation for themselves as an achievement for involving in this part of utilizations. In the bone fracture fixations or orthopedic applications, metallic internal implants have numerous limitations like stress shielding, stress tangibility, leaching of metallic particles, and furthermore resurgery in the wake of recuperating of the broken bone for the extraction of the metal's implants [5,6]. In orthopedic-related issues, when the bone cracks, the patient needs to visit orthopedic specialist, on the off chance that the fracture is minute, the muscular specialist might propose to fix with the scaffolds and on the off chance that the break is serious, the internal implants are embedded to fix and adjust the broken bone through natural remodelling process. The healing of a bone fracture depends upon various factors such as age, gender, location of the bone fracture site, type of bone, etc. The polymeric implants are widely used in various applications in biomedical areas. There are several factors which led to the application of bioabsorbable polymers widely such as tunability, processability, forming, and molding capacity. The combination of polymers along with biobased ceramics, multifunctional elements, and bioactive elements makes the composites more worthful and thus as per the required application, the composites can be fabricated as per the needs. In order to mimic the bone and nearby related biomedical application sites, the nanomaterial, smart material, self-healing material, and self-sustained materials have a great impact for the fabrication of many implants these days [7]. The bioabsorbable polymers consist of many benefits; however, additionally have a few challenges such as less mechanical strength, stability, biocompatibility issues, and processing of the bioabsorbable polymers. The researchers in these areas are engaging themselves to promote or upgrade the materials to fulfill all the limitations that occur these days. The degradation in bioabsorbable polymers is mainly due to thermal exposure, catalytic degradation, mechanical degradation, and hydrolytic degradation [8]. Figure 5.1 shows the factors responsible for degradation of bioabsorbable polymers.

The degradation is taking place due to the stimulus present against the exposure through water, light chemical reaction, types of bonding between the molecules, and the mechanical strength of the composite materials [9]. In the subsequent section, the status of market and their demand for biomedical application is studied in detail.

5.2 STATUS OF MARKET AND BIOABSORBABLE POLYMERIC COMPOSITES IN BIOMEDICAL AREAS

The demand of the bioabsorbable composite is increasing year by year. Bone fracture, knee implant resurgery, hip surgery, cardiovascular application,

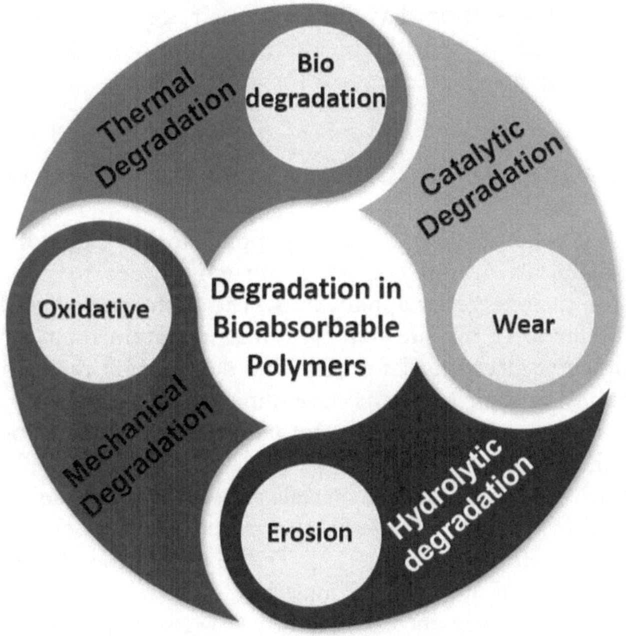

Figure 5.1 The factors responsible for the degradation in bioabsorbable polymers.

dental applications, and other biomedical applications use biobased bioabsorbable composites for their intended applications [10,11]. As per the polymer market research report, the market of the bioabsorbable polymer is estimated to be $1 billion USD and projected to reach $1.6 billion USD by the year 2026, at the CAGR of 10% in the years between 2021–2026 (https://www.marketsandmarkets.com/Market-Reports/bioresorbable-polymer-market-235258717.html). The factors for the huge demands consist of surgical interventions, demands of biomedical devices for internal and external fixations, drug delivery systems, and favorable condition for healing of the wound and fracture in nearby sites. The recent focus on minimal invasive surgeries and lack of quality control in developing countries are the major challenges associated for the bioabsorbable polymers. The environment-friendly and acceptance of regulatory standards by various environment protection norms also led to development of bioresorbable polymeric implants for biomedical applications. As per the report of bioresorbable polymer market report (https://www.marketresearchfuture.com/reports/bioresorbable-polymers-market-969), the market value of resorbable polymers are $3260.07 million USD in 2021 and expected to reach $6463.07 million USD by the year 2028 at CAGR of ~10%. The mainly polymer materials, such as polylactic acid, polycaprolactone, polydioxanone, and others, are widely used in manufacturing of bone plates, bone screws, rods, scaffolds, stents, and other biomedical devices. Thus, it is concluded from this section that the market is huge and it will always be demanding in nature to provide the alternative solution related to various conventional implants.

5.3 BIOABSORBABLE POLYMERIC COMPOSITE MATERIALS

The progress in materials for biomedical applications is expanding rapidly. Currently, materials are in composite form in order to act as multi-functional for their targeted applications. The performance of the bioabsorbable polymeric composite implants must maintain their dimensional stability, sufficient mechanical strength, and degradation time with respect to their stability in their all perspective dimensions of the fixation devices and implants. Figure 5.2 shows the factors influencing the performance of bioabsorbable implants.

The widely used polymeric materials and composites are shown in Table 5.1 below. Table 5.1 shows the polymeric composites, benefits, challenges, and their applications [12]. The clinical grade plastic materials we use for biomedical applications must have high molecular weight, biocompatible, tunable, and nontoxic in nature [13–15]. The role of nanofillers such as hydroxyapatite, cerium particle, zirconia, nano silver, copper, etc. are added in order to make the composite multifunctional as well as

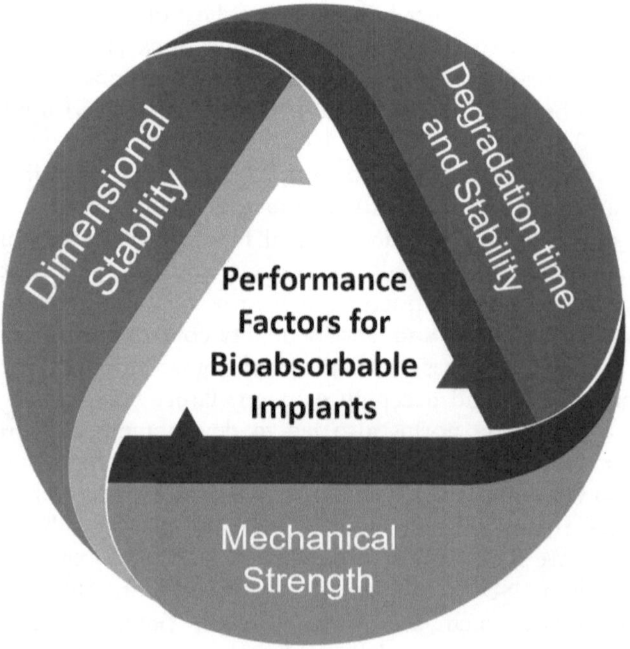

Figure 5.2 Factors influencing the performance of the bioabsorbable implants.

Table 5.1 Polymeric Composites along with Merit, Limitations, and Applications [24–27]

S.no.	*Materials*	*Merits*	*Limitations*	*Applications*
1.	Polyamides	Conjugatable side group, Highly biocompatible, degradation products	Degradation, charges induced, Toxicity	Drug delivery, scaffolds, implants
2.	Polycarbonates	Mechanical properties, surface eroding	Limited degradation, require copolymerization with other polymers	Wound healing cap, surgical devices, knee, artificial limbs
3.	Polylactide	Highly processable, widely available	Limited degradation, highly acidic degradation products	Resorbable implants, scaffolds, patches
4.	Polyurethanes	Mechanically strong, handle physical stresses well	Limited degradation, require copolymerization with other polymers	Artificial support, bone clips, implants and scaffolds
5.	Polycaprolactone	Widely used available	Limited degradation	Ligament repair, tendon regeneration, bone implants, dental, cardiovascular
6.	Polyphosphoesters	Biomolecule compatibility, highly biocompatible degradation products	Complex synthesis	Limbs, knee surgery, wound healing, skin treatment
7.	Poly (ortho esters)	Controlled degradation rates, pH sensitive degradation	Weak mechanical properties, complex synthesis	Skin treatment, wound healing, scaffolds, drug eluting devices
8.	Polyacetals	pH-sensitive degradation, mild degradation properties	Low molecular weight, complex synthesis	Bio patches, skin treatments, wound healing, scaffolds, biofilms
9.	Polyanhydrides	Significant monomer flexibility, controllable degradation rates	Low molecular weights, weak mechanical properties	Biofilms, surgical sutures, skin
10.	Polyphosphazenes	Synthetic, controllable mechanical properties	Complex synthesis	Medical support, surgical devices

antimicrobial [16–18]. The polycarbonates are naturally transparent and majorly used in medical tubing. It has high tensile, shear, and more flexural strength. It is also stronger than acrylic and also heat resistant. The PEEK-based composites are resistant towards chemicals, resist cracking, and can be easily sterilized [19].

The bioactive biodegradable polymer also classified mainly into two types i.e., natural biodegradable polymers and synthetic biodegradable polymers. The natural biodegradable polymers involve collagen, chitosan, fibrin, silk fibroin, etc., whereas the synthetic biodegradable polymers include PCL, PGA, PLA, PLLA, PLGA, PHB, PDS, etc. The biodegradable ceramics also play a vital role in mainly orthopedic fixations, in this context, the bioactive ceramics material include hydroxyapatite (HAp), tricalcium phosphate (TCP), DCPs, calcium sulphate, and silicate-based bio ceramics composite materials. Some of the bioactive glasses and biodegradable metals such as silicate, borate, and phosphate bioactive glasses, iron-based and zinc-based materials, magnesium and its alloys, also enhance neo bone regeneration. The multifunctional elements, like nanosilver, nanocerium particles, zirconia, zinc-based oxide materials, and titanium oxides, are intelligent-based materials with several properties like antifungal and antimicrobial activities in order to sustain the bone growth and development of the organs for performing their intended operations. The polylactic acid has limited mechanical strength but it is FDA-approved polymer widely used in resorbable implants. It was observed that resorbable PLA absorbed within our body within six months and also does not harm our nearby tissues for further development. The application of biocompatible neo materials always marks its presence in tissue engineering applications not only in orthopedics but also in cardiovascular, dental, skin treatment, drug delivery, and wound healing applications [20].

Figure 5.3 shows the application of bioabsorbable polymeric composite in various domains such as scaffold fabrication, soft tissue regeneration, 3D printing, bioimaging, coating, surface modification, and drug delivery applications. In tissue engineering, the growth factors such as nano particles of bone and bone mineral particles have encouraged the formation of neo bones in order to mimic natural bones during the healing phase. The coating of biocompatible materials on the metallic implants also plays an important role for successful proliferation and differentiation of cells for neo bone regeneration [21]. Once the implantation of bioabsorbable implant is inserted into the patients, the patient need not visit again for extraction, which leads to providing additional comfort in both mental and financial ways. In the case of metallic implants, during extraction, there may be a chance infection occurred during the implantation and retraction of the implants [22,23]. In this way, manufacturing of the bioresorbable polymeric implants provides not only an alternate solution but also utilizes safe, environment friendly, biocompatible, and bioactive resorbable implants for almost all widely sought biomedical applications.

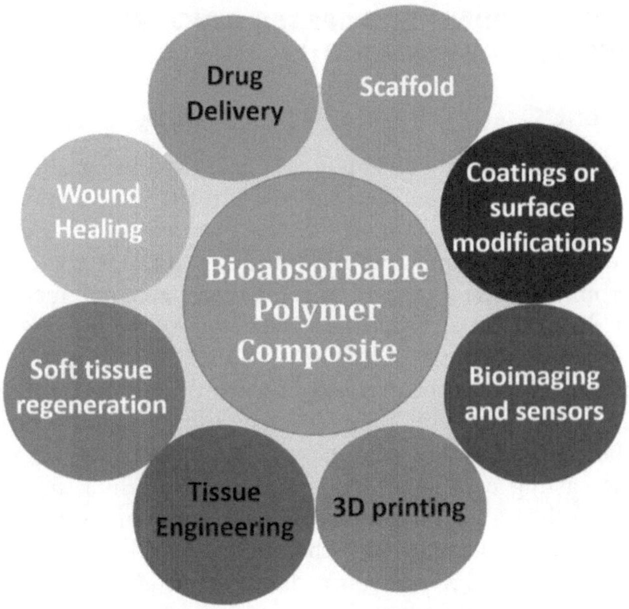

Figure 5.3 Application of bioabsorbable polymer composite materials in various biomedical applications.

The advantages associated with polymeric composites are that they are easier to produce, design to mimic, and cheaper comparatively than metallic biomaterials and implants. Nowadays, polymeric biomaterials are used to fabricate artificial muscles, biosensors, neural interfaces, drug delivery, and overall tissue engineering. The adverse application of tissue engineering involves the application of smart materials, tunable materials, and shape memory materials in order to fulfill the application of the biodegradable composite materials in all aspects of biomedical applications. It was reported that polycaprolactone/collagen fibrous scaffolds by electrospinning. Fibrous scaffolds for tissue engineering were fabricated using collagen extracted from Nile tilapia skin and polycaprolactone (PCL) by electrospinning [28]. It was observed that the ultimate tensile strength of PCL increased with an increase in collagen content. Biocompatibility studies was conducted using circular dichroism and a degradation assay in vitro indicated that the developed composite had good stability and less degradation rate. Thus, it was finally concluded as a promising candidate for tissue engineering. Collagen has been broadly utilized as a biomaterial in the drug and clinical fields for drug delivery, dressings for wound healing, bone matrix, nanocomposites, and for tissue scaffolds [29]. Collagen has several properties that include natural origin, non-immunogenicity, biocompatibility, and biodegradability; thus, the addition of collagen to a bioabsorbable polymer matrix mostly mimics the natural bone and also has wide application as scaffolds for bone fixation devices.

5.4 MANUFACTURING PROCESSES INVOLVED IN BIOMEDICAL APPLICATIONS

5.4.1 Solvent Casting Particulate Leaching

In this process, a proper solvent is selected for the specific polymers; after dissolving the polymers into the solvent, the slurry is composed of bioabsorbable polymers and nano fillers are casted on the glass plate or Teflon plate. After solvent evaporation, the casted polymeric composite is extracted from the plate and used as a bioabsorbable polymer composite sheet for further applications [5]. The schematics of the process are shown in Figure 5.4.

[30] reported biomimetic scaffolds based on hydroxyapatite nanorod/ poly (D, L) lactic acid with their corresponding apatite-forming capability and biocompatibility for bone-tissue engineering. it was mentioned that solvent casting combined with particulate leaching has been very much used for bone scaffold fabrication because of its simplicity and efficiency. It was also mentioned that this method allows to produce highly porous up to 93% and pore diameter of 500 microns by using varying porogen particle size. After the samples of the scaffold are immersed in simulated body fluid for five days, the formation of an apatite layer was observed. Development and characterization of hydroxyapatite/β-TCP/chitosan composites for tissue engineering applications was reported. The physical and mechanical properties, biodegradation, and cytocompatibility of various scaffolds formed form HAp and β-TCP at varying concentrations are formed through solvent casting and then a freeze-drying technique. The mechanical strength was increased when β-TCP was added and the cell culture experiment showed that L929 and Saos-2 cells proliferated on the surface of the composite scaffold and higher ratio of HAp/ β-TCP had a better effect on growth rate [31].

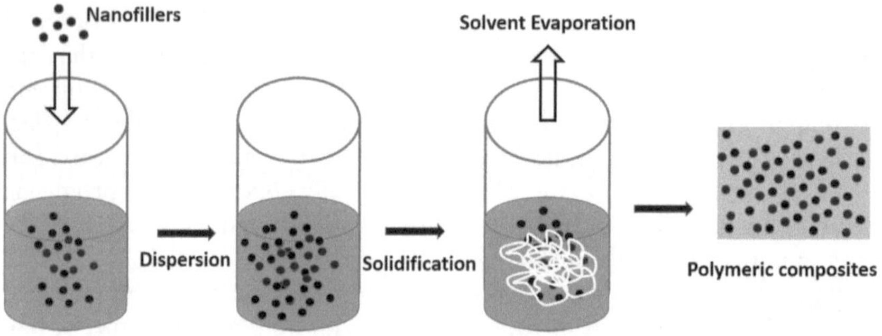

Figure 5.4 Schematics of solvent casting particulate leaching process for the development of polymeric nanocomposite sheet.

5.4.2 Additive Manufacturing Process

The additive manufacturing process utilizes information from the computer-generated computer-aided design programming to further instruct the machine to function their movement of the extruded or bed to create the 3D item as per the required dimensions. The materials are extruded layer upon layer, in exact dimensional shapes. As its name suggests, additive manufacturing adds the materials and fuses them layer by layer to make 3D objects. It is a fast process compared to other processes. Recent advances in 3D printing have the enormous potential to meet the demand of tissues and organ for transplant. The additive manufacturing based on many technologies such as fused deposition modeling, selective laser sintering (SLS), multijet fusion, and material jetting. Polymers are very much used for additive manufacturing process because of their flexibility to various cycles and it very well may be modified to complex shapes with high exactness [32–34] reported additively manufactured macroporous titanium with silver-releasing micro-/nanoporous surface for multipurpose infection control and bone repair. In this work, fabrication of Ti6Al4V scaffolds were fabricated layer by layer by EBM. The method used was cost-effective and practical; the surface obtained of the scaffolds were multifunctional in nature. The biological studies show good growth of the cells. Thus, it was concluded to have an alternate manufacturing process for scaffold fabrication and lead towards directions of biomaterials designing. Researchers are doing various studies on 3D printing technology for medical applications. In their research, It was mentioned that 3D printing technology would be able to solve the donor shortage issue for organ transplantation. It integrates the concept and application of materials science biology and biomedical knowledge related to clinical applications [35]. It was observed that if the structure is not uniform and architecture of the scaffold is not proper, then it makes a challenge to further grow the cells over it, but additive manufacturing manufactured scaffold have no such dimension defects thus it encourages the cell differentiation and proliferation as well. Further a future insight was mentioned about integration of 3D printing with tissue engineering applications [36].

5.4.3 Extrusion Process

Extrusion is basically a continuous production of objects through the extrusion machines. The main elements in extrusion machines consist of hopper, heating champers, screw rod, die, and winding platform. The pellets or granules of the desired plastics is fed through the hopper, which passes through the screw rod inside the heated barrel. There are three zones i.e., feed zone, melting zone, metering zone. The granules were melted down inside the barrel and it is melted mixed with the help of a screw rod, which pushes them towards the nozzle of the extruder.

In this process, the materials are mixed efficiently by the twinning screw rod along the barrel. The PID-controlled heaters are there so that the polymers avoid overheating and thus do not degrade through thermal exposure. Multifunctional nanohydroxyapatite-promoted toughened high-molecular-weight stereo complex poly (lactic acid)-based bio nanocomposite are for both 3D-printed orthopedic implants and high-temperature engineering applications. It was mentioned that the addition of nHAp into stereo complex PLA enhances mechanical properties and makes the composite more tough. From there onward, the representative orthopedic implant was printed using a 3D printer. The filament used for making the implant was made of stereo complex PLA/nHAp extruded composite through a twin screw extruder machine [37]. Thus, additive manufacturing might be used to achieve accurate dimensional implants with the precise dimensions.

5.4.4 Injection Molding Process

The injection molding process is one of the versatile processes, mainly used in a plastic-based industry in order to get the desired product in mass scale. In this process, plastic pellets (thermoplastic or thermoset) are feed through the feeding zone. Inside the barrel of the injection molding machine, various heating zones are there to melt the plastic pellets; then, when it becomes malleable enough, the melt mixed plastic composites are reinjected at high pressure through the nozzle to the desired mould cavity, which after solidification, the final object is formed [38]. During the process, plastic pellets are compressed by ram or screw, heated until molten, and converted into cold; thereafter, the molds are required to be filled with the molten polymers under pressure. The molded object is cooled below glass transition temperature; thereafter, the mold is opened and the final output product is ejected. After cooling to compensate the dimension of the molded object, some excess amount of molten plastic is injected to further fulfill if it is required to maintain the shape of the final molded object. It was observed that crystalline polymers have a regular arrangement of molecules that leads to slow degradation, while in the case of amorphous polymers, it has a random structure and arrangement of the molecules and because of this, the degradation process occurs more easily. Most of the bioabsorbable polymers are semi-crystalline in nature, which shows partial degradation. Manufacturing of medical components and devices using medical-grade plastic material are approachable, reliable, durable, biocompatible, and also meet Food and Drug Administration regulations through the injection molding process [39]. The implant and medical devices produced through the injection molding process are considered cost efficient and also considered as ideal for high-volume production process with tight tolerances. The cost of the medical items, including devices and implants, have been reduced. The numerous numbers of plastics are now molded through an injection molding process in order to create a prototype and artificial limbs

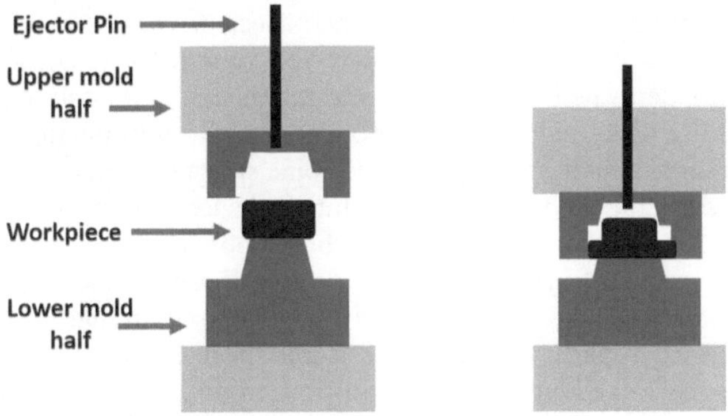

Figure 5.5 Schematics of compression molding process.

and other multi-nanofillers inserted in devices that reduce the risk of inflammatory infections [40]. Thus, the injection molding process related to production of biomedical devices and fixations increased the patient safety and ability to save more and more lives.

5.4.5 Compression Molding Process

Compression molding is a flexible and easy-operating machine for the compression of the polymeric composite materials. There, an upper part and lower part of the mold are attached with the filament coil inside to generate heat. When the polymer pellets spread on the surface of lower plate, then temperature is set to the concerned polymer pellet temperature up to 5 deg C above the glass transition temperature. The specific pressure is allowed to compress by the help of the upper and lower mould so that the molten polymer pellets gets the desired shape. After cooling, the upper and lower molds get opened and thus the compressed mould object is attained in a final shape. It has wide applications for a variety of polymer filler mixtures and also is possible to fabricate polymeric objects including implants and biomedical devices at room temperature for some specific polymers [41]. The challenge associated with this process is the production of flat devices with the mould geometry and poor matrix uniformly depending on the drug, which may result in poor reproducibility in drug release cases. Figure 5.5 shows the schematics of the compression molding process.

5.5 PERSPECTIVES AND CHALLENGES

Bioabsorbable composites have been widely used in the fabrication of various implants and internal fixation devices. The applications, including drug delivery, wound healing, implants, external and internal fixations, sutures, cardiovascular devices, artificial limbs, and knee-related devices, are very in demand. To fulfill the huge demand, the aspect of various manufacturing processes mentioned in this chapter will have a great impact in order to fulfill all the requirements with full accuracy and tolerances. In the coming days, the researchers are using bioabsorbable polymers with multifunctional properties, such as anti-inflammatory, anti-microbial, smart materials, environment friendly, resorbable, and sustain drug-eluting devices and implants [42,43]. The challenges associated with bioabsorbable polymers are mainly lack of mechanical strength, early degradation of the bioabsorbable implants, surface erosion, inadequate shapes, rapid loss of initial strength, and failure during intended applications. To meet as per specific requirements, various nanofillers are mixed in a specific proportion to match the mechanical properties and eliminate the various limitations associated with it. In terms of processing, it was observed that the difficulties were faced during the processing of bioabsorbable polymers in a cost-efficient way. The researchers are engaging themselves to increase the molecular weight so that these bioabsorbable polymers may take place widely in the area of not only biomedical but also in packaging and constructional applications.

5.6 SUMMARY

Bioabsorbable polymers and their composites have wide applications in almost all engineering and some biomedical applications. It provides to fulfill the gap of shortage of the product, which are high in demand with accuracy and tight tolerances. Application of resorbable polymers in various domains of biomedical applications gives a new researcher to select and efficiently use their concerned applications. The manufacturing process for the polymer composite processing are cost effective and flexible. The challenges associated may be overcome by utilization of new smart material, multifunctional materials, and also have some natural resource materials to enhance the mechanical strength, load-bearing capabilities, increase compatibility, and make them reliable for their intended applications. It is also accurate to say that various biomedical devices and implant requirements are increasing day by day; thus, the mass number of productions is essential in order to fill the demand required. Thus, this chapter gives an informative insight related to the use of bioabsorbable polymeric composites and their worthful applications, including their manufacturing processes.

REFERENCES

[1] S. M. Bhasney, K. Mondal, A. Kumar, and V. Katiyar, "Effect of microcrystalline cellulose [MCC] fibres on the morphological and crystalline behaviour of high density polyethylene [HDPE]/ polylactic acid [PLA] blends," *Compos. Sci. Technol.*, vol. 187, p. 107941, 2020.

[2] Y. Zhong, P. Godwin, Y. Jin, and H. Xiao, "Biodegradable polymers and green-based antimicrobial packaging materials: A mini-review," *Adv. Ind. Eng. Polym. Res.*, vol. 3, no. 1, pp. 27–35, 2020.
[3] A. Prasad, "Bioabsorbable polymeric materials for biofilms and other biomedical applications: Recent and future trends," *Mater. Today Proc.*, vol. 44, pp. 2447–2453, 2021.
[4] A. Prasad, S. Bhasney, V. Katiyar, and M. Ravi Sankar, "Biowastes processed hydroxyapatite filled poly (Lactic acid) bio-composite for open reduction internal fixation of small bones," *Mater. Today Proc.*, vol. 4, no. 9, pp. 10153–10157, 2017.
[5] A. Prasad, M. R. Sankar, and V. Katiyar, "State of art on solvent casting particulate leaching method for orthopedic scaffolds fabrication," *Mater. Today Proc.*, vol. 4, no. 2, pp. 898–907, 2017.
[6] A. Prasad, "State of art review on bioabsorbable polymeric scaffolds for bone tissue engineering," *Mater. Today Proc.*, vol. 44, pp. 1391–1400, 2021.
[7] N. Zhao, W. Chen, and T. Yan, "Application of different biomaterials in Achilles tendon repair for exercise injury," *Adv. Mater. Res.*, vol. 886, pp. 329–332, 2014.
[8] M. Li, S. Kim, S. Choi, K. Goda, and W. Lee, "Effect of reinforcing particles on hydrolytic degradation behavior of poly (lactic acid) composites," *Compos.* Part B, vol. 96, pp. 248–254, 2016.
[9] R. babu Valapa, G. Pugazhenthi, and V. Katiyar, *Hydrolytic degradation behaviour of sucrose palmitate reinforced poly(lactic acid) nanocomposites*, vol. 89. Elsevier B.V., 2016.
[10] J. C. Middleton and a J. Tipton, "Synthetic biodegradable polymers as orthopedic devices," *Biomaterials*, vol. 21, no. 23, pp. 2335–2346, 2000.
[11] C. Schulze, R. Morgenroth, R. Bader, M. Habil, D. Kluess, and H. Haas, "Fixation stability of uncemented acetabular cups with respect to different bone defect sizes," *J. Arthroplasty*, pp. 1–9, 2020.
[12] Y. Chen, J. Dou, H. Yu, and C. Chen, "Degradable magnesium-based alloys for biomedical applications *The role of critical alloying* (2019), 33(10), 1348–1372. DOI: 10.1177/0885328219834656
[13] R. Ma et al., "Preparation, characterization, in vitro bioactivity, and cellular responses to a polyetheretherketone bioactive composite containing nanocalcium silicate for bone repair," *ACS Appl. Mater. Interfaces* (2014) 15(6), 12214–12225.
[14] A. T. Neffe, K. K. Julich-Gruner, and A. Lendlein, *Biomaterials for bone regeneration*. Elsevier, 2014.
[15] M. Berni et al., "Tribological characterization of zirconia coatings deposited on Ti6Al4V components for orthopedic applications," *Mater. Sci. Eng. C*, vol. 62, pp. 643–655, 2016.
[16] A. R. Unnithan, R. S. Arathyram, and C. S. Kim, *Scaffolds with antibacterial properties*. Elsevier Inc., 2015.
[17] H. S. Nanda, "Surface modification of promising cerium oxide nanoparticles for nanomedicine applications," *RSC Adv* (2016)., vol. 6, no. 113, pp. 111889–111894.
[18] Z. Liu, Y. Zhang, H. Xia, and L. Zhang, "Biocompatible and biodegradable bioplastics constructed from chitin via green pathway for bone repair,"5(10) 2017, 9126–9135.

[19] R. Ma et al., "Preparation, characterization, in vitro bioactivity, and cellular responses to a polyetheretherketone bioactive composite containing nano-calcium silicate for bone repair," *ACS Appl. Mater. Interfaces*, vol. 6, no. 15, pp. 12214–12225, 2014.
[20] S. Suzuki and Y. Ikada, "Biomaterials for surgical operation," *Biomater. Surg. Oper.*, vol. 9781617795, pp. 1–211, 2014.
[21] A. Liu et al., "Osteogenesis catalyzed by titanium-supported silver nano-particles," *ACS Appl. Mater. Interfaces*, vol. 10, no. 37, pp. 5149–5157, 2018.
[22] Y. Han, X. Wang, H. Dai, and S. Li, "Nanosize and surface charge effects of hydroxyapatite nanoparticles on red blood cell suspensions," *ACS Appl. Mater. Interfaces*, vol. 4, no. 9, pp. 4616–4622, 2012.
[23] A. K. Nasution, M. F. Ulum, M. R. Abdul Kadir, and H. Hermawan, "Mechanical and corrosion properties of partially degradable bone screws made of pure iron and stainless steel 316L by friction welding," *Sci. China Mater.*, vol. 61, no. July 2017, pp. 1–14, 2017.
[24] Bhoi, S., Prasad, A., Kumar, A., Sarkar, R. B., Mahto, B., Meena, C. S., & Pandey, C., "Experimental study to evaluate the wear performance of UHMWPE and XLPE material for orthopedics application," *Bioengineering*, vol. 9, no. 11, p. 676, 2022.
[25] Gupta, Arvind et al., "Multifunctional nanohydroxyapatite-promoted toughened high-molecular-weight stereocomplex poly(lactic acid)-based bio-nanocomposite for Both 3D-printed orthopedic implants and high-temperature engineering applications." *ACS Omega*, vol. 2, no. 7, pp. 4039–4052, 2017
[26] Prasad, A., Chakraborty, G., & Kumar, A., Bio-based environmentally benign polymeric resorbable materials for orthopedic fixation applications. In *Advanced Materials for Biomedical Applications* (pp. 251–266). CRC Press, 2022.
[27] Prasad, A., Biomaterial-based nanofibers scaffolds in tissue engineering application. In *Functional Biomaterials* (pp. 245–264). Springer, 2022.
[28] S. Fu, et al., "In vitro mineralization of hydroxyapatite on electrospun poly(ε-caprolactone)-poly(ethylene glycol)-poly(ε-caprolactone) fibrous scaffolds for tissue engineering application," *Colloids Surf. B. Biointerfaces*, vol. 107, pp. 167–173, 2013.
[29] Q. Zhang, S. Lv, J. Lu, S. Jiang, and L. Lin, "Characterization of polycaprolactone/collagen fibrous scaffolds by electrospinning and their bioactivity," *Int. J. Biol. Macromol.*, vol. 76, pp. 94–101, 2015.
[30] N. K. Nga, T. T. Hoai, and P. H. Viet, "Biomimetic scaffolds based on hydroxyapatite nanorod/poly(D,L) lactic acid with their corresponding apatite-forming capability and biocompatibility for bone-tissue engineering," *Colloids Surf. B. Biointerfaces*, vol. 128, pp. 506–514, 2015.
[31] A. Shavandi, A. E.-D. A. Bekhit, M. A. Ali, Z. Sun, and M. Gould, "Development and characterization of hydroxyapatite/β-TCP/chitosan composites for tissue engineering applications," *Mater. Sci. Eng. C*, vol. 56, pp. 481–493, 2015.
[32] R. Kumar, M. Kumar, and J. S. Chohan, "The role of additive manufacturing for biomedical applications: A critical review," *J. Manuf. Process.*, vol. 64, no. September 2020, pp. 828–850, 2021.

[33] D. Wu, A. Spanou, A. Diez-escudero, and C. Persson, "3D-printed PLA/HA composite structures as synthetic trabecular bone: A feasibility study using fused deposition modeling," *J. Mech. Behav. Biomed. Mater.*, vol. 103, no. August 2019, p. 103608, 2020.
[34] Z. Jia et al., "Additively manufactured macroporous titanium with silver-releasing micro-/nanoporous surface for multipurpose infection control and bone repair – a proof of concept," *ACS Appl. Mater. Interfaces*, vol. 8, no. 42, pp. 28495–28510, 2016.
[35] Q. Yan et al., "Additive manufacturing — review a review of 3D printing technology for medical applications," *Engineering*, vol. 4, no. 5, pp. 729–742, 2018.
[36] D. Puppi and F. Chiellini, "Biodegradable polymers for biomedical additive manufacturing," *Appl. Mater. Today*, vol. 20, 2020.
[37] A. Gupta et al., "Multifunctional nanohydroxyapatite-promoted toughened high-molecular-weight stereocomplex poly(lactic acid)-based bionano-composite for both 3D-printed orthopedic implants and high-temperature engineering applications," *ACS Omega*, vol. 2, no. 7, pp. 4039–4052, 2017.
[38] N. Monmaturapoj et al., "Properties of poly(lactic acid)/hydroxyapatite composite through the use of epoxy functional compatibilizers for bio-medical application," *J. Biomater. Appl.*, vol. 32, no. 2, pp. 175–190, 2017.
[39] L.-T. Lim, R. Auras, and M. Rubino, "Processing technologies for poly(lactic acid)," *Prog. Polym. Sci.*, vol. 33, no. 8, pp. 820–852, 2008.
[40] R. R. Bos et al., "Degradation of and tissue reaction to biodegradable poly(L-lactide) for use as internal fixation of fractures: A study in rats," *Biomaterials*, vol. 12, no. 1, pp. 32–36, 1991.
[41] L. Chen, C. Y. Tang, D. Z. Chen, C. T. Wong, and C. P. Tsui, "Fabrication and characterization of poly-d-l-lactide/nano-hydroxyapatite composite scaffolds with poly (ethylene glycol) coating and dexamethasone releasing," *Compos. Sci. Technol.*, vol. 71, no. 16, pp. 1842–1849, 2011.
[42] Kim, Tiffany, See, Carmine Wang, Li, Xiaochun, & Zhu, Donghui (2020). Orthopedic implants and devices for bone fractures and defects: Past, present and perspective. *Engineered Regeneration*, 1, 6–1810.1016/j.engreg.2020.05.003.
[43] N. Mulchandani, A. Prasad, and V. Katiyar, *Resorbable Polymers in Bone Repair and Regeneration*. Elsevier Inc., 2019.

Chapter 6

Green and Sustainable Manufacturing Processes for Motor Cores in EV

T. Aizawa[1] *and T. Shiratori*[2]

[1]Surface Engineering Design Laboratory, Shibaura Institute of Technology, Japan

[2]Faculty of Engineering, University of Toyama, Japan

CONTENTS

6.1 INTRODUCTION

Motor core was an essential part not only in the electric vehicles (EV) but also in the conventional passenger and commodity cars [1]. Many tiny motors are utilized to improve the comfortability when driving the vehicles. In particular, a high-capacity motor core with low ion loss is highlighted in every automotive company under the worldwide political conditions with an urgent shift to production of EV [2]. This high qualification of motor-cores with resource-saving and energy-saving nature is just corresponding to a future direction to the sustainable society as insisted in the propaganda on the SDGs (sustainable development goals) [3]. Remember that EV has a role to reduce the environmental burden in driving by zero-emission of CO_2 gas in

DOI: 10.1201/9781003436072-6

the green society, but that lots of electrical energy is wasted in driving at the same time. Then, the ion loss, which essentially contributes to the energy loss in EV, must be saved as possible [4]. Remember also that the motor core is made from the magnetic materials, and then their mass must be reduced to advance the light-weight design of EVs. In addition, those have to made from the recycled ones or the raw materials with a big Clarke number in order to promote the resource saving [5]. Under these constraints, these recycled or raw materials are transformed to the magnetic sheet-coils by using the energy-saving process and those sheet coils are net-shaped to a stack of sheared magnetic pieces as a motor-core. Thin Fe–Si electrical steel sheets with high silicon contents were produced by rolling and annealing in the first step [6]. Its sheet-coils were punched out to complex-shaped parts and stacked into a motor-core [7]. The thinner electrical sheet was selected to reduce the weight of motor-core and to increase the specific power generation capacity. Iron and silicon were also selected as its constituent element since they have a large Clarke number. Then, the engineering issue toward high qualification of motor-core is focusing into two items. The first item is a microstructural improvement of electrical steel sheet in the rolling and annealing and in the punching or piercing of rolled sheets to the constituent pieces of motor-core. The second item is a tooling design to punch out the raw electrical steel sheet to complex-shaped constituent pieces of motor core with minimization of iron loss.

In the discussion on the first item, let us consider the relationship between the magnetism and the microstructure in the polycrystalline electrical sheet. In the weak magnetic materials, including the polycrystalline electrical steels, each grain has easy-to-magnetize and difficult-to-magnetize axes. When the high silicon iron alloys are rolled and annealed, their grains are also aligned to the specified orientation. As stated in [8], when these oriented electrical steel sheets are used into a motor core, the anisotropic iron loss is considered in practice. On the other hand, when the easy-to-magnetize direction in each grain is controlled to be randomly oriented, the electrical steel sheet becomes non-oriented one so that the iron loss turns to be isotropic. This crystallographic relation to magnetism also plays a role in the affected microstructure of electrical steel sheets after punching. In parallel to those polycrystalline Fe-Si alloy sheets, the amorphous electrical sheets are also highlighted to improve the capacity of motor core and to reduce its total weight [9]. Higher strength and zero ductility hinders the precise punching and piercing of amorphous sheets [10].

In the discussion on the second item, the tooling design is needed to improve the dimensional accuracy of punched-out sheet piece geometry, to reduce the induced damage and defects by punching, to reduce the losses and to prolong the tool life [11]. In particular, the piercing punch for fabrication of motor core units must be designed from various production engineering and scientific points of view [12]. As stated in [13], the magnetic properties of motor core were often severely deteriorated by

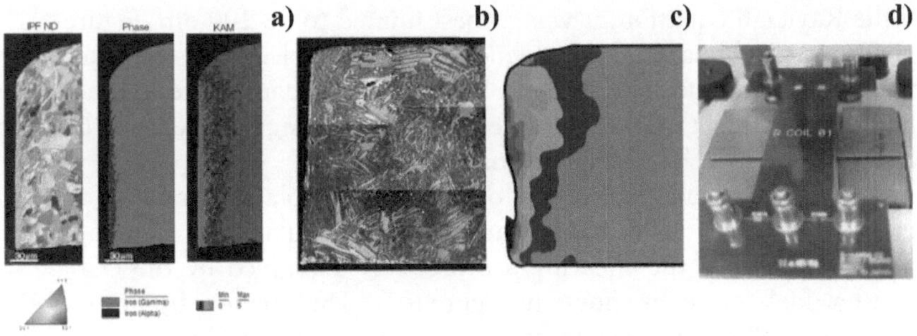

Figure 6.1 Multi-physical evaluation on the deterioration and damage induced into the electrical steel sheets of motor core by punching. a) Microstructure distortion analyzed by EBSD, b) magnetic zone distortion by Kerr-effect microscopy, c) hardness mapping, and d) iron loss in the motor core.

microstructure damage via the punching process. This suggests that a wider perspective view is needed to build up the optimum piercing punch configuration. Reference [14] insisted that a multi-physical modeling is indispensable to make a tailored design on the electro-magnetic parts including the motor cores. In corresponding to this multi-physical model in the theoretical approach, a multi-physical measurement is also necessary in the experimental evaluation on the microstructure and magnetic zones.

In the research and development for the second item, the microstructure distortion and magnetic deterioration induced by shearing the electrical steel sheets must be described and analyzed by this multi-physical measurement. Figure 6.1 depicts a typical multi-physical evaluation method on the deterioration and damage induced into the electrical steel sheets of motor core by punching.

At first in this multi-physical measurement, SEM (scanning electron microscopy) and EBSD (electron back-scattering diffraction) play a role of material science methods to describe the microstructural change during the sharing process. In particular, the crystallographic change, the equivalent plastic strain distribution, and the phase transformation in punching the metallic sheets can be understood by the IPF (inverse pole figure), the KAM (kernel angle misorientation) and the phase mapping as depicted in Figure 6.1a.

In order to visualize the magnetism in the electrical steels, the Kerr effect was utilized, where a rotation of linearly polarized light was caused on reflection from a magnetic material by this magneto-optical interaction [15]. When using the confocal scanning Kerr microscope, the slightly scattered optical change by each magnetic domain was scanned through slits and represented as an image by digitally subtracting the non-magnetic background. Since the first finding of this Kerr effect, many studies have been done to improve the resolution of this optical microscopy. Since the resolution by the constructive interference in the microscope was expressed by the Rayleigh equation, it was at best limited to be 200 nm. Figure 6.1b depicts a typical image on the magnetic domain of electrical steel sheet. The inner straining induced by shearing the sheet significantly affected the zone-alignment so that those affected regions by shearing were detected as the distorted zones in the measured image.

The work hardening behavior or the internal plastic straining state is described by using the hardness mapping. In particular, the residual, plastic strains induced by the shearing are precisely explained by the change in these hardness maps, as shown in Figure 6.1c. The iron loss by shearing the electrical steel sheets was measured by the hysteresis loss of magnetic medium. As shown in Figure 6.1d, the alternative current was applied to the sheared sheet specimen so that the response of the measured voltage and current should have an intrinsic hysteresis curve to each specimen.

Through this multi-physical measurement, a tooling for shearing the electrical steel sheets is optimized to reduce the inner plastic strains as well as the damages in magnetic domains. In this chapter, a new tooling for punching and shearing the electrical steel sheets is proposed to reduce the deterioration and damage toward high qualification of electrical steel sheets of motor cores. This new tooling requires for no lubrication; without additional increase of environmental burdens, this tooling provides a solution to improve the capacity of motor-cores with full usage of original electrical steel sheets.

6.2 TOOLING DESIGN FOR FINE PIERCING PUNCH AND DIE

In the classical treatise on the shearing process in the technology of plasticity, most of the studies have been concerned with optimization of clearance between the die and punch, surface and heat treatment of punch and die substrates, and suitable selection of punch and die materials [16]. In particular, the narrow clearance was believed to be indispensable in the fine piercing and fine blanking processes [17]. Let us start with how to make a tooling design toward the fine piercing of electrical steel sheets with high qualification of products and with engineering durability on the tool life.

Figure 6.2 depicts the platform of tooling design including several engineering items. The punch and die substrate materials have been improved to promote their strength, hardness, and toughness [18]. The tool steel has a family with different content of chromium in its alloying; e.g., SKD11 is selected for cold forging and fine blanking, and SKD61 is utilized for warm and hot forging [19]. WC (Co) with higher cobalt content than 8 mass% is used to improve the fracture toughness for working in severe forging conditions while the cobalt content is lowered to improve its hardness [20].

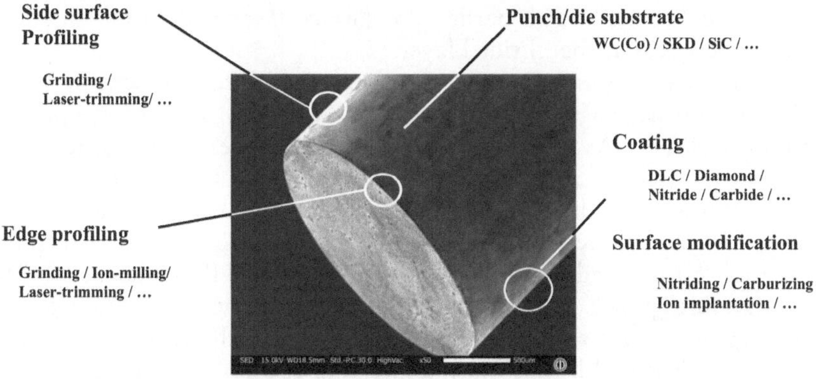

Figure 6.2 Tooling design on punch/die substrate materials, coating and surface modification, and tool edge and side surface profiling for fine punching of electrical steel sheets.

The WC-grain size is more reduced down to μms or sub-μms to protect the drop-out of WC-grains from the punch and die edges during stamping [21]. To be discussed in later, this drop-out and tiny chipping results in the fatal failure of punch and die in severe shearing conditions [22]. In recent, the ceramics, including the alumina, the zirconia, the silicon carbides, and the silicon nitrides, were utilized as a substrate to be working in severe metal-forming conditions [23]. In particular, the PSZ (partially stabilized zirconia) and the silicon carbide are suitable to metal forming, including the deep-drawing [24].

To prevent those pinch and die substrates from fatal failure, the coating was often selected as the first remedy to strengthen those substrates. The binary or tertiary nitride coatings [25] as well as DLC (diamond like coating) and polycrystalline diamond coatings [26] have been applied as a protective layer to prevent the punch and die substrate from chipping and fatal damage during the forging, stamping and punching processes in the cold, warm, and hot conditions even without lubricating oils. In recent, thick CVD (chemical vapor deposition) – DLC, CVD-diamond, and CVD-SiC coatings were highlighted as a coating punch and die substrate to improve the mechanical properties. A thick CVD-DLC coating die was utilized as a mother punch to imprint the fine microtextures into aluminum sheet for thermal radiation control [27]. A thick diamond coating was used as a die to shape the micro-textures into the metallic products [28]. Thick β-SiC coating was also employed as a punch to make galling-free forging of titanium [29].

In parallel with this coating technology, the surface modification played an essential work to significantly prolong the life time, to build up the micro-textured punch and die and to functionalize the die substrates. Among various surface treatments, the plasma nitriding process has grown up as a reliable method to harden the surface layer and to make microstructural control at the nitrided layer.

This plasma nitriding process was categorized into two schemes; e.g., high-temperature plasma nitriding (HT-PN) and low-temperature plasma nitriding (LT-PN). As stated in [30], the former process was characterized by fine CrN (chromium nitride) precipitation into the die matrix. The latter process was characterized by the nitrogen supersaturation into matrix without synthesis of CrN [31]. Due to the relatively thick nitrided layer than 50 μm and higher hardness than 1,200 HV by using LT-PN, the trimming punch life was significantly prolonged in the transfer stamping of copper alloy products [32]. The punch with a multi-head array was fabricated by the plasma nitriding-assisted printing for micro-embossing into the copper products [33]. With the aid of the nitrogen supersaturation with a higher content than 3 mass% into the substrate, the molding stamping dies were shaped and finished to have a mirror-shining surface only by fine machining without grinding and polishing [34]. In later, this LTPN is utilized to yield the nitrided SKD11 punch for fine piercing.

In the similar way, the low-temperature plasma carburizing process works as an innovative method for galling-free forging, fine-blanking and punching of titanium and titanium alloys [35]. Due to the carbon supersaturation into a AISI420-J2 die [35] and SKD11 punch [36], the titanium and titanium alloy work materials were lubricated in solid since the tribofilm was in situ formed by the isolated carbon solutes from the carbon supersaturated die and punch matrices. The friction and work hardening were reduced to continue the galling-free forging with higher reduction of thickness than 50%.

The ion implantation was also an effective method to embed the tailored element into the punch and die matrices with a higher content than 2–3 mass% for reduction of friction and wear. In case of the chlorine implantation into TiN-coated substrates, the original higher friction coefficient than 1.0 was reduced and preserved to be 0.1 to 0.15 in the long-term running by the ball-on-disc testing [37]. The specific wear volume was also reduced in the order of two by the in situ lubrication mechanism of chlorine-implanted TiN coating [38].

In addition to these innovations in the material selection and nanostructuring for punches and dies, their edge and side surface profiling technologies were important to control the stress concentration at the edge and local plastic flow around the edge curvature, to regularize the plastic flow along the side surface, and to preserve the debris particles on the side surface layer and efficiently eject them out of punching front.

The punch-and-die materialization and geometric profiling technologies in Figure 6.2 are considered an innovative tooling concept toward the high qualification of products and the tailored tool-life extension, instead of the classical treatise of piercing punches.

In the following study, four different types of punch-and-die pairs are proposed for piecing experiments to find an engineering direction toward high qualification of motor cores with long-life tooling; e.g., the mechanically ground WC (Co) punch-and-die pair, the ion-milled WC (Co) punch-and-die pair, the diamond-coated WC (Co) punch with and without laser trimming, and the nitrided SKD11 punch with laser trimming and nitrided SKD11 die.

Selection of WC (Co) punch and die is a standard tooling with the use of their high hardness. In particular, the recent technology to reduce the WC-grain size and to precisely control the cobalt binder content enables to enhance the strength and toughness in addition to hardness. This WC (Co) substrate is machined and mechanically ground to sharpen the punch and die edges. The ion-milling becomes a useful method to sharpen the edges of any die materials. In the present ion-milling system, the argon ion beam is ejected from the argon plasmas to irradiate the rotating die sample with the specified skew angle. The edge sharpness with less than 1 μm is attained by optimization the intensity and fluence of argon ion beams [39].

The hard coating methods have been employed to reduce the abrasive wear of WC (Co) punch and die during stamping and forging. Among them, the diamond coating is often deposited onto the pretreated WC (Co) substrate because of its highest hardness [40]. To be noticed, its functional properties are controllable by doping the boron, nitrogen and so forth [41]. As stated in [42], the diamond coating surface has significant roughness due to its three-dimensional nucleation and growth on the WC (Co) substrate. The mechanical rubbing [43], the plasma ashing [44], and the laser trimming [45] have been invented to reduce its surface roughness within the tolerance for tooling. Among them, the femtosecond laser trimming [46–48] provides a way to directly control the punch and die edge and side-surface profiles. In particular, the micro- and nanotexturing during the femtosecond laser trimming process is effective to improve the pierced hole surface quality and to prolong the punch-and-die life [49,50].

The tool steel substrates have been utilized as a punch and die material for fine blanking and press-forging processes because of higher ductility and toughness than WC (Co). Their hardness is much lower than WC (Co); heat treatment and surface modification methods are necessary to improve their hardness enough to reduce the abrasive wear and to prevent them from chipping. In the present approach, the LT-PN is employed as the surface treatment to harden the SKD11 punches by the thick nitrided layer.

In practical operations of punching and forging processes, there are many cases to experience the deteriorations or severe damages of punches and tools. Figure 6.3 summarizes a typical failure mode of punches when punching the metal sheets and plates. WC (Co) punch often suffered

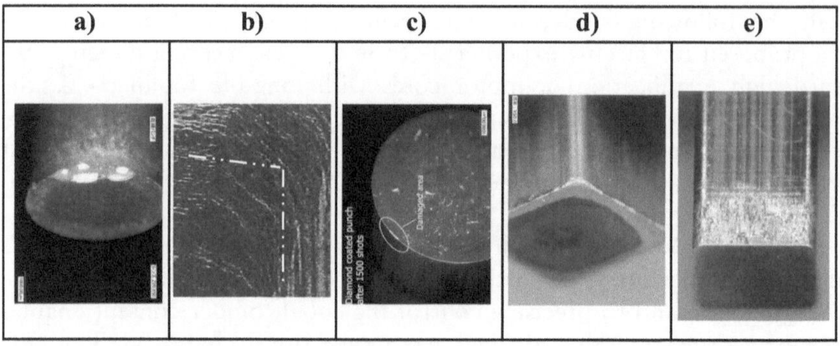

Figure 6.3 Typical failures of the piercing punch to shorten the tool life in piercing the electrical steel sheets. a) Chipping failure of punch substrate, b) abrasive wear of punch substrate, c) chipping failure of diamond coatings, d) abrasive wear of hard DLC coating, and e) adhesion wear of punch substrate.

from the chipping in piercing the electrical steel sheets under narrow clearance. As shown in Figure 6.3a, the punch edge chipped so much and resulted in fatal failure. Figure 6.3b depicts a typical abrasive wear of tool steel punch. Compared to the original geometry of punch, which was represented by two-dot chain lines, a significant amount of edges was worn out when piercing the hard metal sheets. Even when using the diamond-coated WC (Co) punch in piercing under narrow clearance, the diamond coating is difficult to be free from chipping, as shown in Figure 6.3c. Due to the intrinsic high hardness of diamond, the chipped volume of coating is much less than seen in Figure 6.3a; this chipping deteriorates the quality of sheared hole surface. In severe punching operations, DLC coating has a risk of abrasive wear. Figure 6.3d shows the complete worn-out trace of DLC coating around the punch edge in the early stage of punching process. When punching the stainless steel and titanium plates, their debris particles and fragments adhere to the punch side surface. As depicted in Figure 6.3e, the punch suffered from full adhesion wear of debris particles even after punching in a single shot. In the following sections, the importance of a piercing punch-and-die design toward high qualification of punched products is demonstrated through various piercing experiments.

6.3 TOOLING FOR FINE PIERCING OF POLYCRYSTALLINE ELECTRICAL STEEL SHEETS

Most of the motor-cores are made from the polycrystalline electrical steel sheets. In particular, un-oriented high silicon steel sheets are widely utilized for various automobiles including the EV and hybrid cars. During the piercing process, the damages zones are often induced to distort the

Figure 6.4 Three types of the punch edge profile control for piercing the polycrystalline electrical steel sheets. a) Mechanically ground WC (Co) punch, b) ion-milled WC (Co) punch, and c) femtosecond laser trimmed WC (Co) punch.

magnetic domain structure and to increase the iron loss as imagined from Figure 6.1. In this section, three pairs of piercing punch and die are prepared to investigate the effect of piercing process on the pierced hole surface condition, the residual strains, and the iron loss.

Figure 6.4 compares these three WC (Co) piercing punches with each diameter of 5 mm. The first punch was mechanically shaped, ground, and polished after the normal procedure to fabricate the commercial punches in industries [51]. Since the WC-grains dropped out during the mechanical adjustment, the punch edge becomes heterogeneous and the edge curvature increases to 8 mm, as shown in Figure 6.4a. This normal WC (Co) punch was ion-milled to sharpen this edge lines by applying 2 kV with additional DC bias of 500 V. Since the normal punch was setup with a skew angle to the argon ion beam profile, its top and side surfaces were simultaneously ion-milled to clean up both surface conditions. As a result, the punch edge is sharpened to be homogeneous, as depicted in Figure 6.4b. The top and side surfaces of a normal WC (Co) punch was laser-trimmed to reduce each surface roughness and to sharpen the punch edge. The femtosecond laser machining was utilized to advance this trimming process without the thermal effect [52]. In general, with reducing the wavelength of the incident laser beams down to the micro-meter range, the laser irradiation induces the optical interaction between the incidental beam and the scattering beam by the surface roughness [53]. In particular, when using the femtosecond lasers, this interaction forms the 10 nm to sub-μm sized nanotextures onto the irradiated surface [54]. This LIPSS (laser induced periodic surface structuring) forms the rippled surface structure with its pitch of (1/3) to (1/2) of incident laser wavelength [55]. As shown in Figure 6.4c, the punch top surface was polished by this laser trimming; the punch side surface condition was modified to have micro-textures together with the nano-textures with the LIPSS period. As a result, the punch edge line between the top and side surfaces was also sharpened at the same time.

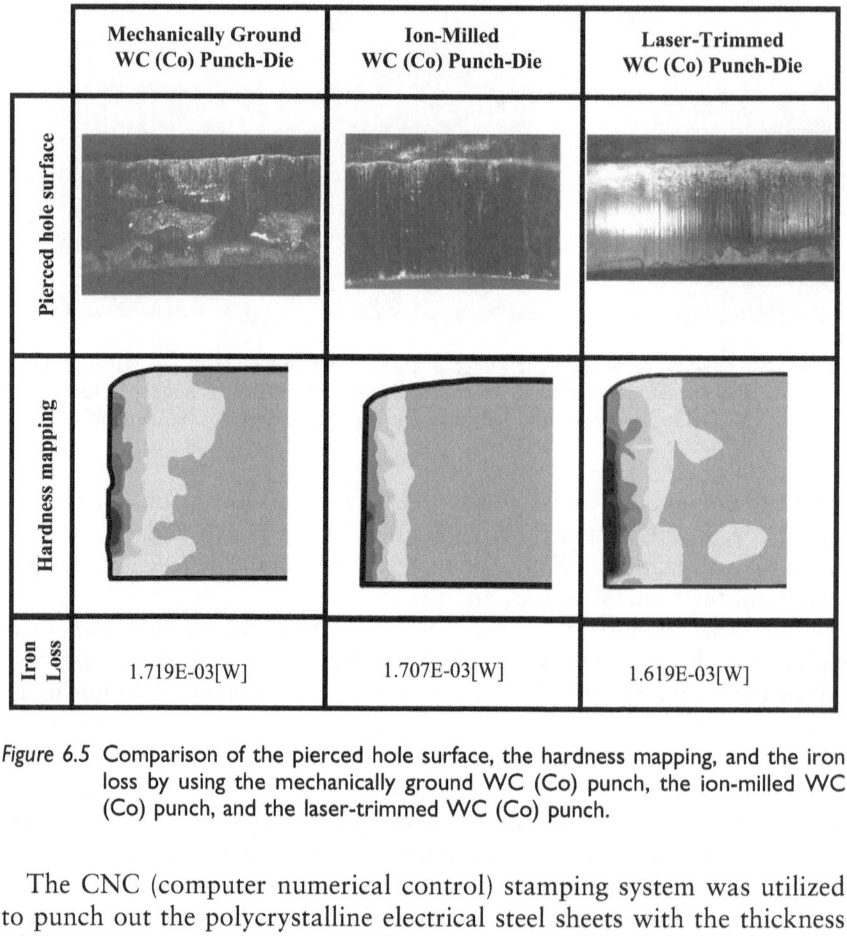

Figure 6.5 Comparison of the pierced hole surface, the hardness mapping, and the iron loss by using the mechanically ground WC (Co) punch, the ion-milled WC (Co) punch, and the laser-trimmed WC (Co) punch.

The CNC (computer numerical control) stamping system was utilized to punch out the polycrystalline electrical steel sheets with the thickness of 0.5 mm by using three punches. A normal WC (Co) die was also prepared and ion-milled to have sharp die edge. This die was used in common in the following punching experiments. Figure 6.5 compares three punched hole surfaces with a diameter of 5 mm. When using the normal WC (Co) punch, the fractured areas were seen on the surface. This reveals that normal WC (Co) punch is never selected to punch out the fine shaped products and to shape the perforated sheet. When using the ion-milled WC (Co), little fractured areas were seen on the punched hole surface. The ratio of burnished surface reached to nearly 100%; the surface seemed to be dull. The laser-trimmed WC (Co) punch was used for this punching process; the punched hole surface was fully burnished with metallic shining. This improvement of punched hole surface by using the ion-milled and laser-trimmed punches suggests that mechanical state and magnetic properties of perforated electrical steel sheets should be improved.

Let us investigate the residual strain profiles in the perforated electrical steel sheets and measure the iron losses in them. The normal, ion-milled, and laser-trimmed WC (Co) punches and dies were prepared to punch out the rectangular hole of 50 mm × 5 mm. Two pieces of punched-out Fe-6Si sheets were set into the circuit unit to measure the iron loss under the alternative current of 1 kA/m. The hardness mapping technique was utilized to measure the residual strain distribution in each punched-out specimen. The work-hardened region area became maximum when using the normal WC (Co) punch, while it minimized when using the ion-milled punch., as depicted in the middle of Figure 6.5. The iron loss decreased with changing the normal WC (Co) punch by the ion-milled and laser-trimmed punches in a series, as shown in Figure 6.5. This multi-physical comparison on the punching effect reveals that high qualification of punch configuration in Figure 6.2 has an essential influence on the accumulated strains and iron loss of the punched-out and perforated electrical steel sheet units.

6.4 TOOLING FOR FINE PIERCING OF AMORPHOUS ELECTRICAL STEEL SHEETS

Recently, the amorphous electrical steel sheet has been highlighted as a new motor-core material to reduce the weight of motors and the hysteresis loss and to improve the magnetic properties. In the literature [56], many trials were reported to make fine piercing these amorphous steel sheets. Most of them failed in piercing without damages in their microstructure and magnetic properties [57,58]. On the other hand, a new amorphous steel sheet was invented to have a higher ultimate strength than 2 GPs; however, it has still no ductility and lower uniform elongation than 1% [59]. These difficulties in punching out the high strength and brittle amorphous sheets to motor-core units still remain to annoy the steel makers, the motor suppliers, and the automotive companies. An innovative procedure is intensely needed to reduce the affected damages into sheets by punching and to preserve their magnetic properties [60].

In this session, the laser trimming pretreatment is utilized to sharpen the punch edge and to nano-texture the punch side surface. The mechanically ground WC (Co) punch is used as a reference to investigate the effect of edge-sharpness and nanotextures on the reduction of induced damages by piercing process.

6.4.1 Amorphous Electrical Steel Work

An amorphous electrical steel sheet has high strength with brittleness. The ultimate strength reaches 2,240 MPa at the nominal strain of 1.09%. The thickness of the sheet was 25 μm. In the piercing experiments, a stack of five sheets was prepared and employed as a work.

6.4.2 Preparation of Four Types of Punches

Four punches were prepared for piercing experiments as listed in Figure 6.6. A normally ground WC (Co) punch with the diameter of 1.986 mm was edge-sharpened to be less than 2 μm by the grinding machine, as depicted in Figure 6.6a. Since several WC grains were mechanically ground during the edge-sharpening process, the edge profile has inhomogeneous angulation. This mechanically ground punch is referred by punch-0. The diamond-coated WC (Co) punch with the diameter of 1.958 mm was laser-trimmed to have the edge curvature radius of 2.75 μm. The skewed nano-grooves with the depth of 100 nm and the pitch of 300 nm were formed from the punch edge onto the side surface with regular alignment as shown in Figure 6.6b. This punch was symbolized as a punch-1. The WC (Co) die with the inner diameter of 2.002 mm was commonly utilized as a die-0 and die-1 in the piercing experiments with the use of punch-0 and punch-1, respectively. The SKD11 substrates were also utilized instead of the WC (Co).

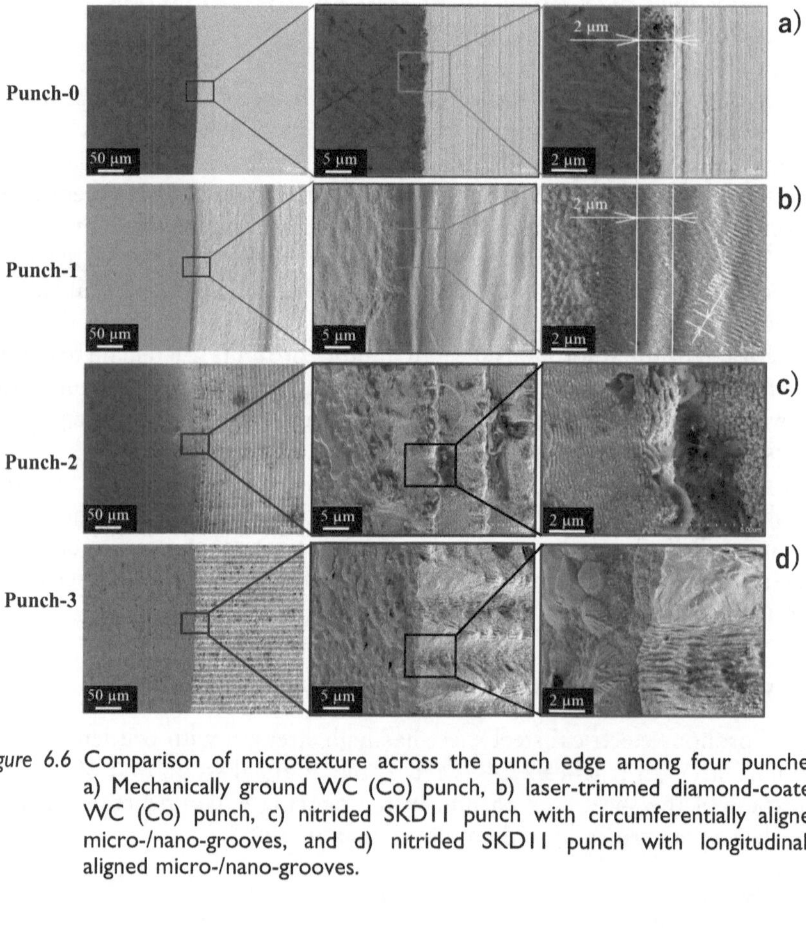

Figure 6.6 Comparison of microtexture across the punch edge among four punches. a) Mechanically ground WC (Co) punch, b) laser-trimmed diamond-coated WC (Co) punch, c) nitrided SKD11 punch with circumferentially aligned micro-/nano-grooves, and d) nitrided SKD11 punch with longitudinally aligned micro-/nano-grooves.

To compensate for the hardness difference between bare WC (Co) and SKD11 substrates, the LT-PN was employed to make solid-solution hardening of SKD11 up to 1,400 HV.

As mentioned, the plasma nitriding process is categorized into two treatments [61]. In the HT-PN at a higher holding temperature than 673 K (or 400°C) and longer duration than 10 hrs, the chromium nitrides precipitate into the substrate matrix and harden it by precipitate strengthening mechanism [62–64]. On the other hand, no iron and chromium nitrides are synthesized in the LT-PN at a lower holding temperature than 673 K and shorter duration than 10 hr. The substrate matrix is hardened by solid-solution strengthening mechanism because of the nitrogen supersaturation with high nitrogen solute content [65–69]. In addition to the different hardening mechanisms between HT-PN and LT-PN, the following four features are essentially different:

1. The chromium content [Cr] is reduced in HT-PN due to the direct synthesis of CrN between Cr and N solutes in the nitrided layer, while [Cr] does not change itself because of none nitride synthesis in LT-PN [70],
2. Most of the nitrogen content is bound with constituent elements in a matrix, especially chromium, as nitrides and the nitrogen solute content [N] is usually lower than the maximum nitrogen solubility in the case of HT-PN. [N] is much higher than this maximum solubility in case of LT-PN [71],
3. [N] exponentially decays from the matrix surface to the depth since the HT-NP is governed by the body-diffusion mechanism, while the nitrogen solute depth profile has a plateau with [N] > 3 mass% [72],
4. After HT-PN, the microstructure in the nitrided layer turns to be a composite of matrix and nitride precipitates without phase transformation and grain-size refinement. While in LT-NP, the microstructure in the nitrided layer is significantly refined to have super-fine granular structure with two-phase structure [73],
5. The microstructure of a matrix below the nitriding front end (NFE) remains the same as before HT-NP, while it is modified by nitrogen supersaturation because of the nitrogen boundary diffusion across NFE [74],
6. No further essential evolution of microstructure takes place with increase the holding temperature in the case of HT-PN, while in LT-PN, the microstructure changes itself with decreasing the holding temperature and with refining the initial crystalline structure [75].

Two types of nitrided SKD11 punch were prepared and laser-trimmed to have different micro-/nano-grooves on each punch side surface, as shown in Figure 6.6c and 6.6d, respectively. The punch-2 with the diameter of 1.981 mm has a circumferential micro-/nano-grooves. These micro-grooves have a pitch of 10 μm and the depth of 4 μm; the nano-grooves have the pitch of 300 nm and the depth of 100 nm. The punch-3 with the diameter of 1.985 mm has a longitudinal micro-/nano-grooves. Their dimension remains to be the same as the punch-2. The nitride SKD11 dies were also prepared to have an inner diameter of 2.003 mm and 2.004 mm, respectively, as the die-2 and die-3 for the punch-2 and punch-3. The clearance (δc) between four punches and dies were fixed to be nearly the same among four punch-die pairs; e. g., δc for punch-0 vs die-0 and punch-1 vs die-1 pairs was 8 μm, δc for punch-2 vs die-2 was 11.0 μm, and δc for punch-3 vs die-3 was 9.5 μm. To be noticed, wider clearance was set up in the latter two punch-die pairs.

6.4.3 Affected Damages to Amorphous Sheet Stack by WC (Co) Punch-Die Pairs

Various damages are induced into the amorphous foils and sheets when piercing them. As depicted in Figure 6.7a, three types of damages were detected by SEM observation at the right-hand side of pierced hole when using the punch-0 and die-0. A-defect denotes for the droop-like damage is induced at the vicinity of pierced hole. B-defect is induced as a periodic peak-to-valley angulation in the circumferential direction, where the circumferential cracks are formed on the convex parts in angulation. This B-defect suggests that amorphous sheet is pushed back by the piercing punch to deform in compression at the valleys and in tension at the peaks. C-defect is a typical inner crack, running in the circumferential direction. These defects are responsible for deterioration of microstructure and magnetic properties in the amorphous electrical steel sheets.

When using the punch-0 and die-0, A-defect width (W_A) reaches 10 μm, B-defect width (W_B) is 45 μm, and C-defect width (W_C) is 30 μm. This huge damage drastically reduces in case of the punch-1 and die-1; e.g.,

Figure 6.7 Comparison of the affected damages to amorphous sheet stack when piercing by piercing two WC (Co) punch-die pairs. a) Characteristic damages induced by piercing with the use of punch-0 and die-0, b) characteristic damages induced by piercing with the use of punch-1 and die-1 pair, and c) comparison of the damage profile among five pierced sheets in a stack by piercing with used of two pairs.

W_A = 2 μm, W_B = 0 μm and W_C = 10 μm in Figure 6.7b. This reveals that amorphous work sheet deformation response in piercing is significantly improved by selection of punches.

Figure 6.7c compares the variation of total damage widths (W_{defect}) with the number of sheets (N) in the stack between two punch and die pairs. In case of the punch-0 and die-0, W_{defect} reaches three times larger than the amorphous sheet thickness for $N > 1$. This implies that those defected zones must be removed from the as-pierced sheets to sustain the original magnetic properties of amorphous steel sheets. This additional treatment is costly and against the material saving policies in SDGs. On the other hand, W_{defect} is bound by nearly the half of sheet thickness to preserve the original amorphous steel microstructure without significant loss of magnetic properties.

6.4.4 Shearing Behavior by Using the Nitrided and Laser-Trimmed SKD11 Punches

Instead of the low toughness WC (Co) and costly diamond coating, the nitrided SKD11 punches and dies were employed to investigate the effect of this material selection on the piercing behavior of amorphous steel sheets. The femtosecond laser trimming was also utilized to form different micro- and nano-textures onto the nitrided SKD11 punches. In the punch-2, they were formed in the lateral directions on its side surface. The depth and pitch of microgrooves were 7 μm and 10 μm, respectively, as seen in Figure 6.6c. The nano-grooves were superposed onto these microgrooves with the period of 300 nm. In the punch-3, the micro- and nano-grooves were formed in the vertical directions along the punch length. The depth and pitch of micro-grooves and the nano-groove period were the same as the punch-2, as depicted in Figure 6.6d.

The amorphous steel sheet stacks with the equivalent thickness of 0.125 mm were pierced by using the punch-2 and die-2 pair and the punch-3 and die-3 pair, respectively. In a similar manner to piercing experiments in 4.3, the punching affected zone width (W) was measured for each constituent sheet in stack. Figure 6.8a compares the W – N variations among the punch-0, the punch-2, and the punch-3. When using the punch-2, W becomes nearly constant irrespective of N and nearly equal to sheet thickness of 25 μm. When using the punch-3, a large initial W of around 50 μm, monotonously decreases with N and reaches 5 to 20 μm for $N > 3$. These W – N variations imply that the induced damage by piercing is controlled by the punch side surface configuration. In the punch design with circumferential micro-/nano-grooves, the constant damage width is reduced by controlling the depth and pitch of micro-grooves as well as the directivity of nano-grooves onto the micro-grooved surfaces. In the punch design with straight micro- and nano-grooves, the monotonously decreasing damage is reduced by controlling the micro- and nano-texture configuration.

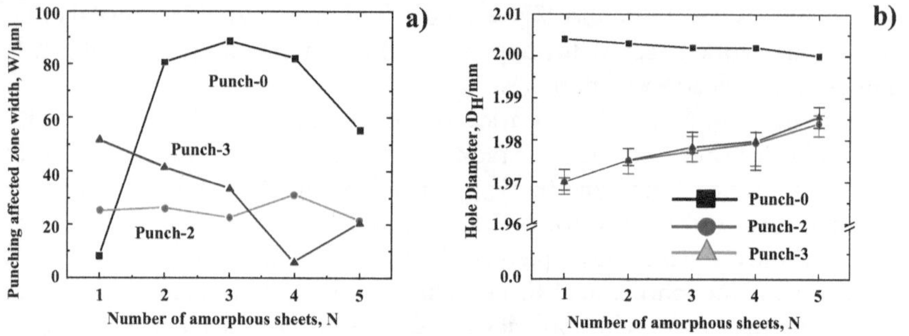

Figure 6.8 a. Comparison of the punching-affected zone width (W) and the pierced hole diameter (D_H) in the sheared amorphous electrical steel sheet stack by using the punch-0, the punch-2 and the punch-3. a) Variation of W with N, and b) variation of D_H with N.

Let us describe the essential difference of piercing behavior among the punch-0, the punch-2, and the punch-3. The pierced hole diameter (D_H) was measured for each pierced amorphous sheet in the stack. Since the punch-0 head diameter (D_P) is 1.986 mm, $D_H > D_P$ is irrespective of N; the pierced hole expands by piercing, as seen in Figure 6.8b. As shown in Figure 6.7a, the amorphous sheet near the pieced hole is bended and compressed to form A- and B-defects. This damage formation accompanies hole expansion during the piercing process. On the other hand, D_P = 1.981 mm for the punch=2 and D_P = 1.985 mm for the punch=3. Since $D_H < D_P$ irrespective of N, the pierced hole shrinks after piercing. This shrinkage corresponds to absence of A- and B-defects at the vicinity of pierced hole in Figure 6.7b. That is, the pierced hole diameter remains to be equal to the punch diameter during piercing; the elastic spring-back occurs just after the piercing process and results in the shrinkage of $D_H < D_P$. This implies that the intrinsic piecing performance without significant damages is reproduced by using the punch-2 and punch-3.

This piercing experiment demonstrates that the nitrided SKD11 punch and die works well in the similar manner to the diamond-coated WC (Co) punch and die and that punch side surface configuration essentially influences on the precise piercing behavior.

6.5 ENGINEERING ENDURANCE ON THE PIECING PUNCH LIFE

The severity of engineering endurance on the piercing punch life is dependent on the various piercing conditions. In each step of the piercing shot, the punch edge profile plays a main role to control the local metal flow around the punch edge and on the punch side surface. When the punch edge has a dull curvature, the metal flows smoothly around the punch edge and

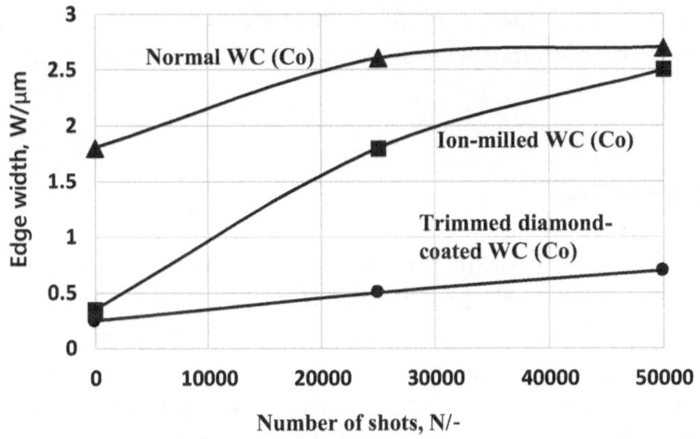

Figure 6.9 Variation of the punch edge curvature radii with increasing the number of shots in piercing the electrical steel sheets with the use of the mechanically ground WC (Co) punch, the ion-milled WC (Co) punch, and the diamond-coated and laser-trimmed WC (Co) punch.

side surface. The punch life prolongs by itself but fatal damages are induced into the electrical steel sheets by piercing. On the other hand, the sharply edged punch easily suffers from the abrasive and adhesive wearing; the punch life shortens by itself. The punch profiling design is also necessary to certify the engineering endurance on the punch life.

Three WC (Co) punches were prepared for too-life evaluation experiments; e.g., the normal WC (Co) punch with mechanically ground edge and side surface, the ion-milled WC (Co) punch with initially sharpened edge width of 0.25 μm, and the laser-trimmed diamond-coated WC (Co) punch with the same initially sharpened edge as the ion-milled one. The ion-milled WC (Co) die was commonly utilized in each piercing experiment. Figure 6.9 depicts the increase of measured edge width (W) with increasing the number of shots (N). The initial edge width of 1.8 μm of normal WC (Co) increases with N; e.g., when N = 50 k-shots, W approaches to 3 μm. The fractured surface area ratio in the polycrystalline electrical steel sheets or the total damage induced into the amorphous electrical steel sheets, significantly increases with this monotonous punch edge with increase. The edge width of the ion-milled WC (Co) significantly increases with N to have nearly the same width as the normal WC (Co) at N = 50 k-shots. When using the laser-trimmed, diamond-coated WC (Co) punch, the increase of punch edge width is suppressed to preserve the sharp edge width less than 1 μm even at N = 50 k-shots.

In addition to the punch edge profile, the adhesion of debris particles to punch head and side surfaces drives to shorten the punch life. When using the normal and ion-milled WC (Co) punches, there are no ways to prevent them from adhesion of debris particles when piercing without lubricating oils. Since the trimmed, diamond-coated WC (Co) punch has a micro-/

nano-textured side surface, fine debris particles are trapped into them and ejected out of the piercing process. This difference also influences the total punch life.

6.6 DISCUSSION

After the classical treatise on the shearing and piercing punch design, the mechanically ground tool steel or WC (Co) punches with the C-cut edge were used under the proper clearance against the die [76]. As pointed out from sections 6.2 to 6.5, the piecing process of electrical steel sheets and stacks is never highly qualified by using this classical standard procedure. In the present study, the punch edge profile is controlled to precisely describe the punching process of electrical steel sheets on the basis of multi-physical materials evaluation. The punch edge sharping and punch side-surface micro-/nano-texturing is essential to lower the inner straining, to suppress the microstructure distortion, and to reduce the iron loss. In the following, new piercing process is reconsidered from three aspects.

First, the effect of the edge profile control on the piercing process is considered to understand the physical model of punch edge as a measure for high qualification in piercing. As depicted in Figure 6.4, the mechanically ground WC (Co) has an inhomogeneous punch edge line. The sheared hole surface has fractured areas, the inner strain accumulates and severe iron loss occurs. Both the ion-milled and laser-trimmed punches have homogeneously sharp edge lines. Their sheared hole surfaces are fully burnished and the accumulated straining state is only seen at the vicinity of pierced hole. To be noticed, a slightly smaller ion loss is attained by using the laser-trimmed punch. This experimental result proves that inhomogeneity in the punch edge line works to disturb the uniform metal flow around the punch edge and to trigger the onset of transition from shearing mode to fracture mode. As demonstrated in comparison of the sheared surfaces in Figure 6.4, the shearing process at the punch edge is stabilized on the micro- and nano-textured punch side surface. This reveals that the punch edge profile within the depth in single μm up to 10 μm from the original edge line plays an important role to control the local metal flow across the punch edge in shearing.

In second, let us consider the physical model on the micro- and nano-textures on the punch side surface in piercing. As seen in Figures 6.7 and 6.8, the damage zones at the vicinity of pierced hole are much reduced at the presence of micro-/nano-textures on the punch side surface. In addition, the difference on the micro-/nano-texturing effect to piercing behavior is also noticed when using two textured punches in the lateral and vertical directions. In parallel to these direct influence of texturing onto the piercing performance, the debris particles from the work sheets are trapped into the micro- and nano-grooves on the punch side surface. In case of the punch-2, those debris particles accumulate in the circumferential micro-/nano-grooves;

while they are also trapped in the vertical micro-/nano-grooves in case of the punch-3. Since no particles are detected on the punch heads, most of them are once trapped into micro-/nano-textures on the side surface. As intensely discussed in [77,78], this trapping mechanism of debris particles is dependent on the orientation of nano-grooves. On the optimally designed micro-/nano-textured side surface, the accumulated debris particles delaminate by themselves to be ejected out of the piercing system. Remember that the debris particles deposit on the punch and die surfaces at the absence of these trapping micro- and nano-grooves. This self-ejection mechanism plays an important role to sustain the integrity of piercing process.

A motor core consists of the electrical steel sheet stacks in hundreds so that the piercing punch must have sufficient life over 1 million shots. The ceramic coating or DLC coating is often necessary to sustain the punch edge sharpness and to prevent the micro- and nano-textured side surface from severe abrasive wear. As seen in Figure 6.3d, the conventional deposited coating onto the flat punch head and side surfaces is easily worn out. As suggested in [79], when a thin DLC coating with the film thickness of 0.1 μm was deposited onto the punch head and micro-/nano-textured side surface, no delamination and chipping occurred between the punch substrate and DLC coating. This reveals that both the punch edge sharpness and the debris-ejection mechanism along the micro-/nano-textured punch side surface can be prevented from wearing damages by proper combination of the thin coating and laser texturing technologies.

Finally, let us reconsider the role of lubrication in the mass production of motor cores through the continuous piercing process. As shown in Figure 6.9, the laser-trimmed diamond-coating prevents WC (Co) punch edge from severe wear without lubrication and sustains the piercing process up to 50-k shots without maintenance. If extrapolating the W – N relation in Figure 6.9, the punch edge width might be more than 1 mm even at 100-k shots. This implies that minimum quantity lubrication (MQL) is necessary to retard the deterioration of punch edge sharpness to termination of mass production and to sustain the integrity of piercing conditions. As suggested by [80], the micro- and nano-textures on the punch side surface is expected to be working as a micro-/nano-reservoir of lubricating oils.

6.7 CONCLUSION

Various kinds of motor core with wide spectrum on its size and structure, are utilized in EVs, moving vehicles, conveyor robots, medical systems, and so forth. In recent, high-power motors requires high-frequency magnetic components without an increase of hysteresis losses [81]. High qualification of the punching die technology provides a way to significantly reduce the energy and materials losses in fabrication of the motor-cores. The minimization of damaged zones in the pierced electrical steel sheets straightforwardly results in low energy loss. In addition to recycling the skeletons of punched-out steels, a mount of debris is also reduced by fine-piercing tools. Under high integrity of punches and dies in the long-term punching operations, little intermissions for punch and die maintenance is needed to improve the cost-competitiveness without increase of environmental burdens.

Two engineering issues for the fine piercing process are considered to foresee the future perspective of tooling for fabrication of motor-cores. How to fabricate a miniature motor core to be working instead of permanent magnet is the first issue to respond to the essential needs in the micro-motor market. The next issue is a cooling of generated heat by the energy loss in motor-cores during running.

In most miniature motors, a permanent magnet is widely utilized as a magnet core. Its 3D-geometry and structure is difficult to be controlled in manufacturing; a flexible design of magnet cores is nearly impossible to perform. As proposed in [82–84], the plasma-assisted 3D printing enables the buildup of the multi-head punch and multi-cavity die without significant increase of tact time and cost. With the aid of the screen printing and the maskless lithography, the original CAD (computer-aided data) for micro-punch and micro-cavity arrays is transferred onto the micro-pattern on the die substrates. The unprinted areas are plasm nitrided by LT-PN; the un-nitrided parts are chemically etched or mechanically blasted. As-blasted multi-head punch and multi-cavity die are directly utilized for the fabrication of miniature electrical steel sheets and stacks. In fact, T-shaped electrical steel sheet unit was fabricated by one-shot piercing with the use of as-blasted punch and die [85].

The electric energy losses, including the iron loss, finally transform to heat. In particular, the increase of energy losses in the high-frequency operation of motors generates more heat even in the large volume. Thermal spreading system is needed to transfer the generated heat flux to coolant channel in the rotating motor cores. As discussed in [86,87], the micro-/nano-textured sheet is effective to enhance the convection heat transfer under the forced air blow. Higher cooling capacity than a pure copper is attained by using the acicular-textured copper films. In addition, a heat radiation mechanism is enhanced by using the tailored micro-textures. Since the height of these textures is usually smaller than 10 μm, the generated heat flux is transferred in convection and radiation through the air blowing channels between the electrical steel sheet stacks with micro-textured films.

ACKNOWLEDGEMENTS

The authors would like to express their gratitude to Mr. Y. Komori (Graduate School of Engineering, University of Toyama), Mr. Kira (Mazac, Co., Ltd.), Mr. E. Katsuta (Nippatsu, Co., Ltd.), and Mr. S. Kurozumi (Nano-Film Coat, LLC.) for their help in experiments. This study was financially supported in part by the METI-program on the supporting Industry from 2019 to 2022, and by Grant-In-Aid from MEXT with the project number of 21K03789, and by JKA Social association founding of 2021 year, and by Amada foundation at 2021 year.

REFERENCES

[1] Bouiabadi M. M., Aliabad A. D., Mousavi S. M., Ebrahim A. E., Design and analysis of E-core PM-assisted switched reluctance motor. *IET Electr. Power Appl.* 14 (5) (2020) 859–864.

[2] Enomoto Y., Soma K., Development of the amorphous motor balancing both resource saving and energy saving. *J. Soc. Mech. Eng.* 117 (2014) 753–756.

[3] www.mofa.go.jp/mofaj/gaiko/oda/sdgs/about/index.html. (Cited at 2022/3/21).

[4] Krings A., Soulard J., Overview and comparison of iron loss models for electrical machines. *J. Elect. Eng.* 10 (3) (2010) 162–169.

[5] Zhang G., Ni H., Li Y., Lie T., Wang A., Zhang H., Resource-saving production of Fe-based amorphous alloys from carbothermal reduction of high-phosphorous oolitic iron ore. *J. Noncrystalline Solids.* 579 (1) (2022) 121365.

[6] Sanjari M., He Y., Hilinski E. J., Uue S., Kestens L. A. I., Texture evolution during skew cold rolling and annealing of a none-oriented electrical steel containing 0.9 wt% silicon. *J. Mater. Sci.*, 52 (2017) 3281–3300.

[7] Kurosaki Y., Mogi H., Fujii H., Kubota T., Shiozaki M., Importance of punching and workability in non-oriented electrical steel sheets. *J. Magnetism Magnetic Mater.* 320 (20) (2008) 2474–2480.

[8] Ferrara E., Appino C., Ragusa C., de la Barriere O., Fiorillo F., Anisotropy of losses in grain-oriented Fe-Si. *AIP Adv.* 11 (2021) 115208.

[9] Enomoto Y., Suzuki K., Okita S., Eto K., Evaluation of a motor with an amorphous iron core punched by a die. *IEEJ J. Industry Appl.* 10 (6) (2021) 785–792.

[10] https://www.hitachimetals.com/infrastructure-energy/. (retrieved at 2022/4/11).

[11] Kumaresh A. K., Balaji B., Kumar M. R., Design and analysis of punching die. *Int. J. Res. Eng. Technol.* 5 (2016) 249–255.

[12] Martin F., Aydin U., Sundaria R., Rasilo P., Belahcen A., Arkkio A., Effect of punching the electrical sheets on optimal design of a permanent magnet synchronous motor. *IEEE Trans. Magnetics* 54 (39) (2018) 8102004.

[13] Kaido C., Mogi H., Fujikura M., Yamasaki J., Punching deterioration mechanism of magnetic properties of cores. *IEE-Japan Trans. FM* 128 (8) (2008) 545–550.

[14] Itoh S., *Multi-physics modeling on the magnetism of electrical steels.* PhD Thesis, Kyoto University (2017).

[15] Mansuripur M., The magneto-optical Kerr effect. Ch. 13 In *Classical Optics and its applications.* Cambridge University Press (2011).

[16] Maeda T., *Introduction to the technology of plasticity.* Nikkan-Kougyo-Shinbun (1966).

[17] Osada S., Yanagimoto J., Fundamental and applications of the technology of plasticity. (2010) Korona.

[18] Dieter G. E., Materials selection and design. *ASM Handbook* 20 (1997). doi: https://doi.org/10.31399/asm.hb.v20

[19] Semiatin S. L., Metalworking: Bulk forming. *ASM Handbook* 14A (2005).

[20] Miyoshi A., Hino K., Hara A., Kurata S., A study on the Co-phase composition contained in WC-Co base cemented carbides. *Jpn. J. Metals* 30 (1966) 603–606.

[21] Imasato S., Sakaguchi S., Okad T., Hayashi Y., Effect of WC grain size on corrosion resistance of WC-Co cemented carbide. *Jpn. J. Powder and Powder Metall.* 48 (7) (2001) 609–615.

[22] Tsujita Y., Improvement of product-life and chipping resistance of a cemented carbide rotary cutter by cemented carbide-coated (CC) anvil roll. *Nippon Tungsten Rev.* 40 (2016) 44–51.

[23] Kataoka S., Murakawa M., Aizawa T., Ike H., Tribology of dry deep-drawing of various metal sheets with used of ceramic tools. *Surf. Coat. Technol.* 178 (2004) 582–590.

[24] Tamaoki K., Manabe K-I., Kataoka S., Aizawa T., Electro-conductive ceramic tooling for dry deep drawing. *J. Mater. Process. Technol.* 210 (2010) 48–53.

[25] Dohda K., Aizawa T., Funzuka T., Process Tribology of PVD Nitride Coated Dies in Hot Metal Forming. *J. Friction* (2022) (In press).

[26] Dohda K., Aizawa T., Tribo-characterization of silicon doped and nano-structured DLC coatings by metal forming simulators. *Manufacturing Letts.* 2 (2014) 82–85.

[27] Aizawa T., Micro-texturing onto amorphous carbon materials as a mold-die for micro-forming. *Appl. Mech. Mater.* 289 (2013) 23–37.

[28] Yunata E. E., Aizawa T., Micro-grooving into thick CVD diamond films via hollow cathode. *Manufacturing Letters.* 4 (2016) 17–22.

[29] Aizawa T., Yoshino T., Shiratori T., Dohda K., Anti-galling β-SiC coating dies for fine cold forging of titanium. *J. Physics: Conference Series 1777* (2021) 012043.

[30] Novi G., Aizawa T., Kuwahara H., Multi-stripe pattern formation in the microstructure of plasma nitrided Fe-14Cr alloy. *J. Surface Coating.* 27 (2004) 134–142.

[31] Aizawa T., Fukuda T., Itoh K., Duplex coating of AISI-316 and 420 Dies for Hot Mold-Stamping. *Steel Research Int. J.* (2012) 933–936.

[32] Aizawa T., Morita H., Dry progressive stamping of copper-alloy snaps by the plasma nitrided punches. *Mater. Sci. Forum* 920 (2018) 28–33.

[33] Shiratori T., Aizawa T., Saito Y., Dohda K., Fabrication of micro-punch array by plasma printing for micro-embossing into copper substrates. *J. Mater.* 12 (16) 2640 (2019) 1–12.

[34] Aizawa T., Fukuda T., Microstructure and micro-machinability of plasma nitrided AISI420 martensitic stainless steels at 673 K. Chapter 1 in Top 5 contributions in materials sciences: 6th Edition. *Avid Science* (2019) 2–23.

[35] Aizawa T., Yoshino T., Suzuki Y., Shiratori T., Anti-galling cold dry forging of pure titanium by plasma carburized AISI420J2 dies. *J. Appl. Sci.* 11, 595 (2021) 1–12.

[36] Aizawa T., Yoshino T., Suzuki Y., Shiratori T., Free-forging of pure titanium with high reduction of thickness by plasma carburized SKD11 dies. *J. Mater.* 14 (10) 2356 (2021) 1–14.
[37] Aizawa T., Mitsuo A., Yamamoto S., Muraishi S., Sumitomo T., Self-lubrication mechanism via the in-situ formed lubricious oxide tribofilm. *Wear* 259 (2005) 708–718.
[38] Aizawa T., Akhadejdamrong T., Mitsuo A., Self-lubrication of nitride ceramic coating by the chlorine ion implantation. *Surf. Coat. Technol.* 178 (2004) 573–581.
[39] Wu W., Xu Z., Fang F., Lie B., Xiao Y., Chen J., Wang X., Lie H., Decrease of FIB-induced lateral damage for diamond tool used in nano cutting. *Nucl. Inst. Meth. Phys. Res. Sec. B* 330 (2014) 91–98.
[40] Reineck I., Sjoestrand M. E., Karner J., Pedrazzini M., Diamond coated cutting tools. *Int. J. Refractory Metals Hard Metals* 14(1996) 187–193.
[41] Zhao G., Li Z., Hu M., Li K., He N., Jamil M., Fabrication and performance of CVD diamond cutting tool in micro milling of oxygen-free copper. *Diamond & Related Mater.* 100 (2019) 107589.
[42] Koga N., Usu K., Xu C., Deep drawing of thin stainless steel sheets using diamond tools without lubricant. *J. JSTP* 53 (2012) 74–78.
[43] Yokozawa T., Takagi J. Kataoka S., Tanaka S., Study on polishing of CVD diamond film (3rd Rep.); An attempt of polishing flat surface with non-abrasive ultrasonic vibration polishing. *Jpn. JPME* 72 (8) (2006) 1018–1024.
[44] Yamauchi K., Aizawa T., High density plasma ashing of used diamond coated short-shank tools without damage to WC (Co) teeth. *Proc. 11th ICOMM* 50 (2016) 1–5.
[45] Aizawa T., Inohara T., Geometric adjustment and sizing of CVD-diamond coatings via oxygen plasma etching and laser machining. Proc. 7th SEATUC Conference (Bandon, Indonesia) (2013) 141–146.
[46] Roy S., Das M., Malik A. K., Balla V. K., Laser melting of titanium–diamond composites: Microstructure and mechanical behavior study. *Mater. Lett.* 178 (2016) 284–287.
[47] Aizawa T., Inohara T., Pico- and femtosecond laser micromachining for surface texturing. Ch. 3. In *Micromachining*. IntechOpen, London, UK (2019): 35–62.
[48] Aizawa T., Inohara T., Yoshino T., Shiratori T., Laser treatment of CVD diamond punch for ultra-fine piercing of metallic sheets. Ch. 1. In *Engineering Applications of Diamond*. Intech Open, London, UK (2021) 1–24.
[49] Aizawa T., Shiratori T., Inohara T., Short-pulse laser precise trimming of CVD-diamond coated punch for fine piercing. Proc. 2nd Asian Pacific Symposium on Technology of Plasticity (APSTP) (2019) 123–128.
[50] Aizawa T., Shiratori T., Yoshino T., Inohara T., Femtosecond laser trimming of CVD-diamond coated punch for fine embossing. *Mater. Trans.* 61 (2) (2020) 244–250.
[51] Martin F., Aydin U., Sundaria R., Rasilo P., Belahcen A., Arkkio A., Effect of punching the electrical sheets on optimal design of a permanent magnet synchronous motor. *IEEE Trans. Magnetics.* 54 (3) (2018) 2768399.
[52] Chen M-Q., He T-Y., Zhao Y., Review of femtosecond laser machining technologies for optical fiber microstructures fabrication. *Optics Laser Technol.* 147 (2022) 107628.

[53] Fuentes-Edfuf, Y., Sánchez-Gil J. A.; Garcia-Pardo M., Serna R., Giannini V., Solís Céspedes J., Siegel Jan, The role of surface roughness in the formation of LIPSS on metals. Proc. USTS (Madrid, Spain; Nov. 7th, 2019).
[54] Gečys, P., Vinčiunas, A., Gedvilas, M., Kasparaitis, A., Lazdinas, R., Račiukaitis, G., Ripple formation by femtosecond laser pulses for enhanced absorptance of stainless steel. *J. Laser Micro/Nanoeng.* 10 (2015) 129–133.
[55] van Driel H. M., Sipe J. E., Young J. F., Laser-induced periodic surface structure on solids: A universal phenomenon. *Phys. Rev. Lett.* 49 (1982) 1955–1958.
[56] Koga N., Okada S., Yamaguchi T., Effect of various blanking conditions on properties of cut surface of amorphous alloy foil and tool life. *J. JSTP* 59 (2018) 176–180.
[57] Takahashi M., Murakoshi Y., Sano T., Matsuno K., Stacked punching of amorphous alloy foil. Proc. 40th Jap. Joint Conf. Technol. Plast. (Ehime, Japan, 16–18 October 1989) 531–534.
[58] Aoki I., Suzuki K., Nakagawa T., Shearing characteristics of amorphous alloy foils. *J. JSTEP* 27 (1986) 860–867.
[59] Hitachi Metals. News Release. 24 October 2018. Available online: https://www.hitachi-metals.co.jp/e/press/pdf/2018/20181024en.pdf (accessed on 4 January 2022).
[60] Komori Y., Suzuki Y., Abe K., Aizawa T., Shiratori T., Fine piercing of amorphous electrical steel sheet stack by micro-/nano-textured punch. *J. Materials* 15, 1682 (2022) 1–13.
[61] Aizawa T., Low temperature plasma nitriding of austenitic stainless steels. *Chapter 3 in Stainless steels and alloys*. Intech Open, London, UK (2019) 31–50.
[62] Kuwahara H., *Surface modification of iron alloys by plasma nitriding and carburizing* [PhD Thesis]. Kyoto University (1992).
[63] Granito N., Kuwahara H., Aizawa T., Normal and anormal microstructure of plasma nitrided Fe-Cr alloys. *J. Mater. Sci.* 37 (4) (2002) 835–844.
[64] Aizawa T., Sugita Y., High density RF-DC plasma nitriding of steels for die and mold technologies. *Res. Rep. SIT.* 57 (1) (2013) 1–10.
[65] Bell T., Surface engineering of austenitic stainless steel. *Surface Engineering.* 18 (2002) 415–422.
[66] Dong H., S-phase surface engineering of Fe-Cr, Co-Cr and Ni-Cr alloys. *Int. Mater. Rev.* 55 (2) (2011) 65–98.
[67] Borgioli F., Galvanetto E., Bacci T., Low temperature plasma nitriding of AISI 300 ad 200 series austenitic stainless steels. *Vacuum* 127 (2016) 51–60.
[68] Katoh T., Aizawa T., Yamaguchi T., Plasma assisted nitriding for micro-texturing onto martensitic stainless steels. *Manufacturing Review.* 2 (2) (2015) 1–7.
[69] Farghali A., Aizawa T., Nitrogen supersaturation process in the AISI420 martensitic stainless steels by low temperature plasma nitriding. *ISIJ International.* 58 (3) (2018) 401–407.
[70] Aizawa T., Functionalization of stainless steels via low temperature plasma nitriding. In: Proc. 7th Annual Basic Science International Conference (Malang, Indonesia; 2017) 1–16.

[71] Aizawa T., Yoshihara S-I., Inner nitriding behavior and mechanism in stainless steels at 753 K and 623 K. SEATUC J. Sci. *Eng. (SJSE)* 1 (2019) 13–20.

[72] Aizawa T., Yoshino T., Shiratori T., Yoshihara S-I., Grain size effect on the nitrogen supersaturation process into AISI316 at 623 K. *ISIJ Int.* 59 (2019) 1886–1892.

[73] Farghali A., Aizawa T., Yoshino T., Microstructure/mechanical characterization of plasma nitriding fine-grain austenitic stainless steels at low temperature. *J. Nitrogen* 2 (2021) 244–258.

[74] Aizawa T., Yoshino T., Suzuki Y., Nitrogen supersaturation of AISI316 base stainless steels at 673 K and 623 K for hardening and microstructure control. Ch. 1. In *Stainless Steels*. IntechOpen, London, UK (2022) (in press).

[75] Aizawa T., Shiratori T., Komatsu T., Micro-/nano-structuring in stainless steels by metal forming and materials processing. Ch. 5. In *Electron Crystallography*, IntechOpen, London (2020) 101–122.

[76] Ikumapati Omolayo M., Afolalu Sunday A., Fatoba Olawale S., Kazeem Rasaq A., Adetunla Adedotun A., Ongbali Samuel O., A concise study on shearing operation in metal forming. Proc. 3rd Int. Conf. Design and Manufacturing Aspects for Sustainable Energy. 309 (WEB; October 7th, 2021) 1–8.

[77] Aizawa T., Shiratori T., Kira Yoshihiro, Inohara T., Simultaneous nanotexturing onto a CVD-diamond coated piercing punch with femtosecond laser trimming. *J. Appl. Sci.* 10 (2674) (2020) 1–12.

[78] Aizawa T., Development of high functional die-technology by micro- and nano-texturing. *J. Die Mold Technol* 5 (2022) (in press).

[79] Suzuki Y., Yoshino T., Shiratori T., Aizawa T., Progress report on the supporting industry projects. *METI-Report* (2022).

[80] Inohara T., Aizawa T., Friction control by micro-dimple laser texturing onto the tool surface. *J. Tribologist* 65 (5) (2020) 300–307.

[81] Kazimierczuk M. K., *High-frequency magnetic components* (Second ed.). Wiley, Chichester (2014).

[82] Aizawa T., Wasa K., Plasma printing of micro-nozzles with complex shaped outlets into stainless steel sheets. *J. Micro-Nano-Manuf.* 7 (2019) 034502.

[83] Shiratori T., Aizawa T., Saito Y., Wasa K., Plasma printing of an AISI316 micro-meshing punch array for micro-embossing onto copper plates. *J. Metals* 9, 396 (2019) 1–15.

[84] Aizawa T., Saito Y., Hasegawa H., Wasa K., Fabrication of optimally micro-textured copper substrates by plasma printing for plastic mold packaging. *Int. J. Automation Technol.* 14 (2) (2020) 200–207.

[85] Aizawa T., Suzuki Y., Yoshino T., Shiratori T., Fabrication of punch and die using plasma-assisted 3D printing technology for piercing sheet metals. *J. Manuf. Mater. Pros* 6 (49) (2022) 1–15.

[86] Aizawa A., Ono N., Nakata H., Boiling heat transfer by micro-textured interfaces. Ch. 1 In: *Heat Transfer*. IntechOpen, London, UK (2022).

[87] Aizawa T., Nakata H., Nasu T., Manufacturing and characterization of acicular Fe-Ni micro-textured heat-transferring sheets. Proc. 5th WCMNM (Sep., 2022; Leuven, Belgium) (2022) (in press).

Chapter 7

Microwave Energy for Joining of Dissimilar Metals

Siddharth Tamang[1] *and S. Aravindan*[2]
[1]Indian Institute of Technology Kharagpur, Kharagpur, West Bengal, India
[2]Indian Institute of Technology Delhi, Hauz Khas, New Delhi, India

CONTENTS

7.1 INTRODUCTION

Joining technology is a study of the joining of materials to fabricate/manufacture a product that is needed. It has its challenges, and the challenges are more when the materials to be joined are dissimilar. Industry requirements dictate the need for higher-strength materials. The usage of such materials certainly necessitates the joining technology for these newer materials. Thus, the flexibility of the materials used in a product can be incorporated from the functionality point of view. This is one-way industries can cut costs while maintaining the product performance needed. In addition, the industry is slowly but surely moving towards sustainable joining processes. Many processes are available for dissimilar material joining. Every joining process has its own merits and demerits. The mechanical joining processes are limited to use for only

DOI: 10.1201/9781003436072-7

thick metals with high strength and ductile metals. Joining by adhesion is limited not only by the strength of the bond, but also by the service temperature. The wide dissimilarity between the dissimilar material combinations, such as variation in thermal expansion, melting points and crystallographic nature, induce problems during dissimilar material joining. Thus, it is challenging to join dissimilar materials with the help of conventionally available welding processes like gas metal arc welding (GMAW), shielded metal arc welding (SMAW), gas tungsten arc welding (GTAW), and submerged arc welding (SAW), etc. Solid-state bonding technologies like friction, explosive welding, and diffusion bonding can be used for dissimilar joining. Dissimilar joints are also accomplished by utilizing concentrated energy beams such as laser and electron beam. Low temperature joining methodologies such as brazing and soldering can also be used since the interlayer employed accommodates the residual stress produced, owing to the difference in thermal expansion. These methods are expensive, time-consuming, and undergo stringent processing parameters. A quick joining process is required of today's industries to achieve improved productivity and energy saving. One goal is to find a dissimilar material joining process that can solve a joining problem faster/cheaper/more efficiently than is presently possible.

The role of joining dissimilar metals is highly pronounced in technological applications such as nuclear, rail, automobile, aviation, chemical, power, and electronics industries. In plants of such industries, materials are applied as per their specific characteristic and properties. However, in many cases, these materials usually are expensive making the installation and operational cost of plants and machinery very high. In an ideal scenario, such high-cost materials have to be replaced by a low-cost/readily available material without compromising the machine's function and maintaining the product performance. This is where the role of joining dissimilar metals comes into play. In the case of nuclear power plants, the ferritic steel (SA508Gr.3Cl.1) water pressure vessel needs to be joined with stainless steel (SS304LN) gas pipelines [1]. These materials are used due to their characteristic properties such as the retained strength at high temperatures of ferritic steel and the corrosion resistance of SS304LN. Thus, by integrating high-performing/high-cost materials with easily available/low-cost materials, the total assembly cost can be reduced. However, joining these dissimilar metals also faces challenges since most materials have different crystallographic structure, elemental composition, and thermal expansion coefficient. For example, in the above case, due to the difference in composition, carbon diffuses to the SS304LN to form chromium carbide. As a result, the chromium oxide layer deteriorates to form chromium carbide, leading to joint failure. There are many such applications where dissimilar materials are required to be joined, such as in automotive, aerospace industries, electric industries, etc. In this chapter, we will focus on joining low density, low melting alloys, and joining of high density, high melting alloys.

The automotive and aerospace sectors have a shared requirement of decrease in weight. Future cars will contain more lightweight metals like aluminum (Al) and magnesium (Mg) alloy [2]. The aerospace industry will also be using more and more aluminum, magnesium, and their composites. The aluminum alloy exhibits excellent corrosion resistance, thermal conductivity, and strength-to-weight ratio [3]. In particular, 6xxx series is preferred over 2xxx and 7xxx series aluminum alloy due to their low cost and superior weldability and formability [4]. 6xxx series are alloys of Mg and Si and can be heat-treated and can undergo precipitation hardening. One of the commonly available aluminum alloys is the AA6061, which contain magnesium plus silicon 1.5% or more. This alloy can be easily extruded into a wide variety of shapes and can also be formed into sheets and plates [5]. Similarly, magnesium has a lower density than aluminum, with a good strength-to-weight ratio, good castability, and machinability [6]. Due to the above-mentioned properties, magnesium-based alloys are widely used in structural applications. Magnesium alloys have very good damping properties that make them ideal for vibration reduction applications, especially for automotive and aerospace applications. AZ31B is a vastly used magnesium alloy with good formability. The major alloying elements are aluminum and zinc for AZ31B. They vary in purity, with AZ31C being low purity and used for lightweight structures where corrosion resistance is not needed. AZ31B is widely used in housing of equipment with the chassis. The joining of these lightweight alloys is necessary to reduce the weight of vehicles.

Another popular dissimilar material combination is the joining of copper with stainless steel. Copper has very high thermal and electrical conductivity, whereas stainless steel is recognized for its high corrosion resistance property. Due to these special properties, they are used together in refrigeration industries. Heat exchangers require qualities, like high thermal conductivity and resistance to corrosion by cooling fluids [7–9]. In electrical industries, the stainless steel grounding plates (placed inside the ground) can be joined with copper rods to provide earthing to a power supply. Copper may contain some oxygen which may reduce its properties. Thus, a high purity form of copper with oxygen less than 0.0005% is used, and the same is termed *oxygen-free copper* (OFC) [10]. Due to its high purity, it has the highest thermal conductivity and finds its use in bus bars. However, the same property of high conductivity of Cu hinders its joining since the heat utilized for joining is rapidly dissipated to the base metal.

SS304 is the most commonly available stainless steel, which lies in the austenitic stainless steel category corresponding to AISI 300 series. Most other types of stainless steel are some variants of this stainless steel SS304. They have a very high modulus of elasticity, and high strength at room and high temperatures [11]. Stainless steel has good corrosion, oxidation, abrasion, and erosion resistance.

7.1.1 Microwave Heating

Microwaves are electromagnetic waves of frequency 300 MHz to 300 GHz that travel at the speed of light. Microwaves have a wavelength from 1 mm to 1 m and are used in communications, security and RFID, cancer treatment, and microwave heating. A frequency of 2.45 GHz is employed for commercial microwave processing globally. Microwave heating is the most commonly used method for heating and cooking food in a domestic microwave oven and in the vulcanization of rubber [12]. Microwave heating behavior of materials is not identical to conventional heating, as seen in Figure 7.1. In conventional heating, the material heats from the surface to the core [Figure 7.1(a)], whereas with microwave heating, the material is hotter in the core [Figure 7.1(b)]; thus, having a reverse thermal gradient.

Due to the lack of waste of materials and full utilization of energy without any losses, this technology is one of the sustainable alternative to existing processes. In addition, no harmful gases are emitted during this process.

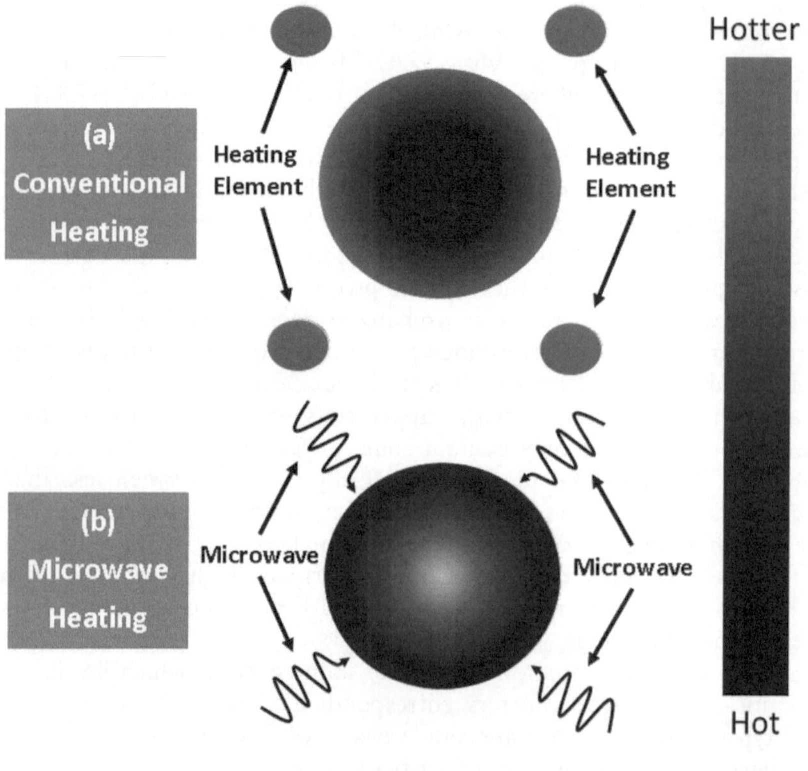

Figure 7.1 Schematic of denoting the behavior of material to (a) Conventional Heating and (b) Microwave Heating [13].

7.1.2 Microwave Interaction with Metals

Metals in bulk form act as reflectors to microwave at room temperature. This reflective property of bulk metals is utilized for manufacturing microwave applicators such as multimode and single-mode cavities. Microwaves can enter into the material up to a certain depth, termed *penetration depth* or *skin depth*. It is the depth of penetration of the microwave from the outer layer, such that the microwave field strength reduces to $1/e$ times [14], where e is Euler's number.

Skin depth [14],

$$d = \sqrt{\frac{\rho}{\pi f \mu_0 \mu'}} \tag{7.1}$$

where ρ = electrical resistivity (Ω.m), f = frequency of microwave (Hz), μ' = material permeability (H/m), and μ_0 = permeability of free space (H/m).

Metals have skin depth ranging between 1 and 5 μm. This means that the microwave penetration in metals is negligible compared to its bulk dimensions. This small penetration induces a current on the metal, which in turn exerts a force on the electrons. These electrons form an electron cloud and accumulate on the edges of the bulk metal to form sparks.

On the contrary, when the metal has a diameter less than or equal to the skin depth, it heats uniformly in a microwave [15]. The microwave heating of powdered metals was first observed by Walkiewicz et al. [16] in 1988. Microwave heating was observed for metals such as Cu and Ni to temperatures up to 228°C and 384°C, respectively. However, high temperature was not achieved. If the metal powder has a larger particle size than the skin depth, microwave heating is only up to its skin depth, leading to non-uniform heating of powder. However, microwave hybrid heating can be applied to heat such metals to high temperatures. Later, Sheinberg et al. [17] in 1990 proved and patented the heating of metal powders with an oxide layer by microwave radiation, provided the powder particles were lower than or equal to the skin depth of the microwave.

7.1.3 Microwave Hybrid Heating

This is a process in which the parts are heated up directly and indirectly by microwave. A susceptor surrounds the part in the microwave hybrid heating process. Susceptors are materials that absorb microwave energy readily and convert it to thermal energy. On exposure to microwave, the susceptor rapidly heats up and increases the temperature of the surface of the material adjacent to it. The microwave will thus directly heat a sample from the core and well as indirectly by transferring the heat of the susceptor from the surface. Thus, the materials' uniform heating can be obtained contrary to the reverse temperature gradient, as seen in Figure 7.1(b).

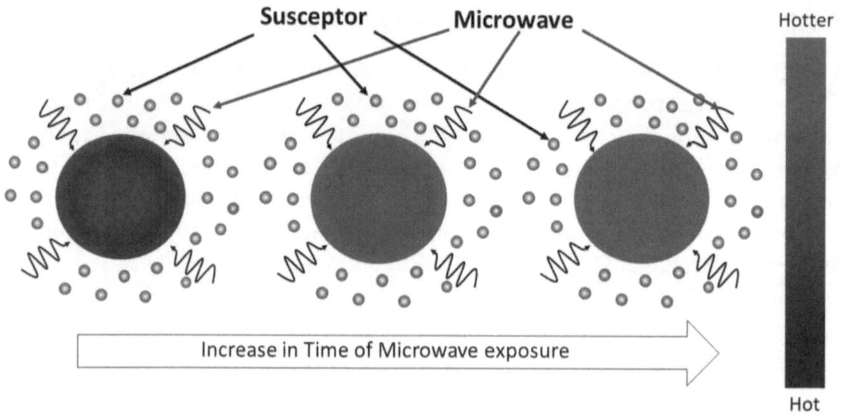

Figure 7.2 Schematic diagram of temperature distribution across a material in microwave hybrid heating [13].

Thus, uniform heating is observed as the microwave exposure time increases (Figure 7.2). Microwave hybrid heating was originally developed as a method of heating microwave transparent materials like alumina and zirconia [18]. Susceptors are used in such a way that they heat easily on exposure to the microwave, and the heat is transferred to the adjacent materials. After reaching a certain temperature, the transparent materials start coupling directly with the microwave and heat rapidly. They were used in the sintering of zirconia-toughened alumina and later in the sintering of metals.

Microwave hybrid heating is also capable of heating metal powders larger than the skin depth. A susceptor is placed near the metal powder such that the susceptor directly heats by microwave, and the heat is transferred to the metal powder via conduction, convection, or radiation. After reaching a critical temperature of 400–600°C [18], the metal powder starts coupling directly with microwave and heats up to its melting point. The main reason for this is that for metals, as the temperature increases, the resistivity increases, thus increasing the skin depth (equation (7.1)). This hybrid heating was first used as a method to fully sinter metal powder compacts by Roy et al. [19] in 1999. Since then, microwaves have been used extensively in fully sintering metal powders [20–23] and metal composites [24,25]. In the last seven years, several researchers have studied bulk metals joining using microwave hybrid heating [26–30]. To conclude, microwave hybrid heating is used due to the following advantages:

- Uniform heating
- Faster heating
- Heating microwave transparent materials
- Melting microwave reflective materials

7.2 MICROWAVE JOINING OF BULK METALS

7.2.1 Joining of Similar Metals by Microwave

The joining of metals was first reported by Siores and Rego [31], who utilized the arc formation to join 0.1–0.3 mm thin sheets to form a joint. Arcing between the surfaces caused the heating and melting of sheet locally and the pressure was required for the fusion of the specimen. However, this repeatability of the process was not possible. The proper joining of bulk metals in a multimode microwave furnace was initially reported by Srinath et al. [28]. The successful joining of 4 mm copper plates utilizing microwave hybrid heating by applying a ~5 µm copper powder in the form of paste. The microwave was reported to be absorbed by the copper powder with additional heating by the microwave interaction with epoxy and the hybrid heating of the bulk metal. X-ray diffraction (XRD) study disclosed the significant formation of about 26% oxides of copper (CuO), that increase the coupling with microwave. The scanning electron microscope image confirmed the successful continuous joint with less porosity (1.92%) and circular pores. The Vicker's micro-hardness of the joint cross-section was reported to be 84% of the base metal. The tensile test reported good elongation of 29.21% and ultimate tensile strength of 164.4 MPa, signifying metallurgical bonding of the sample.

SS316 was joined in another study by microwave hybrid heating, using SS316 powder [27]. An epitaxial grain growth (grain growth having crystallographic orientation same as substrate) with chromium carbide forming predominantly on the grain boundary was reported due to affinity of Cr to C at high temperatures. Furthermore, high hardness was observed at the grain boundaries preventing high-order plastic deformation due to these carbides. Iron-nickel and austenite matrix was also observed in the fusion zone. The tensile strength of the joint was reported to be good enough (82% of the base metal). A similar study was carried out for joining SS316 [32] by nickel base powder of 40 µm average size. The intermetallic compounds observed in the joint zone were $(FeNi)_{23}C_6$, nickel carbides, and chromium carbides. Due to the formation of carbides at the interface, high micro-hardness (420 HV) was reported at the interface, and the average micro-hardness of the fusion zone was reported to be 290 HV. The SEM images showed very good crack-free joints having columnar grains towards the stainless steel side and equiaxed grains at the center. The electron probe microanalysis results reported that the iron was the dominating element towards the base metal and chromium and carbon in the center. The joint showed good tensile strength.

Gupta et al. [29] joined SS316 plates using nickel-based alloy powder. The microstructure revealed crack-free joints. The fusion zone was formed of FeNi, NiSi, Cu_2Si, and $FeSi_2$ phases that were facilitated by microwave heating. The average hardness of the joint was found to be 145.3 HV, tensile strength of 323 MPa, and percentage elongation of 11.30%.

Inconel 625 was joined successfully by using microwave hybrid heating [26]. Nickel-based powder as an interlayer produced a good fusion zone without interfacial cracks. The fusion zone was found to have carbides and oxide of nickel, molybdenum, etc. The presence of chromium carbides at the joint interfaces caused the rise in hardness and consequently resulted in tensile strength of 35% of the Inconel 625.

The strength of Inconel 718 joints by applying the same metal powder as an interlayer was also observed to be about half of the parent material due to the formation of Laves phase by segregation of Nb and Mo [$(Ni, Cr)_2$ (Nb, Mo)] and carbides [33]. However, the post-weld solution treated and aged joints exhibited similar strength and hardness values as the parent material due to the dissolution of laves phase.

Aluminum alloy (6061) was joined using microwave hybrid heating by incorporating aluminum powder as an interlayer with a processing time of 10 minutes. The presence of aluminum and its oxides at the fusion zone leads to the forming of Al-α-Al_2O_3 composites. Micro-hardness was reported to be 45.2 ±10 HV in the base metal, 50.3 ±10 HV in the interface, and highest at the fusion zone at 72.4 ± 10 HV.

Another nickel-based alloy – Hastelloy was studied for joining through microwave applying nickel-based powder. These joints exhibited a tensile strength of 82% of Hastelloy [34].

Bagha et al. [35] studied the effect of powder nickel size (20 to 50 μm) of the interface on the microwave joining of SS304. It was reported that there was an observable increase in hardness of the bead as well as heat-affected zone with a smaller Ni powder. The ultimate tensile strength was reported to increase with a reduction in Ni powder size, with the highest tensile strength achieved by 20 μm Ni powder. It was concluded that this was due to the best homogeneity for smaller Ni powder.

7.2.2 Dissimilar Metals Joining by Microwave Energy

The study of the joining of stainless steel (SS) and mild steel by microwave heating was reported by Srinath et al. [36] and Gupta et al. [37]. Successful metallurgical joining was observed with full melting of the interface (nickel-based powder). The ultimate tensile strength of 340 MPa of the microwave joining was reported. The joint was successfully fabricated at 450 seconds [36]. The XRD study confirmed the formation of nickel-chromium oxide, chromium carbide, chromium oxide and cementite (Fe_3C). The presence of chromium in the joint region was further confirmed by EDS studies. The hardness of the joint region was reported to be almost half of the SS316, thus justifying the elongation obtained in the tensile test as 13.58%.

Bansal et al. performed dissimilar joining of Inconel 718 to SS-316L with Inconel 718 powder as an interlayer [38]. Different intermetallic compounds like NbC, $Cr_{23}C_6$, Cr_3C_2, TiC, and δ-Ni_3Nb were observed to form in the fusion zone along with the Ni-Cr matrix. Niobium carbides were observed to be near the Inconel side of the joint zone. An average microhardness of 230±5 HV was reported in the weld zone with some Laves particles. The tensile strength of the joint was reported to be almost similar to that of the SS-316 base metal.

NiTi shape memory alloy brazing with Ag–Ti and Ag–Cu–Ti [39] through microwave heating was also reported. NiTi to Hastelloy C-276 and A240 stainless steel was also joined by microwave heating. A good metallurgical joint was observed between all the joint combinations having a thickness of less than 8 microns.

7.3 JOINING OF ALUMINUM TO MAGNESIUM ALLOY

7.3.1 Joining with Other Processes

Energy economy and energy-saving dictate the usage of lightweight materials for automotive and aerospace industries. Al and Mg alloys possess low density; hence, the joining of Al alloy to Mg alloy is inevitable for such automotive applications. The joining of Al and Mg alloy is challenging because of the high solubility between them and the formation of intermetallic compounds (IMCs), namely Al_3Mg_2 (β phase) and $Al_{12}Mg_{17}$ (γ phase) at temperatures above ~450 °C [40].

The two intermetallic compounds are formed due to the activation energy difference between the β and γ phases. The $Al_{12}Mg_{17}$ phase has higher activation energy than Al_3Mg_2 [41]. The solubility of magnesium in aluminum is around 17.1 wt% [42] for equilibrium conditions.

The hard and brittle IMCs were reported in the fusion zone when Al and Mg alloy were joined by TIG [43]. The reactivity of Mg with Al to form these intermetallic compounds is high due to the low activation energy of diffusion of Mg in Al [44]. Solid state welding processes like friction stir welding (FSW) and diffusion bonding that have low operation temperature during joining have been used to study the joining of Al to Mg alloys. It was reported that the FSW [45] as well as diffusion bonding [46,47] of Mg alloy to Al alloy, cause the formation of hard IMCs in the reaction layer. The IMCs, MgAl, Mg_3Al_2, and $Al_{12}Mg_{17}$, in the fusion zone result in deterioration of the strength of the joint [46,47]. Similarly, both the processes of laser welding [48] and laser weld bonding [49] reported the presence of Al_3Mg_2 and $Al_{12}Mg_{17}$ on the fusion layer; however, the IMCs were reduced for laser weld bonded workpiece due to the gasification of the adhesive used in laser weld bonding. In order to prevent the IMC formation, researchers have used an interlayer having metallurgical compatibility with the workpieces, and that prevents the Al-Mg IMC formation. Studies have been conducted using nickel [50–54], iron [55,56], silver [57,58], tin [59,60], titanium [61], and zinc [62–67] as an interlayer to join various Al alloy to Mg alloy. Ni has been used as an interlayer to successfully prevent Al-Mg IMCs formation in resistance spot welding [50], laser welding [51], and diffusion bonding [52] since it prevents the diffusion of Mg-Al acting as a separator between the two. However, the development of Mg_2Ni and Al_3Ni has been observed [51,52] in the reaction layer towards Mg and Al workpieces [51,53]. The inter-diffusivity of Mg in Ni is observed to be greater than Al in Ni since the diffusion coefficient is higher for Mg_2Ni than Al_3Ni [53]. The failure of the joint was reported to be due to Mg_2Ni reaction layer in the diffusion-bonded joint [52]. The Mg_2Ni reaction layer exhibited high hardness. In resistance welding of Al5754 and AZ31B using Ni as an interlayer, weld strength of 90% of AZ31B was reported [50]. Lamellar structures of (Mg–Mg_2Ni eutectic) + α-Mg + AlNi were reported to be formed at the center of the nugget, which was the main cause of the failure of the joint [54]. Another reason for the failure of the joint was the crack formation due to the variation in thermal expansion of Ni and Mg. Towards the Al-Ni interface, the formation of a very thin Al_3Ni reaction layer along with α-Al + Al_3Ni was observed. Similarly, other researchers reported that Al to Ni interface is stronger than Mg to Ni interface [54]. The intermetallic transition zone is reported to increase with the increase in temperature. The Ni interlayer prevented Al-Mg intermetallic since the diffusion activation energy of Ni is more than that of Mg and Al in the Ni-Mg and Ni-Al interface [53].

Zn also successfully prevented the formation of Al-Mg brittle IMCs, and has been studied by many researchers [62–67] while maintaining acceptable joint strength. One reason is the similar crystal structure of Zn and Mg. Secondly, Zn and Al are negligible soluble in each other. On applying Zn as an interlayer very little solid solution was observed between Zn and Al substrate [63,68]. For diffusion bonding Zn interlayer results in joints having good strength and no Al-Mg IMCs [62]. Thus, the strength improves with Zn interlayer on joining Al alloy to Mg alloy. It was also reported that with the decrease in thickness of Zn interlayer, the joint shear strength increased since the thickness of the Zn interlayer is directly proportional to the thickness of the reaction layer formed along with its hardness for diffusion-bonded Al to Mg alloy [62]. A liquid phase was formed between Mg and Zn on brazing Al and Mg alloy by Zn based braze alloy [68] with dispersed intermetallic particles. Joint strength was reported to be highest for Zn-5Al interlayer compared to pure Zn and Zn-8Al [69]. Some researchers have reported that Zn interlayer was successful in preventing the formation of Al-Mg IMCs even with the use of conventional welding processes like gas metal arc welding (GMAW) [64] and gas tungsten arc welding (GTAW) [65]. It was reported that TIG welded joints have exhibited a tensile strength of 93 MPa (without Zn 28 MPa), and MIG welding has 64 MPa lap tensile strength, with the fusion zone consisting of mostly Mg-Zn intermetallics, mainly $MgZn_2$. This work was carried forward by using pre-roll assisted A-TIG [70] and resistance spot welding [66] by applying Zn as an interlayer. Xu et al. [67] reported four times enhancement on tensile shear strength when Mg alloy was friction stir spot welded to Zn-coated Al alloy instead of uncoated Al alloy. A brazed zone consisting of Mg-Zn along with Al- Zn diffusion zone along the shoulder prevents Al-Mg intermetallics formation.

7.3.2 Joining by Microwave

Both nickel and zinc have been observed to prevent the formation of Al-Mg IMCs. The strength of the joint obtained by applying nickel and zinc interlayer is far higher than the joints without any interlayer.

Microwave has been utilized to join Al6061 to AZ31B by employing various powder interlayers like TiCuSil (alloy of Cu, Ag, and Ti), Ni, and Zn.

The commercially available TiCuSil paste has been utilized to join Al6061 and AZ31B with microwave hybrid heating [71]. A reaction layer was observed to form between the two parent metals. Hard and brittle intermetallic compounds were formed across the reaction layer as seen in Figure 7.3. The formation of the large reaction zone and the intermetallic compounds was reported due to the enhanced diffusion of elements across the joint zone.

Though the use of nickel as an interlayer has been able to mitigate the formation of Al-Mg intermetallic alloy in diffusion welding and other welding processes, in the case of joining by microwave, it was observed that the Al-Mg intermetallic formation was not prevented, since Ni was used in the powder form. Since Ni has a high melting point compared to Al and

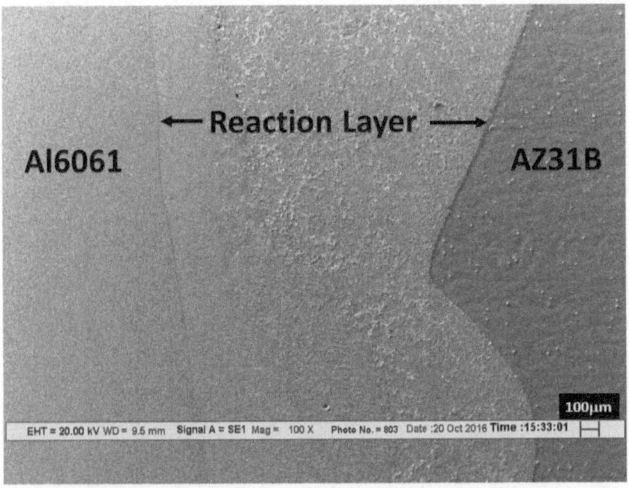

Figure 7.3 SEM microstructure of the reaction layer formed on joining Al6061 to AZ31B by TiCuSil paste using microwave heating.

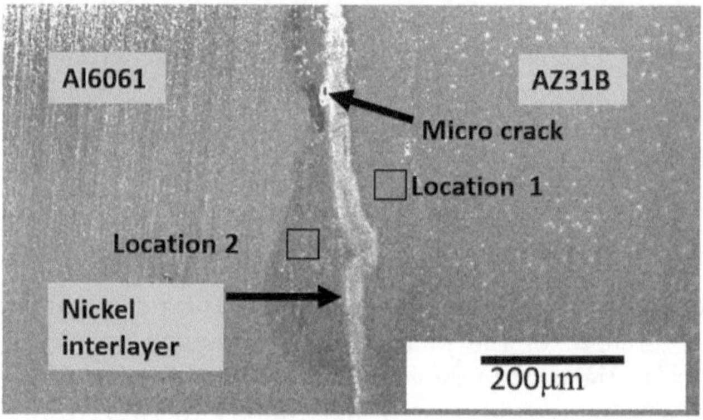

Figure 7.4 Microstructure of the joint formed by using Ni interlayer to join Al6061 to AZ31B.

Mg alloy, the temperature achieved during microwave heating was not able to melt the Ni but sufficient enough the melt the faying surfaces and cause the formation of Al-Mg intermetallic compounds as observed in locations 1 and 2 of Figure 7.4 [72].

Al-Mg intermetallic evolution was mitigated when zinc powder was used as an interlayer [73]. A reaction layer was developed only adjacent to the magnesium alloy. The shape of the reaction layer was observed to be random when graphite was used as a susceptor (Figure 7.5) and uniform when charcoal was used as a susceptor. It was concluded that the graphite susceptor produced joints with Zn interlayer in a shorter time in comparison to the charcoal susceptor. The reaction zone consisted of Mg solid solution and MgZn phase. The % of epoxy that gives the best results was observed to be 19 wt%.

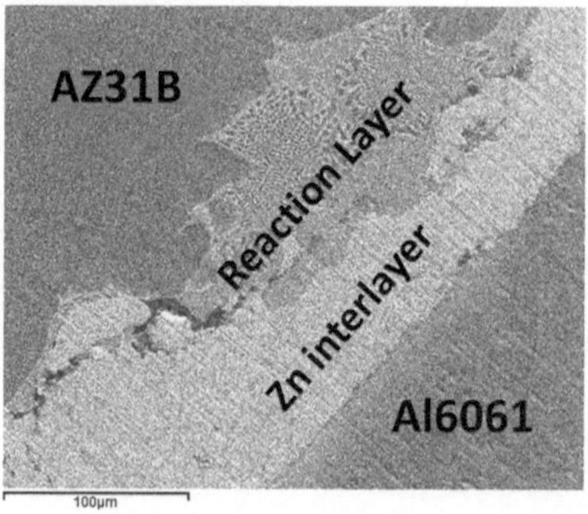

Figure 7.5 Microstructure of using Zn as interlayer to join Al6061 and AZ31B.

7.4 COPPER TO STAINLESS STEEL JOINING

7.4.1 Joining by Other Processes

Copper to stainless steel joints have high technology applications in power industries and heat exchangers due to their exotic properties. However, it is tough to weld copper to stainless by fusion welding due to differences in melting point, electrical resistivity, specific heat capacity, thermal conductivity, and other properties [74]. During arc welding, the heat is dissipated rapidly from the arc zone due to the large thermal conductivity of copper. In addition, copper and iron are immiscible with each other at equilibrium [75]. Thus, they do not form any intermetallic compounds during welding. However, the molten solution of Cu and Fe undergoes metastable phase transformation during solidification [76]. Magnabosco et al. [77] have reported that the formation of these non-equilibrium phases has porosities also. These porosities are formed due to the shrinkage of copper surrounded by a steel matrix. In addition, the penetration of Cu into the grain boundary of unmelted stainless steel was also reported. Sabetghadam et al. [78] studied diffusion bonding of Cu to SS410 using nickel as an interlayer and reported the presence of a corrugated pattern at the interface along with voids. The intermetallic compounds were reported to have formed. The intermetallic compounds were observed to be increased with an increase in temperature. In order to mitigate the uniform heating problem, in the case of Cu to SS joints, Yao et al. [79] assembled the butt joint in the form of a scarf and focused the laser on the steel. Defect-free joints having a limited amount of dissolved copper towards the steel melt were reported.

Explosion welding in a solid-state joining process, of Cu to SS, exhibited successfully joined with improved strength [80]. A wavy structure was observed at the interface between the two samples. However, the hardness was reported to be increased due to collision between two plates i.e., it increases with the stand-off distance and rate of explosion. This is due to the severe plastic deformation of the material on collision, which leads to work hardening [81]. No melting of the parent material and no intermetallic compound formation were reported in this process. However, due to high velocity during explosive welding, the grains were subjected to plastic deformation that resulted in elongated grains [81]. Shanjeevi et al. [82] during friction welding, reported that the high friction pressure and low upset pressure increased the tensile strength of the joint (highest of 2.52% higher than copper) and decreased tensile strength with low friction pressure and high upset pressure. The joining of Cu to stainless steel was studied by varying the electric field supplied to the samples undergoing friction welding. When the samples were connected to the cathode in DC, a homogeneous larger grain of copper was reported during friction welding with uniform distribution under a negative electrostatic field [9].

The highest torsional strength was observed when the AC electric field was applied. Explosion welding and friction welding although reported good joint but lacked in variation of joint configuration.

The literature clearly reports that joining Cu to SS is not feasible by most welding processes.

7.4.2 Microwave Joining

The joining of pure copper and stainless steel was studied using microwave hybrid heating with different interlayers.

A nickel interlayer of size [83] 45 μm and 200 nm was observed to successfully work as interlayer to join Copper to SS304. The small Ni powder resulted in joints at a shorter time with a lower thickness of interlayer, producing better metallurgical bonding, as shown in Figure 7.6. Ni-Cu solid solution was observed in the interlayer by elemental analysis.

Copper interlayer was also reported to form metallurgically sound joints of oxygen-free copper (OFC) and SS304 [84]. It was reported that the interlayer was not observed between the pure copper interlayer and OFC due to the seamless joint between the OFC and Cu interlayer, as seen in Figure 7.7. The shear strength of 82 MPa was observed for a time of 14 minutes. The authors have successfully simulated the joining of these metals by microwave energy using finite element analysis and reported the temperature across the joint zone with respect to change in time. It was observed by simulation and experiment that the joints formed for a microwave exposure time of 12 minutes and above.

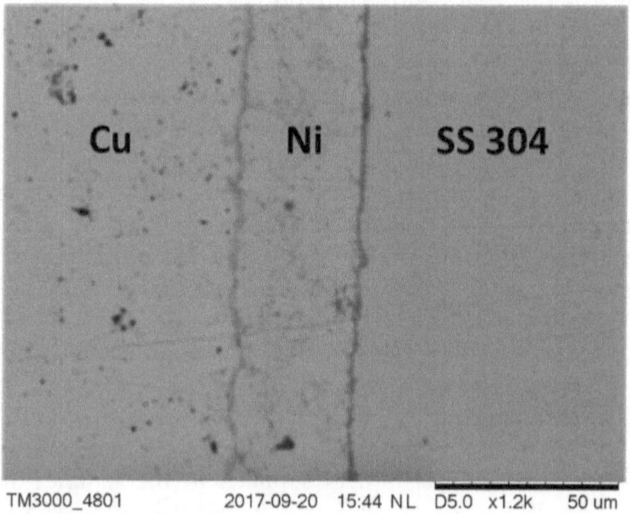

Figure 7.6 Microstructure of the joint formed for Cu to SS304 using Ni interlayer.

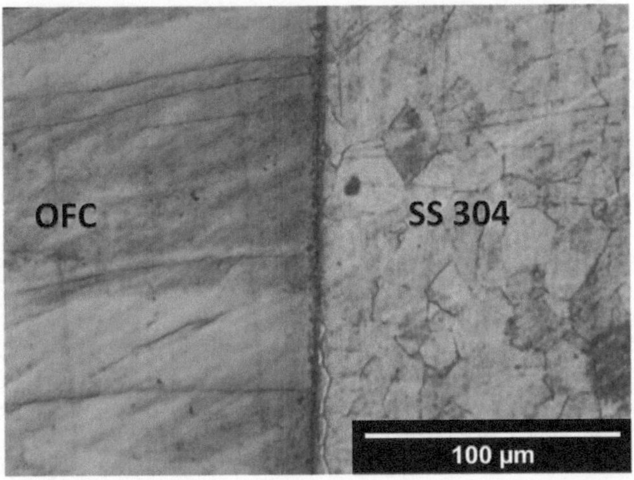

Figure 7.7 Optical microstructure of the joint formed between OFC to SS304 using Cu powder interlayer.

7.5 CONCLUSION

The study of dissimilar metal joining and welding is being carried out for many metals and alloys. Various conventional arc welding and other advanced welding processes are used for welding dissimilar metals. Nowadays, the use of microwave hybrid heating to weld bulk metals have been reported by many researchers. This hybrid heating mechanism is utilized to reduce the time taken to form joints since the material heats at the molecular level by coupling with the alternating microwave and converting the microwave energy directly to heat. The use of a susceptor will firstly help in increasing the surrounding temperature. This increased temperature leads to improved penetration of microwave to the powder interlayer. The powder interlayer thus will couple with microwave and heat internally and will additionally be heated by the externally supplied susceptor heat. Hence, we can say that the powder interlayer will start directly heating by microwave with an increase in microwave penetration as well as is indirectly heated by microwave-heated surrounding susceptors. This has been observed to form successful metallurgical joints for welding aluminum alloy to magnesium alloy and by applying TiCuSil and Zn powder as an interlayer. Copper and stainless steel (SS304) have also been observed to be joined, utilizing Ni and Cu powder interlayer and microwave energy.

REFERENCES

[1] Rathod DW, Singh PK, Pandey S, Aravindan S (2016) Effect of buffer-layered buttering on microstructure and mechanical properties of dissimilar metal

weld joints for nuclear plant application. *Mater Sci Eng A* 666:100–113. 10.1016/j.msea.2016.04.053

[2] Berger L, Lesemann M, Sahr C (2009) SuperLIGHT-CAR – the Multi-Material Car Body. In: 7th European LS-DYNA Conference. pp 1–10.

[3] Schubert E, Klassen M, Zerner I, et al. (2001) Light-weight structures produced by laser beam joining for future applications in automobile and aerospace industry. *J Mater Process Technol* 115:2–8. 10.1016/S0924-0136(01)00756-7

[4] Troeger LP, Starke EA (2000) Microstructural and mechanical characterization of a superplastic 6xxx aluminum alloy. *Mater Sci Eng A* 277:102–113. 10.1016/s0921-5093(99)00543-2

[5] Davis JR (2001) Aluminum and aluminum alloys introduction and overview. *ASM Int* 351–416. 10.1361/autb2001p351

[6] Mordike BL, Ebert T (2001) Magnesium Properties - applications - potential. *Mater Sci Eng A* 302:37–45. 10.1016/S0921-5093(00)01351-4

[7] Chen S, Huang J, Xia J, et al. (2013) Microstructural characteristics of a stainless steel/copper dissimilar joint made by laser welding. *Metall Mater Trans A Phys Metall Mater Sci* 44:3690–3696. 10.1007/s11661-013-1693-z

[8] Suga T, Murai Y, Kobashi T, et al. (2015) Laser brazing of a dissimilar joint of austenitic stainless steel and pure copper. *Weld Int* 7116:1–9. 10.1080/09507116.2014.921090

[9] Fu L, Du SG (2006) Effects of external electric field on microstructure and property of friction welded joint between copper and stainless steel. *J Mater Sci* 41:4137–4142. 10.1007/s10853-006-6224-5

[10] Copper Development Association Oxygen-free High Conductivity Copper. In: *Coller Alliance*. https://copperalliance.org.uk/. Accessed 12 Feb 2019.

[11] ASM International (2000) Introduction to Stainless Steels. In: *Alloy Digest Sourcebook: Stainless Steels*. 172:1–6.

[12] Clark DE, Sutton WH (1996) Microwave processing of materials. *Annu Rev Mater Sci* 26:299–331. 10.1146/annurev.ms.26.080196.001503

[13] Tamang S (2020) Joining of dissimilar materials through microwave hybrid heating. *Indian Institute of Technology Delhi.*

[14] Metaxas AC, Meredith RJ (1983) *Industrial Microwave Heating*. Peter Peregrinus Pvt Ltd, London.

[15] Mishra RR, Sharma AK (2016) Microwave-material interaction phenomena: Heating mechanisms, challenges and opportunities in material processing. *Compos Part A Appl Sci Manuf* 81:78–97. 10.1016/j.compositesa.2015.10.035

[16] Walkiewicz JW, Kazonich G, McGill SL (1988) Microwave heating characteristics of selected minerals and compounds. *Mineral Metallurgy Proceedings* 5:39–42.

[17] Sheinberg H, Meek TT, Blake RD (1990) Microwaving of Normally Opaque and Semi-Opaque Substances. https://www.osti.gov/biblio/7167604

[18] Sutton WH (1992) Microwave processing of ceramics – An overview. *MRS Proc* 269:3.

[19] Roy R, Agrawal D, Cheng J, Gedevanlshvili S (1999) Full sintering of powdered-metal bodies in a microwave field. *Nature* 399:668–670. 10.1038/21390

[20] Whittakera AG, Mingos DMP (1995) Microwave-assisted solid-state reactions involving metal powders. *J Chem Soc Dalt Trans* 12:2073–2079. 10.1039/DT9950002073
[21] Luo J, Hunyar C, Feher L, et al. (2004) Potential Advantages for Millimeter-Wave Heating of Powdered Metals. *Int J Infrared Millimeter Waves* 25: 1271–1283. 10.1023/B:IJIM.0000045137.68600.13
[22] Mondal A, Agrawal D, Upadhyaya A (2009) Microwave Heating of Pure Copper Powder with Varying Particle Size and Porosity. *J Microw Power Electromagn Energy* 43:5–10.
[23] Rodiger K, Dreyer K, Gerdes T, Willert-Porada M (1998) Microwave sintering of hardmetals. *Int J Refract Met Hard Mater* 16:409–416. 10.1016/S0263-4368(98)00050-X
[24] Rajkumar K, Aravindan S (2009) Microwave sintering of copper-graphite composites. *J Mater Process Technol* 209:5601–5605. 10.1016/j.jmatprotec.2009.05.017
[25] Sankaranarayanan S, Hemanth Shankar V, Jayalakshmi S, et al (2015) Development of high performance magnesium composites using Ni50Ti50 metallic glass reinforcement and microwave sintering approach. *J Alloys Compd* 627:192–199. 10.1016/j.jallcom.2014.12.009
[26] Badiger RI, Narendranath S, Srinath MS (2015) Joining of nconel-625 alloy through microwave hybrid heating and its characterization. *J Manuf Process* 18:117–123. 10.1016/j.jmapro.2015.02.002
[27] Bansal A, Sharma AK, Kumar P, Das S (2014) Characterization of bulk stainless steel joints developed through microwave hybrid heating. *Mater Charact* 91:34–41. 10.1016/j.matchar.2014.02.005
[28] Srinath MS, Sharma AK, Kumar P (2011) A new approach to joining of bulk copper using microwave energy. *Mater Des* 32:2685–2694. 10.1016/j.matdes.2011.01.023
[29] Gupta P, Kumar S (2014) Investigation of stainless steel joint fabricated through microwave energy. *Mater Manuf Process* 29:910–915. 10.1080/10426914.2014.892975
[30] Singh S, Suri NM, Belokar RM (2015) Characterization of joint developed by fusion of aluminum metal powder through microwave hybrid heating. *Mater Today Proc* 2:1340–1346. 10.1016/j.matpr.2015.07.052
[31] Siores E, Rego DD (1995) Microwave applications in materials joining. *J Mater Process Technol* 48:619–625. 10.1016/0924-0136(94)01701-2
[32] Srinath MS, Sharma AK, Kumar P (2011) A novel route for joining of austenitic stainless steel (SS-316) using microwave energy. *Proc Inst Mech Eng Part B J Eng Manuf* 225:1083–1091. 10.1177/2041297510393451
[33] Bansal A, Sharma AK, Kumar P, Das S (2015) Structure–property correlations in microwave joining of Inconel 718. *JOM* 67:2087–2098. 10.1007/s11837-015-1523-4
[34] Singh S, Singh R, Gupta D, Jain V (2017) Preliminary metallurgical and mechanical investigations of microwave processed hastelloy joints. *J Manuf Sci Eng* 139:064503. 10.1115/1.4035370
[35] Bagha L, Sehgal S, Thakur A, Kumar H (2017) Effects of powder size of interface material on selective hybrid carbon microwave joining of SS304–SS304. *J Manuf Process* 25:290–295. 10.1016/j.jmapro.2016.12.013

[36] Srinath MS, Sharma AK, Kumar P (2011) Investigation on microstructural and mechanical properties of microwave processed dissimilar joints. *J Manuf Process* 13:141–146. 10.1016/j.jmapro.2011.03.001
[37] Gupta P, Kumar S, Kumar A (2013) Study of joint formed by tungsten carbide bearing alloy through microwave welding. *Mater Manuf Process* 28:601–604. 10.1080/10426914.2013.763966
[38] Bansal A, Sharma AK, Das S, Kumar P (2015) On microstructure and strength properties of microwave welded Inconel 718/stainless steel (SS-316L). *Proc Inst Mech Eng Part L J Mater Des Appl* 0:1–10. 10.1177/1464420715589206
[39] Eijk C van der, Sallom ZK, Akselsen OM (2008) Microwave brazing of NiTi shape memory alloy with Ag-Ti and Ag-Cu-Ti alloys. *Scr Mater* 58:779–781. 10.1016/j.scriptamat.2007.12.017
[40] Baker H (1998) *Alloy Phase Diagrams*. ASM International.
[41] Wang L, Wang Y, Prangnell P, Robson J (2015) Modeling of intermetallic compounds growth between dissimilar metals. *Metall Mater Trans A Phys Metall Mater Sci* 46:4106–4114. 10.1007/s11661-015-3037-7
[42] Bahari Z, Elgadi M, Rivet J, Dugué J (2009) Experimental study of the ternary Ag-Cu-In phase diagram. *J Alloys Compd* 477:152–165. 10.1016/j.jallcom.2008.10.030
[43] Liu P, Li Y, Geng H, Wang J (2007) Microstructure characteristics in TIG welded joint of Mg/Al dissimilar materials. *Mater Lett* 61:1288–1291. 10.1016/j.matlet.2006.07.010
[44] Jafarian M, Khodabandeh A, Manafi S (2015) Evaluation of diffusion welding of 6061 aluminum and AZ31 magnesium alloys without using an interlayer. *Mater Des* 65:160–164. 10.1016/j.matdes.2014.09.020
[45] Somasekharan A, Murr L (2004) Microstructures in friction-stir welded dissimilar magnesium alloys and magnesium alloys to 6061-T6 aluminum alloy. *Mater Charact* 52:49–64. 10.1016/j.matchar.2004.03.005
[46] Peng L, Yajiang L, Haoran G, Juan W (2005) A study of phase constitution near the interface of Mg/Al vacuum diffusion bonding. *Mater Lett* 59:2001–2005. 10.1016/j.matlet.2005.02.038
[47] Li Y, Liu P, Wang J, Ma H (2007) XRD and SEM analysis near the diffusion bonding interface of Mg/Al dissimilar materials. *Vacuum* 82:15–19. 10.1016/j.vacuum.2007.01.073
[48] Borrisutthekul R, Miyashita Y, Mutoh Y (2005) Dissimilar material laser welding between magnesium alloy AZ31B and aluminum alloy A5052-O. *Sci Technol Adv Mater* 6:199–204. 10.1016/j.stam.2004.11.014
[49] Liu L, Wang H, Zhang Z (2007) The analysis of laser weld bonding of Al alloy to Mg alloy. *Scr Mater* 56:473–476. 10.1016/j.scriptamat.2006.11.034
[50] Penner P, Liu L, Gerlich A, Zhou Y (2013) Feasibility study of resistance spot welding of dissimilar Al/Mg combinations with Ni based interlayers. *Sci Technol Weld Join* 18:541–550. 10.1179/1362171813Y.0000000129
[51] Wang H, Liu L, Liu F (2013) The characterization investigation of laser-arc-adhesive hybrid welding of Mg to Al joint using Ni interlayer. *Mater Des* 50:463–466. 10.1016/j.matdes.2013.02.085
[52] Zhang J, Luo G, Wang Y, et al. (2012) An investigation on diffusion bonding of aluminum and magnesium using a Ni interlayer. *Mater Lett* 83:189–191. 10.1016/j.matlet.2012.06.014

[53] Zhang J, Luo QG, Shen Q, Zhang LM (2014) Diffusion mechanism and kinetics of diffusion bonded Mg/Ni/Al Joint. *Key Eng Mater* 616:286–290. 10.4028/www.scientific.net/KEM.616.286

[54] Sun M, Niknejad ST, Zhang G, et al. (2015) Microstructure and mechanical properties of resistance spot welded AZ31/AA5754 using a nickel interlayer. *Mater Des* 87:905–913. 10.1016/j.matdes.2015.08.097

[55] Qi X, Liu L (2012) Fusion welding of Fe-added lap joints between AZ31B magnesium alloy and 6061 aluminum alloy by hybrid laser-tungsten inert gas welding technique. *Mater Des* 33:436–443. 10.1016/j.matdes.2011.04.046

[56] Wang HY, Zhang ZD, Liu LM (2013) The effect of galvanized iron interlayer on the intermetallics in the laser weld bonding of Mg to Al fusion zone. *J Mater Eng Perform* 22:351–357. 10.1007/s11665-012-0260-x

[57] Wang Y, Luo G, Zhang J, et al. (2013) A microstructure and mechanical properties of diffusion-bonded Mg – Al joints using silver film as interlayer. *Mater Sci Eng A* 559:868–874. 10.1016/j.msea.2012.09.035

[58] Wang Y, Luo G, Li L, et al. (2014) Formation of intermetallic compounds in Mg-Ag-Al joints during diffusion bonding. *J Mater Sci* 49:7298–7308. 10.1007/s10853-014-8440-8

[59] Sun M, Niknejad ST, Gao H, et al. (2016) Mechanical properties of dissimilar resistance spot welds of aluminum to magnesium with Sn-coated steel interlayer. *Mater Des* 91:331–339. 10.1016/j.matdes.2015.11.121

[60] Patel VK, Bhole SD, Chen DL (2012) Improving weld strength of magnesium to aluminium dissimilar joints via tin interlayer during ultrasonic spot welding. *Sci Technol Weld Join* 17:342–347. 10.1179/1362171812Y.0000000013

[61] Gao M, Mei S, Li X, Zeng X (2012) Characterization and formation mechanism of laser-welded Mg and Al alloys using Ti interlayer. *Scr Mater* 67:193–196. 10.1016/j.scriptamat.2012.04.015

[62] Liu LM, Tan JH, Zhao LM, Liu XJ (2008) The relationship between microstructure and properties of Mg/Al brazed joints using Zn filler metal. *Mater Charact* 59:479–483. 10.1016/j.matchar.2007.02.005

[63] Zhao LM, Zhang ZD (2008) Effect of Zn alloy interlayer on interface microstructure and strength of diffusion-bonded Mg-Al joints. *Scr Mater* 58:283–286. 10.1016/j.scriptamat.2007.10.006

[64] Zhang HT, Song JQ (2011) Microstructural evolution of aluminum/magnesium lap joints welded using MIG process with zinc foil as an interlayer. *Mater Lett* 65:3292–3294. 10.1016/j.matlet.2011.05.080

[65] Liu F, Zhang Z, Liu L (2012) Microstructure evolution of Al/Mg butt joints welded by gas tungsten arc with Zn filler metal. *Mater Charact* 69:84–89. 10.1016/j.matchar.2012.04.012

[66] Zhang Y, Luo Z, Li Y, et al. (2015) Microstructure characterization and tensile properties of Mg/Al dissimilar joints manufactured by thermo-compensated resistance spot welding with Zn interlayer. *Mater Des* 75:166–173. 10.1016/j.matdes.2015.03.030

[67] Xu RZ, Ni DR, Yang Q, et al. (2016) Influence of Zn coating on friction stir spot welded magnesium-aluminium joint. *Sci Technol Weld Join* 1718:1–8. 10.1080/13621718.2016.1266735

[68] Liu L, Tan J, Liu X (2007) Reactive brazing of Al alloy to Mg alloy using zinc-based brazing alloy. *Mater Lett* 61:2373–2377. 10.1016/j.matlet.2006.09.016

[69] Liu LM, Zhao LM, Xu RZ (2009) Effect of interlayer composition on the microstructure and strength of diffusion bonded Mg / Al joint. *Mater Des* 30:4548–4551. 10.1016/j.matdes.2009.04.040

[70] Zhang HT, Dai XY, Feng JC (2014) Joining of aluminum and magnesium via pre-roll-assisted A-TIG welding with Zn interlayer. *Mater Lett* 122:49–51. 10.1016/j.matlet.2014.02.008

[71] Tamang S, Aravindan S (2017) An investigation on joining of Al6061-T6 to AZ31B by microwave hybrid heating using active braze alloy as an interlayer. *J Manuf Process* 28:94–100. 10.1016/j.jmapro.2017.05.027

[72] Tamang S, Aravindan S (2018) Microstructural investigation on joining of Al6061-T6 to AZ31B using nickel interlayer by microwave hybrid heating. In: Proceeding of the International Symposium on Joining Of Materials

[73] Tamang S, Aravindan S (2021) Effect of susceptors on joining of AA6061-T6 to AZ31B by microwave hybrid heating. *J Mater Eng Perform* 31:1130–1139. 10.1007/s11665-021-06220-2

[74] Chen S, Huang J, Xia J, et al. (2015) Influence of processing parameters on the characteristics of stainless steel/copper laser welding. *J Mater Process Technol* 222:43–51. 10.1016/j.jmatprotec.2015.03.003

[75] Yavari AR, Desre PJ, Benameur T (1992) Mechanically driven alloying of immiscible elements. *Phys Rev Lett* 68:2235–2238. 10.1103/PhysRevLett.68.2235

[76] Turchanin MA, Agraval PG, Nikolaenko IV (2003) Thermodynamics of alloys and phase equilibria in the copper-iron system. *J Phase Equilibria* 24:307–319. 10.1361/105497103770330280

[77] Magnabosco I, Ferro P, Bonollo F, Arnberg L (2006) An investigation of fusion zone microstructures in electron beam welding of copper-stainless steel. *Mater Sci Eng A* 424:163–173. 10.1016/j.msea.2006.03.096

[78] Sabetghadam H, Hanzaki AZ, Araee A (2010) Diffusion bonding of 410 stainless steel to copper using a nickel interlayer Shear strength. *Mater Charact* 61:626–634. 10.1016/j.matchar.2010.03.006

[79] Yao C, Xu B, Zhang X, et al (2009) Interface microstructure and mechanical properties of laser welding copper – steel dissimilar joint. *Opt Lasers Eng* 47:807–814. 10.1016/j.optlaseng.2009.02.004

[80] Durgutlu A, Gülenç B, Findik F (2005) Examination of copper/stainless steel joints formed by explosive welding. *Mater Des* 26:497–507. 10.1016/j.matdes.2004.07.021

[81] Bina MH, Dehghani F, Salimi M (2013) Effect of heat treatment on bonding interface in explosive welded copper/stainless steel. *Mater Des* 45:504–509. 10.1016/j.matdes.2012.09.037

[82] Shanjeevi C, Satish Kumar S, Sathiya P (2013) Evaluation of mechanical and metallurgical properties of dissimilar materials by friction welding. *Procedia Eng* 64:1514–1523. 10.1016/j.proeng.2013.09.233

[83] Tamang S, Aravindan S (2019) Joining of Cu to SS304 by microwave hybrid heating with Ni as interlayer. In: 17th International Conference on Microwave and High Frequency Heating. Editorial Universitat Politecnica de Valencia, Valencia, Spain, pp 98–104.

[84] Tamang S, Aravindan S (2022) Joining of dissimilar metals by microwave hybrid heating: 3D numerical simulation and experiment. *Int J Therm Sci* 172:107281. 10.1016/j.ijthermalsci.2021.107281

Chapter 8

Physics Behind High Strain Rate Powder Compaction

Tanuj Vishwakarma, S Janakiraman, and Ashish Rajak
Department of Mechanical Engineering, Indian Institute of Technology Indore, Madhya Pradesh, India

CONTENTS

8.1 INTRODUCTION

Powder metallurgy (PM) is a highly advanced method of manufacturing with the ability to produce near net shape, high strength, and durable compacts at a low cost. It involves consolidating powder particles together by compressing it in a product-shaped die and sintering it if required, to directly produce

DOI: 10.1201/9781003436072-8

a finished product with minimum material lost in its production. Apart from this, it also provides many other advantages over its competing metal-forming technology, which also makes it fit to be used in various automobile, aerospace, and structural applications.

One of the steps involved in PM is powder compaction, which is a crucial step as it largely determines the final properties and strength of product, and needs to be studied well so as to pick up optimum process parameters and precises powder morphological requirements to create a product with the property as close to that generated from its competing processes, which is not achieved to date but day by day with the extensive study's going on, it is very likely to be achieved, and perhaps can go even better than that, so choices we make in every step are to be chosen wisely to achieve optimum results.

For example, depending on the rate at which powder is compacted, the mechanism of adhesion between particles differs. Compaction at a lower rate strain rate, such as that done in the case of conventional methods like isostatic pressing and die compaction, particle undergoes plastic deformation, due to which mechanical interlocking takes place, which further requires sintering for achieving maximum strength but in case of doing the same compaction at a high strain rate, like that in the case of hydraulic impact, explosives, magnetic or spring-loaded hammers technic allows no time for heat to flow out, and that trapped heat eventually melts the metal at the powder particle interface, which leads to welding of particles together at their interface itself, eradicating the need for sintering. Some of the melts also occupy the void between powder particle hence reducing porosity, increasing strength, and all that produces denser product; and with the increase in density, everything improves as the tread suggests, and hence compaction at a high strain rate turns out to be advantageous and therefore preferred, and that is also the reason why there was a shift of research interest toward dynamic compaction that can be seen in its timeline.

Power compaction is itself a complex phenomenon and its dynamic elements add more elements to it. Here one needs to deal with the pre-processing factors, such as powder material and morphological properties, processing parameters like strain rate, pressure and temperature characteristics and post-processing parameters like sintering temperature, all in one set such that it gives out density, strength, hardness homogeneity, porosity, inclusion, cost as output, which needs to be tailored to our needs by picking up the right optimum parameter to be executed. Researchers have come up with various theory and relations depicting governing of process, such as those that are also included in this literature content.

Different attempts have been made toward it to understand the cause and effect of varying parameters on the results obtained and, through this literature, we have tried to compile some of these findings and their justifications to generate a good understanding of high strain rate compaction or high velocity compaction or explosive compaction process.

8.2 DYNAMIC COMPACTION SETUP

There can be many ways through which powder can be compacted, within which also setup arrangement can differ and modified accordingly as per requirements and convenience. It is not possible to list out all of them, but for illustration purposes, some of the most common variants of hydraulic, electromagnetic and gas explosive methods are discussed below.

8.2.1 Hydraulic Press

In this setup, a heavy hammer weighed in tons is dropped on a punch fixated to a die filled with powder that is needed to be compacted. The hammer strikes a punch with an impact velocity and whole momentum is transferred to the punch to compact the powder Figure 8.1.

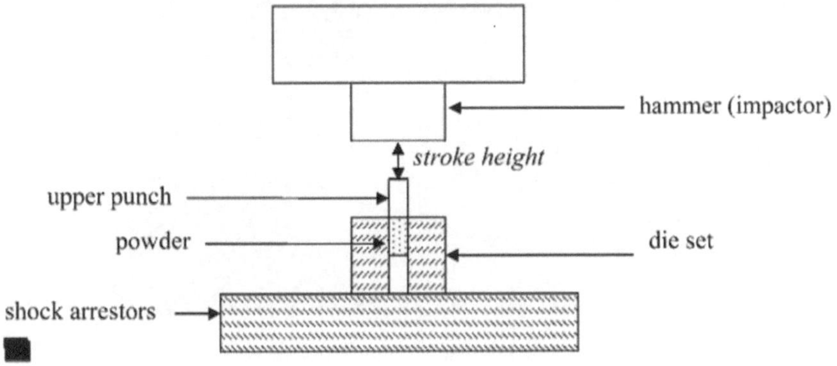

Figure 8.1 Schematic of high-velocity compaction hydraulic press [1].

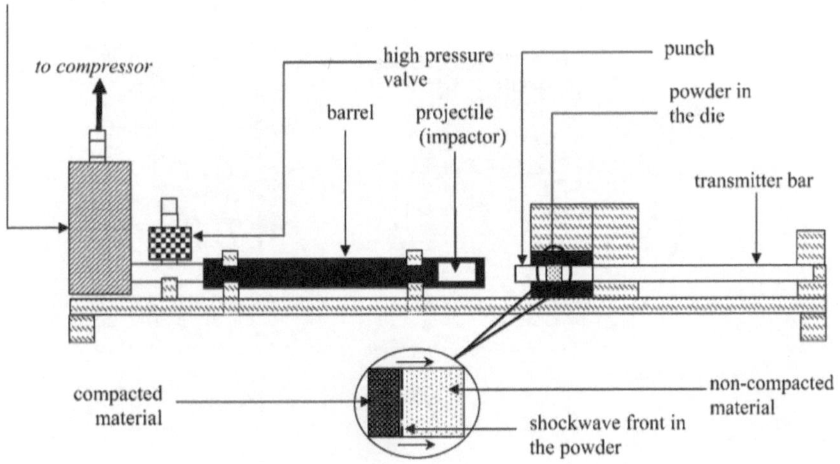

Figure 8.2 Schematic of high-velocity compaction using a gas gun [1].

8.2.2 Gas Gun

In a gas gun setup, compressed air is used to fire a projectile from a barrel targeting a punch fixed to a die containing powder. When a projectile hits a punch, then momentum transfer takes place, which is utilized to compress the powder in a die, as shown in Figure 8.2.

8.2.3 Electromagnetic Powder Compaction

This process utilizes a Lorenz force generated through electromagnetism phenomena for compacting power, one of such set up for it is shown in Figure 8.3. Here powder is filled in easily deformable ductile material like

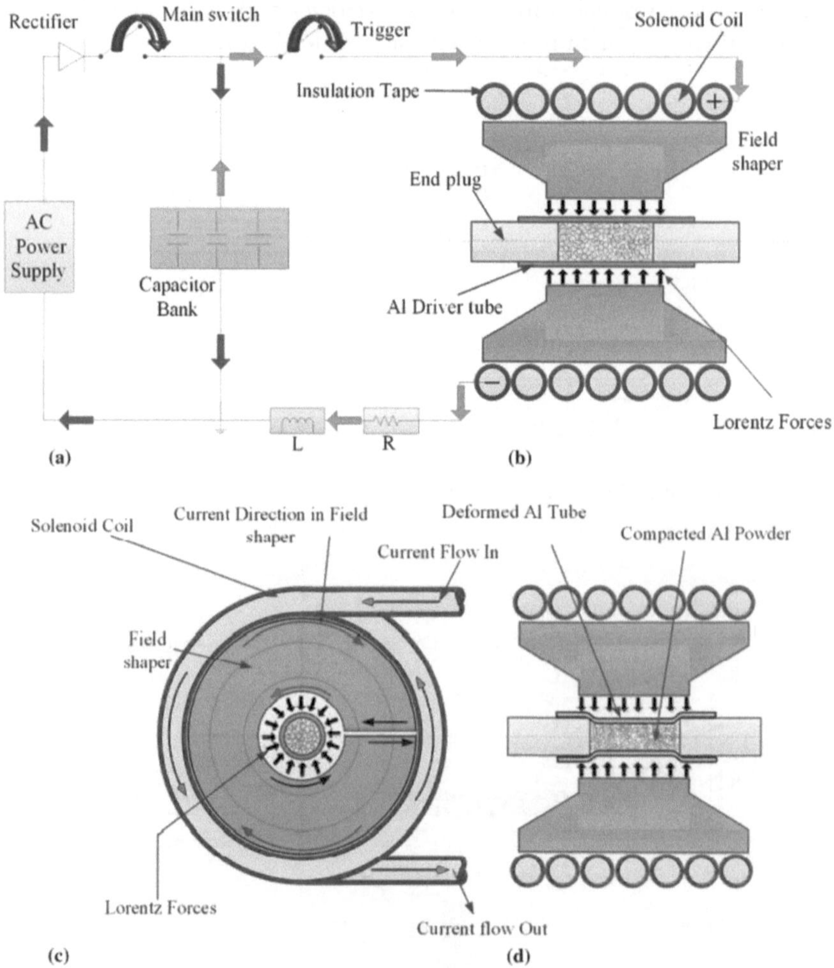

Figure 8.3 Schematic of high-velocity compaction using Electromagnetics Compaction Technics [2].

aluminum tube and packed with mild steel corks at the end of the tube, and this is further placed inside fields sharpener for strengthening magnetic field surrounded by solenoid coil connected to circuitry. When the solenoid coil is powered, then fluctuating magnetic fields are set up, inducing currents on the surface of the conductor placed inside the coil, and due to its repealing nature, generates a magnetic pressure on the aluminum tube that deforms it in radial direction, compacting powder inside.

8.3 COMPACTION PHENOMENA

8.3.1 Compaction Stages

8.3.2 General Compaction Mechanisms

Initially, when powder is just put in container, it is in the stage where one can decrease the bulk volume just by shaking it, this will lead to rearrangement of powder particle where powder is set in another mechanical equilibrium configuration that is more compact; this happens due to slipping and adjusting of powder grain into the spacings that is big enough to occupy powder grains but closed earlier to provide entrance to powder particles but by shaking temporary gaps are generated between particles through which other particles can slip into the void spacings. This will lead to an increase in coordination in the number of particles and density of powder bulk (Figure 8.4). The amount of final densification obtained is also dependent on

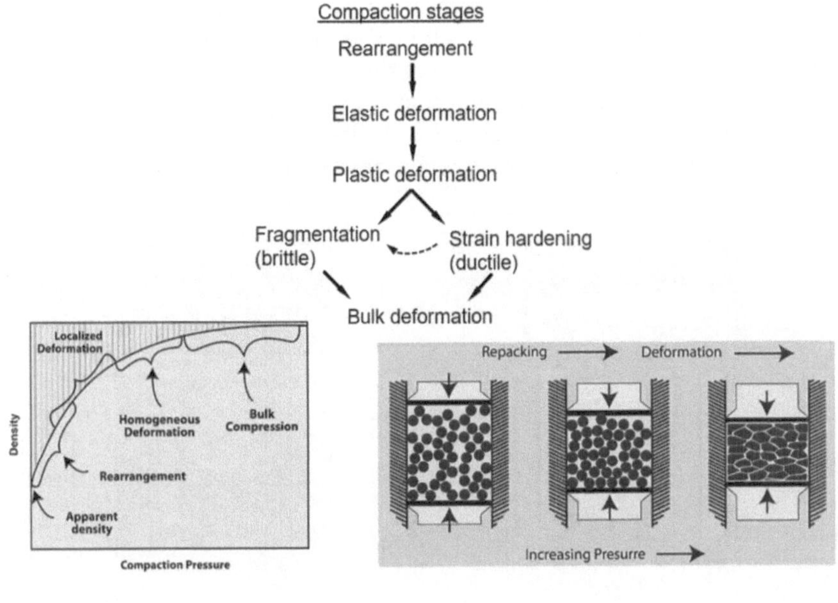

Figure 8.4 Stages of compaction.

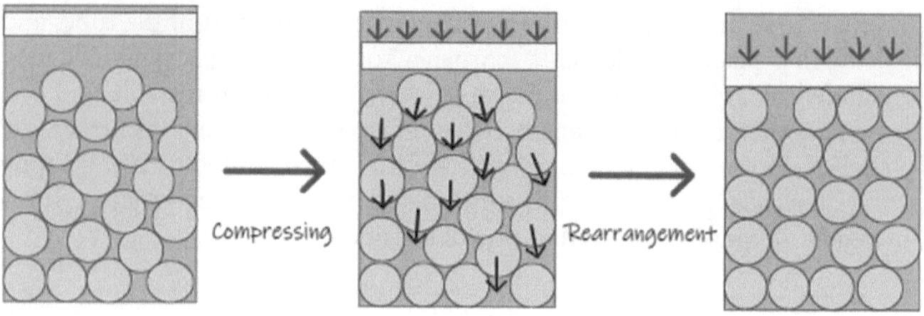

Figure 8.5 Stages of compression and rearrangement.

initial density by which the compaction process has started so thereby also dependent on amount of initial densification [3].

As shown in Figure 8.5, at the start of compaction there will be more rearrangement, but this time some powder particles are pressed at the junction of two particles in contact, leading to side pushing of two particles, inserting a third particle in between if there is space available for that. This will further increase the density.

After nearly all of the spacing is filled and there is no more room left for rearrangement of particles, elastic deformation of particles begins and powder grains will be compressed from the sides, where they are in contact with surrounding powder grains but simultaneously expand towards sides where it is free to expand i.e. in the void spaces and that along with volumetric compression of powder particles also leads to partially filling up of void spaces, leading to further densification. Perhaps this process should not be stopped here; otherwise, there will be a spring-back effect that will lead to a decrease of density as we are still in the elastic zone, and sometimes the density obtained could be less than that what achieved before compaction in elastic zone, as shown in Figure 8.6. If we successfully cross the elastic zone, then comes the plastic zone, where power grains will undergo plastic

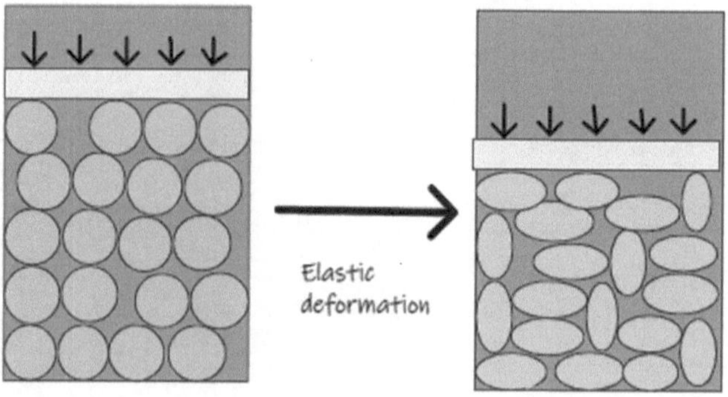

Figure 8.6 Stages of elastic deformation in powder compaction.

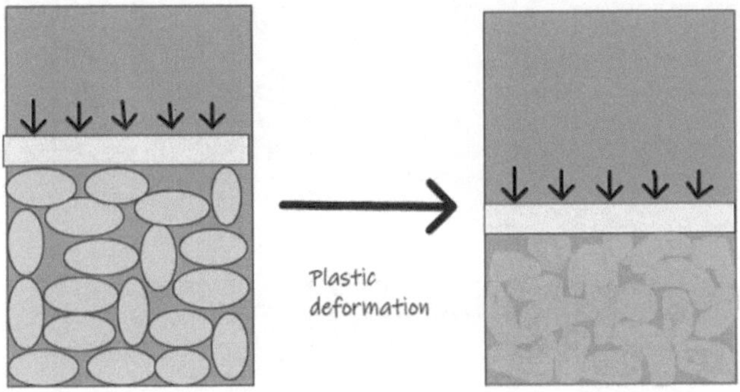

Figure 8.7 Stages of plastic deformation in powder compaction.

deformation. Note that it is possible that while one grain is undergoing plastic deformation mean while another powder grain at some other location in compaction might be still undergoing elastic deformation as the distribution of pressure intensity is generally not uniform.

Plastic deformation gives the same results of densification by further filling up void spacings as that in case of elastic deformation, but here that took place in an irreversible manner as there is no risk of a spring-back effect, as shown in Figure 8.7.

When grains run out of options of ways to adjust with compaction, they undergo fragmentation or strain hardening depending on whether it is made of brittle or ductile material. Sometimes even ductile material can behave as brittle if the compaction process is happening at a very high strain rate or at a very low temperature [4] so that is also needed to be accounted.

In the overall process, the powder is flowing and grains are slide slipping each other, generating heat due to friction. Also, as energy is supplied, and due to the adiabatic nature of high strain rate power compaction, this energy, apart from getting stored in as a deformation, will also appear as heat, which will raise the bulk temperature sufficient enough to cause interface melting and softening [5], which leads to welding at the interface of powder particles and some of the melt will also fill up the void spacing, bringing down porosity [6]; hence, enhancing densification leading enhanced strength of green compact as shown in Figure 8.8 [7].

After this, the compact is going to be cooling down, and depending on the rate of cooling governed by thermal properties of materials, and environmental conditions, change in microstructure may be observed at a powder grain interface conforming partial melting giving different hardness and microstructure for different locations in the compact [8], after removal of load, some spring-back effect can be observed due to release of residual stress induced during loading, as shown in Figure 8.9 [9].

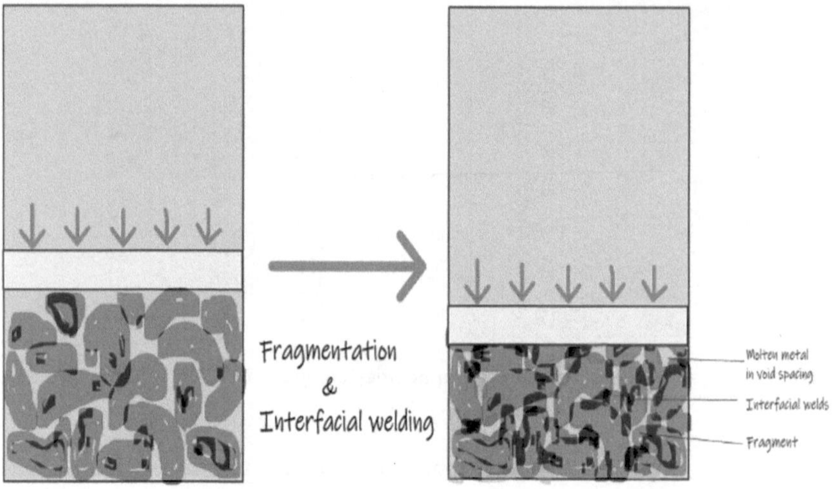

Figure 8.8 Stages of fragmentation and interfacial welding in powder compaction.

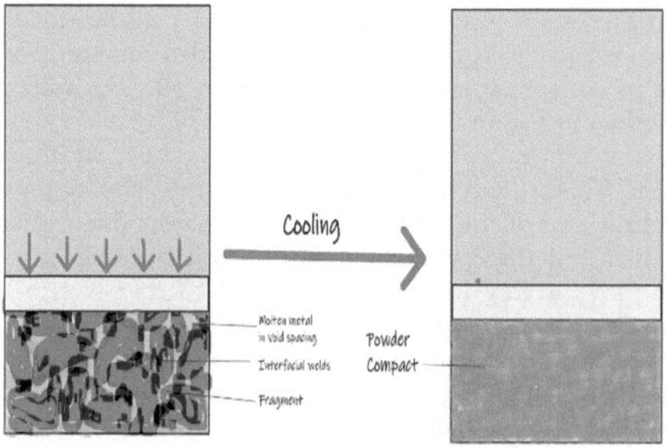

Figure 8.9 Stages of cooling.

Green strength of material is directly related to final density achieved after compaction [10], so more and more attempts are being made to obtain parameters that can take us to achieve density as close to that of a pure metal with near to zero porosity or inclusion.

8.3.3 Multifurcation of Mechanisms Under Different Conditions

A major part of densification occurs during the dynamic compaction of powder. Dynamic compaction is associated with shockwave propagation, which will only happen when the velocity of compaction is greater than the

region/curve	description
A	pressing in absence of shock wave
B	particle deformation close to static, inspite of the shockwave character
C	strength loss and plastic flow of the softened volume
D	localization of energy and melting of interfaces
E, F, G	similar to B-D respectively; incomplete densification
2	separation of plastic flow & incipient melting
3	intensified plastic flow linked with loss of strength
4	time of pressure impulse

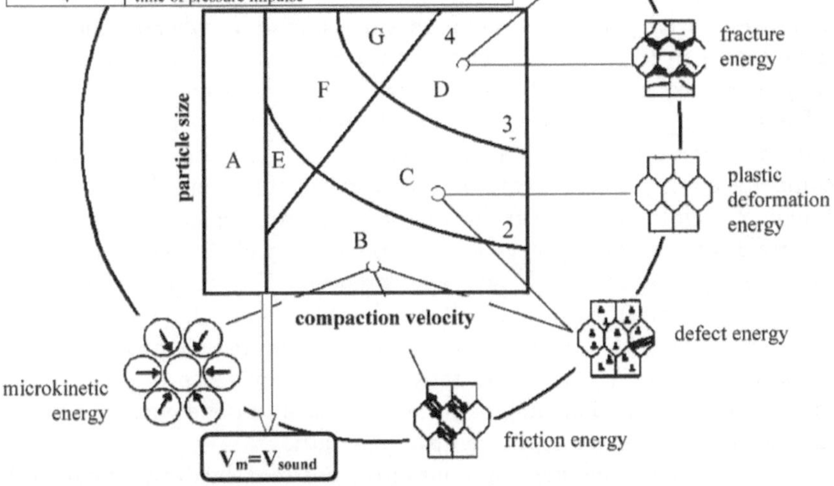

Figure 8.10 Major densification mechanisms occurring during dynamic compaction. [11,12].

velocity of sound in powder; thus, mechanisms differ for low and high compaction rates. At a low loading rate, mechanisms are governed by arrangement and friction dominating effect, whereas at higher loading rates, mechanisms will be taken over by interparticle melting and a strengthening mechanism.

Figure 8.10 describes the various densification phenomena happening during dynamic compaction under a set of particle size and compaction velocity. It can be observed from Figure 8.10 that controlling phenomena of dynamic compaction changes from a region of microkinetic energy to interparticle surface melting as particle size increases.

In Region A, compaction velocity is less than the velocity of sound; therefore, no generation of shock wave and follows regular quasistatic compaction. At the boundary line of Region, A, the material speed of the powder is equal to the velocity of sound in the powder and this line differentiates quasistatic compaction from dynamic compaction, therefore acting as a starting line of dynamic compaction [13].

As we enter Region B with lower particle size, but greater compaction velocity than the sound velocity in powder, there is the generation of the shock wave, but its presence doesn't make much of a change and the process behaves much like in quasistatic compaction in Region A. This happens due to poor packing of smaller size particles causing high dispersion of compressive stress waves and much of wave energy is dissipated in overcoming interfacial friction between particles.

As the particle size increases from Region B to Region C and to Region D in a higher compaction velocity region, greater compaction energy is input, and this leads to plastic deformation and work hardening of material for moderately sized particles without thermal softening, whereas for larger particle sizes in Region D, the effect of thermal softening is also observed along with greater plastic deformation and work hardening mechanism.

The softening was caused in Region D by the localized temperature rise in heat trap zones due to the adiabatic nature of the process and this localized temperature rise is much greater in the particle contact interface, sufficient enough to melt the material at the interface, that further leads to welding at powder grain interface. Some of this molten state also flows into porous void spacings improving density and strength. This molten metal after solidification transforms into different phases than the parent metal, microstructural analysis of which suggested that this molten phase could go up as high as to 15% to 20%, thus serving as evidence of mechanism. Thermal softening and melting of particles cause it to bond with each other, leading to enhancement of the compact strength.

The controlling phenomena are put into a partition by drawing curves 2, 3, and 4 in Figure 8.10 each representing change in mechanisms of compaction. As one goes across curve 2 from Region B to C, densification mechanisms make a sudden change from microkinetic energy to plastic deformation mechanisms However, this transition is smooth across curve 3, where the influence of thermal softening gradually increased to limits where it should be considered.

Curve 4 in Figure 8.10 is the situation where particle deformation time plus freezing time of interparticle melting is equal to the time duration of acting pressure and is represented by a straight line as deformation time is directly proportional to particle size and inversely proportional to speed. The positions of all three curves depend on the thermal properties, density, and hardness of the particles.

8.4 POWDER CHARACTERISTICS

Powder attributes like powder material, powder bulk property, powder shape and size morphology, process parameters, and part geometry are all going to play a key role in steering the path of the process to land on point with certain properties at the end of compaction. If one wants a process to land on the desired set of properties, then one needs to know how the process behaves under what set of parameters, to control and direct it to yield the desired results. Some of the behaviors that have been found by various researchers under specific parameters are described below to illustrate the dependence of the process on such parameters.

8.4.1 Powder Material Properties

The basic material properties that affect the compression process in dynamic compaction are:

i. **Yield, tensile and fracture strength of power material**
ii. **Strain hardening behavior**
iii. **Strain, strain rate and temperature sensitivity** (from Johnson-Cook equation)
iv. **Thermal properties** (such as specific heat, thermal conductivity, diffusivity, latent heat of fusion and melting temperature)
v. **Crystal structure**
vi. **Ductility/brittleness behavior**

Compressibility is inversely proportional to yield strength, tensile strength, strain hardening and strain sensitivity, and directly proportional to temperature sensitivity [1,14].

The deformation behavior of the materials is also affected by the crystal structure because the dislocation behavior is dependent on the same [1,15–17]. Strain hardening of body-centered cubic (BCC) metals is usually much more sensitive to strain rate than face-centered cubic (FCC) metals that remain ductile over a wide range of strain rates. Body-centered cubic metals tend to lose their ductility with increasing strain rates and generally undergo a brittle transition [1,4].

Thermal diffusivity accounts for the different thermal properties affecting the process as these determine temperature rise on an interface of powder grains and extent of interfacial melting taking place for welding between power particles and amount of melt generated to fill up voids spacing in powder.

For brittle materials like metal-oxide and ceramic materials, limited dislocation mobility increases the threshold for yielding, promoting the fracture of individual grains during densification. Therefore, the mechanisms controlling consolidation in brittle materials can be quite different from ductile materials like metals where greater plastic deformation is allowed before fracture.

8.4.2 Powder Bulk Properties

8.4.2.1 Velocity of Sound in Powder Material

High strain rate compaction is generated by creation of shock, for that compaction velocity should be greater than equal to velocity of sound in powder to establish shock waves in the powder. This shock wave propagates and compacts powder along its way. The velocity of sound in a powdered

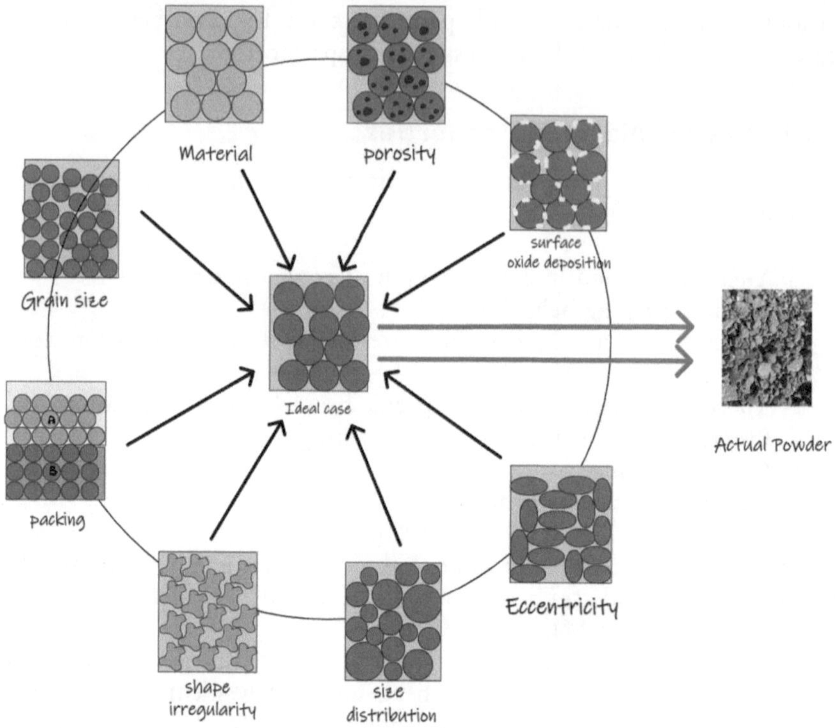

Figure 8.11 Different powder attributes as an ingredient to depict actual powder.

material is at least a hundred orders of magnitude lower than its corresponding velocity in the solid form (or the bulk sound velocity) therefore an important parameter to be considered. This velocity is independent of the powder morphology and size but is dependent on the internal lubricant amount, in addition to material properties such as young modulus [1,18–20], as shown in Figure 8.11.

8.4.2.2 Powder Apparent Density

The ranges of sound velocity in powder are based on the corresponding relative apparent density of the powder and change with time as the experiment progresses.

8.4.2.3 Average Coordination Number of Powder Grain in Bulk

The coordination number is the number of grains with which one grain is in direct contact. In bulk, it's taken on an average for bulk and density will increase as the coordination number increases [21].

8.4.2.4 *Interfacial and Wall Friction*

The compact possesses non-uniform density distribution, which is mainly due to friction among the particles and the mold wall [22]. Analyzing the friction coefficient curve of the iron powder under different pressures, it was found that friction coefficient increases with the increasing compressing force at the beginning of the compression. With the increase of the compact density and the gradual stability of the contact interface, the friction coefficient asymptotically decreases [23]. Improving the friction conditions can reduce the loss of compaction force.

8.4.2.5 *Void Space and Packing Fraction*

As powder mixture consists of particles of all shapes and sizes and overall arrangement can't be free of space voids, therefore apparent density of powder is less than that of pure constituents of powder and depends on powder size and roundness distribution.

8.4.2.6 *Powder Fill*

The amount of powder filled in the die plays a critical role in transference of the shock wave through the powder. The specific energy (compaction energy per unit mass) follows a parabolic trend with powder fill height to achieve the same density in hydraulic press setup for powder compaction [24].

8.4.3 Powder Size, Morphology and Porosity

8.4.3.1 *Particle Size Distribution*

The study shows significant influence of the particle size distribution on the mechanical behavior in powder compaction. A smaller-sized powder fraction will result in a uniform energy distribution over the volume at high pressure, but that can also lead to de-compaction of the compact due to the relief or unloading waves. A larger-sized particle will result in higher adiabaticity, but that may only lead to partially deformation of powder particles, leaving scope of porosity and weak mechanical bonding. A powder can consist of a particle having a range of sizes, varying from very stepped to greatly humped distribution and this will have an effect on deciding process parameters to set up for optimum quality and needed to be taken care of.

8.4.3.2 *Particle Morphology Distribution*

The study suggested that irregular powder (air atomized) suffers greater bending stresses and has greater points of stress concentration (energy

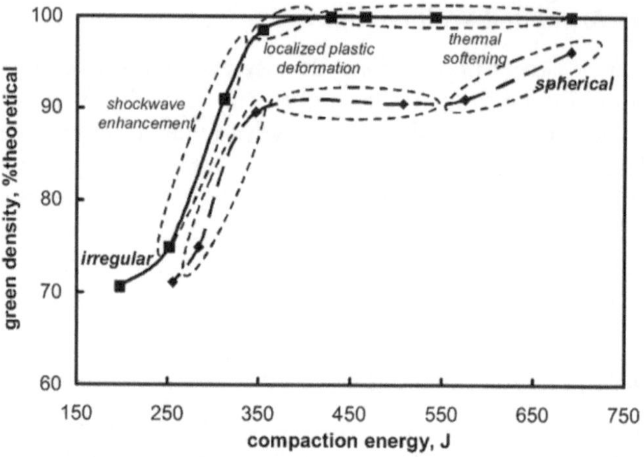

Figure 8.12 Effect of powder shape, irregular and spherical in Al alloy powder [25].

deposition is more) that cause enhanced localized thermal modifications (loss of strength), and hence lead to better densification than spherical powders (inert gas atomized) [13]. Also, after the localized plastic deformation stage, thermal softening stage occurs due to the adiabatic nature of process, as shown in Figure 8.12. This stage occurs earlier in irregular-shaped particle powder as compared to regular-shaped particle powder. Irregular particles suffer more straining and bending moments leading to generation of larger number of thermal softening points supporting densification and strengthening compact [25].

8.4.3.3 Purity in the Powder Particle

As shown in Figure 8.13, the powder produced from different processes has different levels of purity and porosity. For example, the sponge powder that is made from the atomic spraying technic possesses less purity than one produced from electrolytic methods, and in experiments, it was found that electrolytic iron powder is more compressible than sponge iron powder; hence, increasing purity can lead to higher densification [26].

8.4.3.4 Surface Condition

Usually, in the atomic spraying technic, molten drops at higher temperatures undergo surface oxide formation; therefore, surface oxide is inherently there on the powder surface, which is less in the case of powder prepared from an electrolytic process. As shown in Figure 8.14, in the progress of the compaction process, also there is generation of oxide film

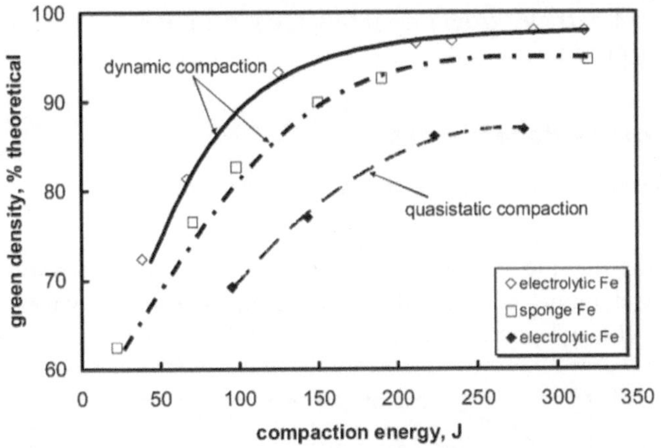

Figure 8.13 Green density of different powder types of iron powder, which are dynamically and quasi- statically compacted [26].

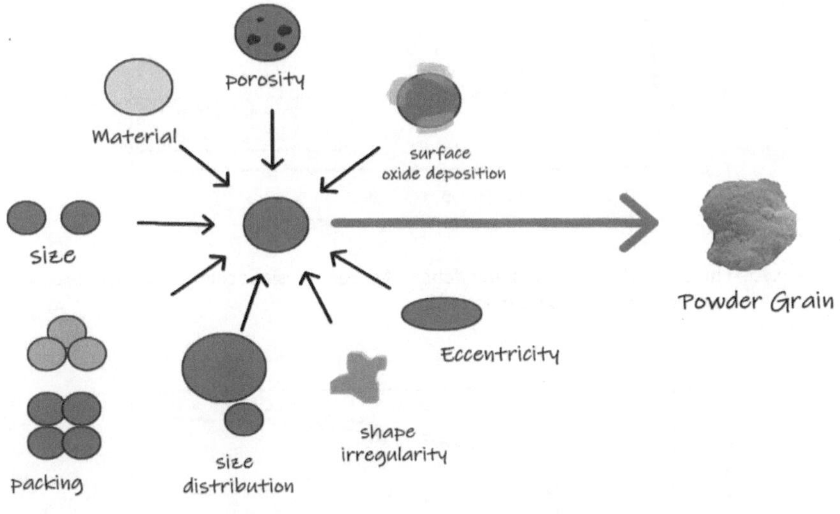

Figure 8.14 Describes different powder attributes as an ingredient to depict actual powder particle.

that is comparatively less when electromagnetic compaction is opted [3]. This oxide layer's formation mainly depends on the compaction pressures, i.e., compaction energy. The oxide layer breaks due to mechanical friction interaction at the interface at a higher energy compaction process; therefore, the oxide layer doesn't appear at a higher compaction pressure [27] and the extent and nature of this surface oxide describes adhesion interaction of particles from one another, contributing its effect on strength of compact produced.

8.4.4 Process Parameters

8.4.4.1 Compaction Pressure Builds Up Characteristics

The compaction pressure is usually measured using a pressure transducer placed in the die and this pressure builds up and varies with location and time as the dynamic compaction happens and the shock wavefront flows away from the compaction end, as shown in Figure 8.15 and Figure 8.16.

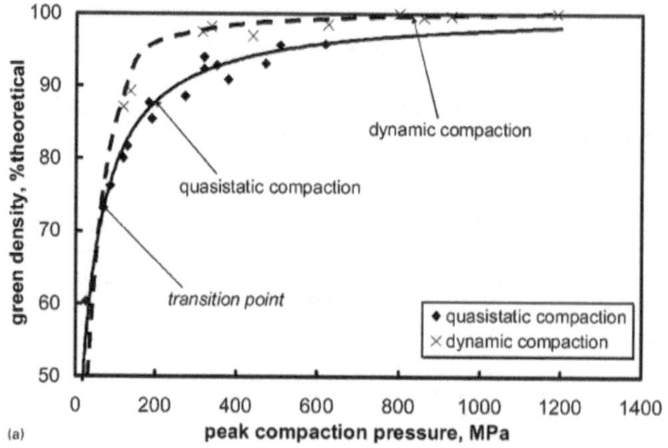

Figure 8.15 The curve depicts the dependence of green density achieved to peak pressure noticed for iron powder (a) [28].

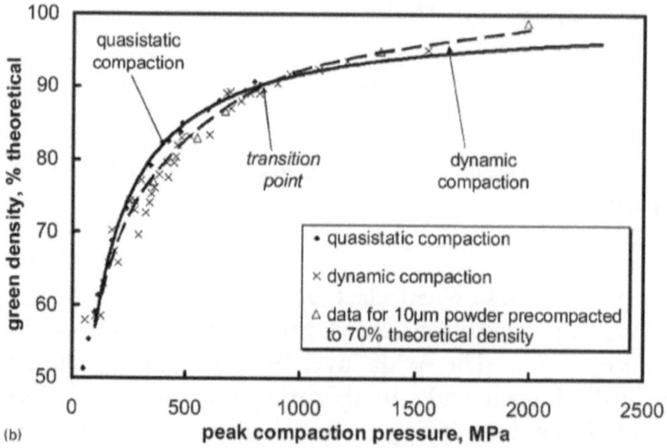

Figure 8.16 The curve depicts the dependence of green density achieved to peak pressure noticed for aluminum powder (b) [28].

The more compaction energy you put in for a shorter period, the higher the pressure rise peak will be observed in general, which could have positive impact of getting higher compact density [28].

8.4.4.2 *Compaction Energy Supplied*

Compaction energy in dynamic compaction per unit mass of powder plays a critical role in framing the properties of powder compact. For iron, aluminum and copper powder, it is observed that greater the compaction energy per unit mass of powder added in the process, the greater the green density it achieved at the end of compaction [28,29] due to greater melting state of powder particle appearing in process available to fill the void spacing between particles, but that effect dissipates as we move closer to achieve maximum theoretical possible density and nature of curve goes asymptotically towards maximum theoretical possible density, as illustrated in Figure 8.17.

8.4.4.3 *Loading Rate*

As the shock waves lead the compaction end, the powder between those two has been compacted by both pressing effect and due to shock wave, whereas powder in front has only been compressed by pressing, so the rate of compaction coupled with velocity and location of shock will decide the nature of compaction and green density achieved, as shown in Figure 8.18 [29,30].

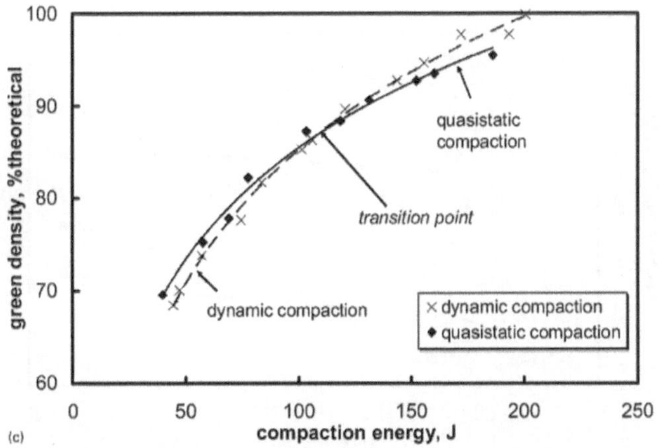

Figure 8.17 E–D (energy–density) curves for copper powder [29].

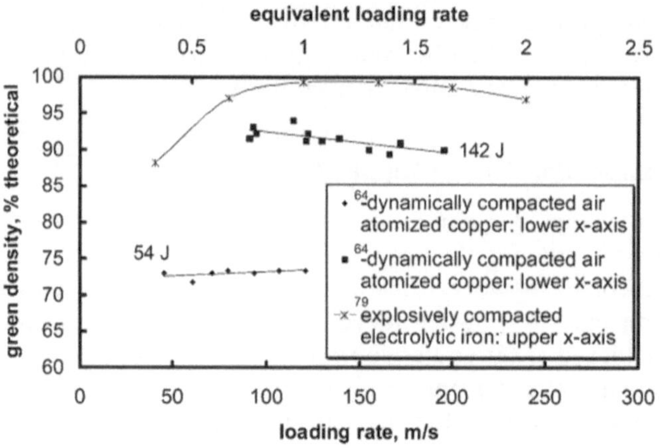

Figure 8.18 Effect of loading rate at constant energy in air atomized copper powder and explosively compacted pure electrolytic iron powder [29,30].

8.4.4.4 Part Geometry and Dimensions of the Container

It is noticed that geometry and dimension of container used for containment of the powder also play a role (mainly height to diameter ratio in case of cylindrical geometry) as pressure distribution suggests an exponential decline of pressure with height with dependence on height to diameter ratio of the compactness of powder. The lower height-diameter ratio makes the transfer effect of compaction force better. Also, the specific energy (compaction energy per unit mass) follows a parabolic trend with powder fill height to achieve the same density in the hydraulic press setup for powder compaction, as shown in Figure 8.19 and Figure 8.20 [1].

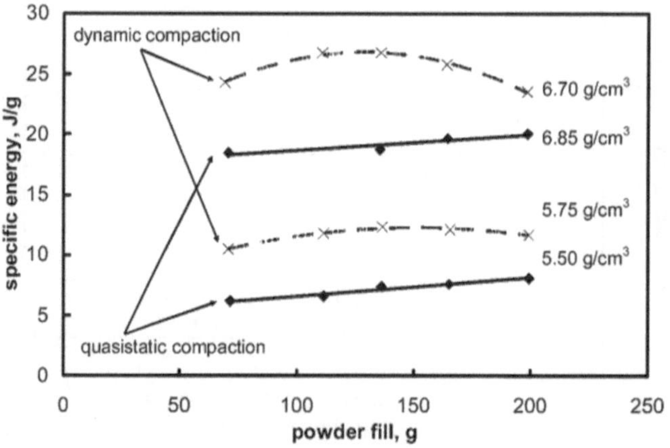

Figure 8.19 Specific energy required to produce the same density with different powder fills (fill heights) in quasi-static and dynamic compaction for iron powder [24].

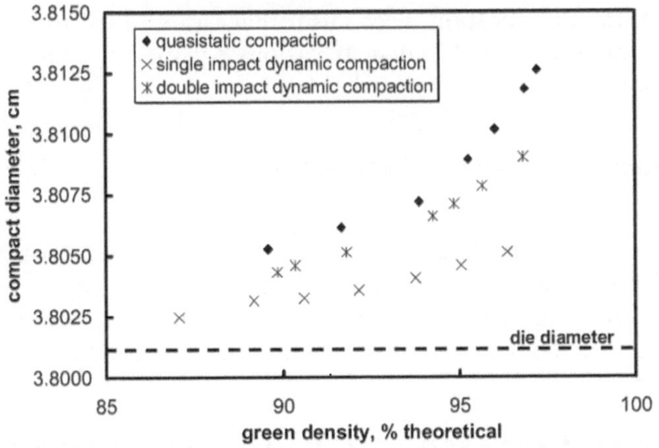

Figure 8.20 Dimensional control in dynamic and quasi-static compaction [1,31].

8.4.4.5 Lubrication Effect

The increase in temperature can improve the lubricating performance of the lubricant, reduce the frictional resistance between the inner wall of the mold and the powder particles, and facilitate the displacement and deformation of the powder particles but larger amount of internal lubricant hinders the densification at higher compaction pressures. In quasi-static compaction, the limitation is due to the space occupied by the internal lubricant. This must also apply to dynamic compaction and is also detrimental to wave propagation through the powder [32].

8.5 THEORETICAL BASIS OF HIGH-VELOCITY COMPACTION (HVC)

The high-velocity compaction process is a complex phenomenon, analysis of which includes nonlinearity problems such as material, geometric and boundary conditions nonlinearity, powder deformation and flow, friction considerations, localized-welding, inertia effect, stress wave propagation, rheological characteristics and thermomechanical couplings. Based on their research work, scholars have put forward various theoretical analysis methods, all of which describe it approximately and not exactly under some range of conditions. Two such theories are described below.

8.5.1 High-Velocity Compaction Equation

A high-velocity compaction equation is an experimental approximation equation obtained by fitting in experimental results by regression and is valid on the specific range of conditions. It is a quantitative analysis derived considering the densification mechanism that can be used to improve the theoretical models. Table 8.1 depicts various equation fittings under mentioned conditions.

8.5.2 Classical Theory of High-Velocity Compaction

Classical theories of high-velocity compaction are sets of theories valid under different sets of conditions, as they are relatively complicated processes, with branching out phenomena so there is no single compaction theory depicting the governing path. Some of those theories, which have tried to describe the phenomena, are presented in Table 8.2.

8.6 SIMULATION-BASED STUDY AND EXPERIMENTAL VALIDATIONS

The numerical simulation of HVC has gained a lot of attention and gained research focus in the field of powder compaction. As in this case, a complete experimental setup requires a lot more investment of time and resources, and a large number of such experiments are needed to be carried out to investigate conditional insights of phenomena; in this case, conducting a numerical simulation run for matching the expectations using already known knowledge makes very much sense. Some of such attempts are described below.

During simulation studies, it is vital to have knowledge of which model gives closer results with experimental results thus in the simulation study for comparisons between using 2D or 3D MPFEM simulations in modeling uniaxial high-velocity compaction behaviors of Ti-6Al-4V powder, it was observed that the 3D MPFEM model generated closer results than the 2D model in following the experimental green density achieved at the end of the compaction and its relationship with true strain and energy per unit mass (E_m) [45].

Zheng Zhou-shun et al. applied the discrete element method to powder compaction processes to establish a contact model of particle and formulate an equation of motion of each particle. The process of powder flow and density distribution during HVC is stimulated based on computing software PFC-2D. The result obtained for density distribution appears to follow a certain law in which density decreases from top to bottom. The highest density is achieved at the top surface, while the lowest density is achieved at the bottom corner. Also, by keeping the height-to-diameter ratio minimal the difference of density on different parts reduces and hence becomes more uniform in density [46].

In order to illuminate the effect of particle grain size and its distribution on force transmission through powder grain contact points, Zhang Wei et al. utilized the concept of force chain, and here in counting the length,

Table 8.1 High-Velocity Compaction Equation [33]

Name of Equation	*Scope of Application*	*Theoretical Characteristics*	*Conclusion Formula*
Kawakita's compaction equation [34]	Suitable for soft metal powder under small pressing force.	It takes into account the plastic deformation and yield limit of powder particles, and is often used as the compression curve of various pharmaceutical powders.	$C = \frac{abP}{1 + bP}$
Bal'shin equation [35]	Only suitable to calculate the medium hardness green body density in the range of hard powder or medium compaction pressure.	It regards the powder as an ideal elastomer, and does not consider the influence of work hardening and friction, pressing time and fluidity.	$\log\left(\frac{P_{max}}{P}\right) = L(\beta - 1)$
Heckel suppression equation [36]	It is used to describe the plastic deformation of metal powder and its stress "hardening effect'.	It is summarized after experiments on a large number of metal powders.	$\log\left(\frac{\epsilon_0}{\epsilon}\right) = KP$
Huang Peiyun suppression equation [37]	It is suitable for soft and hard powders at low, medium and high pressures.	It adopts the concept of natural strain, considering factors such as stress and strain relaxation during powder compaction, work hardening agent, large degree of strain, etc.	$\frac{\left(\frac{d_m}{d} - 1\right)}{\left(\frac{d_m}{d_0} - 1\right)} = e^{\frac{P^n}{M}}$
Johnson-Cook constitutive equation [38]	In view of the large strain, high strain rate and high temperature of ductile materials, it is proposed to describe the behavior characteristics of metallic materials at high deformation rate and high temperature.	It is an empirical equation describing the thermal-viscous plastic constitutive relationship. The material parameters have clear physical meaning and relatively simple, easy to fit experimental data, and have strong versatility. It is usually used as a material constitutive model for numerical calculations.	$\sigma = \alpha \times \beta \times \gamma$ Where: $\alpha = A + B\epsilon^n$ $\beta = 1 + C\ln(\epsilon^*)$ $\gamma = 1 - \frac{T - T_t}{T - T_m}$
Thermal softening shear densification mechanism [39]	After pre-compaction, the powder is struck again by an impact hammer, resulting in instantaneous local high-temperature welding.	It is assumed that the green density is related to the amount of heat softening zone and that the shear zone temperature increases exponentially with the applied impact energy.	$\ln(\varphi) = \frac{K_1 \Delta H_L}{e^{K_2\left(1 - \frac{P}{\Delta H_L}\right)} + c}$

Table 8.2 Classical Theory of High Velocity Compaction [33]

Suppression Theory	*Model Hypothesis*	*Theoretical Basis*	*Theoretical Characteristics*
Impact molding theory [40]	The density of dynamic compaction does not depend on the impact kinetic energy, but on the maximum impact force and impact time.	When the impact kinetic energy is constant, if the hammer mass, powder material, particle deformation and initial loose density are different, the impact stress and action time in different States will be produced, and the corresponding compact density will be completely different.	The compaction effect of a hammer with certain kinetic energy can be seen as the superposition of smaller impact force and longer stroke, or the superposition of larger impact force and shorter stroke.
Inertia effect [40]	Powder particles have inertia in the process of high velocity compaction molding.	The green body starts from the state where the initial velocity is zero and reaches the impact velocity and completes the deformation in a very short time. A high acceleration will be generated in the green body, and then inertia will appear.	There are three inertia effects: global inertia, strain inertia and internal inertia.
Stress wave theory [41,42]	Real stress wave form	All solid materials have inertia and deformability, and the deflection caused by external load gradually propagates from near to far in the medium to form stress waves.	The stress wave has a sawtooth waveform, and its duration is affected by the loading rate. Each loading waveform has several extreme points, which will cause tensile stress after being reflected by the free end surface, resulting in delamination and spalling of the green body surface.
Rheology theory [43,44]	Nonlinear viscoelastic model, plastic model	It regards powder as rheological fluid and considers the influence of time on powder compression process.	It has rheological properties such as strain delay, stress relaxation, compression creep, elastic aftereffect and powder internal friction.

and directional deviation of these force chain was reported for various range of particle size and it was found out that when the particle size distribution range is small, force chain distribution is more uniform, and the anisotropy is not significant, but as the range spreads keeping lower bound same then the fraction of larger size grain increase, and with that fluidity of particle also increases [47]. Analysis of force, such chain formation and its distributions, are only possible through simulation only.

Another simulation-based study was conducted by David J. Benson attempting to calculate shock velocity – particle relation, where hundreds of discrete particles were analyzed with Eulerian finite element program and based on a comparison by least square fitting of experimental data, a reasonably accurate relation for shock velocity and particle relation was established using the model [21].

Numerical simulation turns up to be a reasonable and effective way for investigations, which can largely reduce the experimental cost and time investment in high-velocity compression, with the only limitations being that it is based on a theoretical framework carrying assumptions that might not apply exactly to the real-life world, and what is obtained is an approximate numerical solution, and not the exact one; therefore, it cannot completely replace the experiment procedure and validation methods but still possess a lot of potentials to sort out directions.

8.7 PROCESS LIMITATIONS AND PROBLEMS IN DEVELOPMENT

The high-velocity compaction process is successfully put to use in applications in automobile, aerospace and structural applications, but there are still some limitations to overcome and challenges to tackle to make best use of technic; some of the known ones are described below.

1. There are still a larger number of contradictory views among scholars about the densification mechanisms which is been followed. For example, many scholars [43,48] have raised their doubt about the existence of the stress wave phenomenon. Lack of concrete and convincing evidence is still missing in the knowledge of HVC and needs to be addressed.
2. The parts produced by the high compaction process still lag to deliver properties such as strength, wear resistance and fatigue life over the competing metal forming process, so extensive work is still needed to be done to uplift its shortcomings to achieve the best possible results.
3. So far researchers have mainly focused on the development and experimentation on mainly convectional materials like iron, aluminum and titanium powder, so that needs to change, and new material should also be introduced and tested for the process.

4. Maturity of setup and need for dedicated molds used for the experiments have been noticed, as high energy and high frequency impact characteristics of high-velocity compaction cannot be noted using conventional compaction molds; therefore, a need for smartly designed mold and setup dedicated for the purpose must be produced to extract more and more incites on process.
5. Due to the complexity of the process, a large number of experimental observations are required to investigate the phenomena, but most colleges, universities and enterprises do not have the dedicated setup for conducting HVC, which slows down its development progress and could take longer time to make it fully mature for use extensively.

8.8 NEW DEVELOPMENT TRENDS

New development trends and applications prospects of high-velocity compaction have emerged, some of which are described here.

1. Peng Ni et al. proposed a new method of powder compaction using laser shock for compaction of powder in a small die for creating small parts using powder metallurgy. The compact size of Φ 2.5 × 0.5 mm of aluminum powder using 1,800 mJ laser energy is achieved. Here, 99.72% of green compact density is reached and the microhardness between 50 and 60 HV with an increase of 60% is found with this method [49].
2. While performing experimentation with the electromagnetic powder compaction method on aluminum 6061 alloy powder, it was found out that the result of density variation with input voltage and compaction energy per unit mass needed was coming out to be in good agreement with the simulation study performed by them with less than 2% error [47]. Thus, in some cases like this, a simulation study not only turns to be cost-effective but also possesses broader reach for implementation of sophisticated new-age processes like electromagnetic powder compaction.
3. K. Sajun Prasad et al. in their paper presented a new and rapid throughput experimental setup to do impulsive and shock physics experiments. Setup involves accelerating a projectile mass to a very high velocity using plasma pressure. This plasma is generated by instant vaporization of aluminum foil with the passage of a current of around 200 KA via capacitor discharge. Using this method, CP-Ti Powder was successfully compacted. Along with that shock, velocity can be evaluated using a free surface velocity of the compact-loaded barrel [50].

8.9 SUMMARY

High-velocity compaction (HVC) technology is a key technology to realize high density, high precision, high performance and low cost for P/M parts. High-velocity compaction involves propagation of a compressive stress wave through a powder to cause its densification. Its densification starts with rearrangement, followed by elastic and plastic deformation and melting and welding. This melting and welding are due to its adiabatic nature of process, which causes a heat trap localized melting zone in the regions at high stress concentration zones, which leads to welding of particles together at their interface, providing strength to compact. Some of the melts also occupy the void between powder particles and hence reduces porosity, increasing strength. Mechanisms through which it occurs are greatly governed by powder characteristics and process parameters.

The factors such as material parameters, part shape, molding equipment and process conditions play a role that is discussed in this literature. This also showed that the selection of different pre-processing and process parameters is critical for a high-performance dynamic compaction system is essential. Due to its own technical characteristics, high-velocity compaction has found the optimal solution between the performance and cost of powder metallurgy parts.

Based on their research work, scholars have put forward various theoretical analysis methods, which describe governings of processes approximately if not exactly under some range of conditions. In this various simulation approach there has also been put results of which are quite consistent with experimental; hence, proving potential for the simulation approaches.

High-velocity compaction technology also has the advantages of energy savings, environmental protection, high efficiency and so on, and is widely applied in the field of powder metallurgy industry. However, there are also some problems that cannot be ignored, such as large volume of high-velocity compaction molding equipment, high precision requirements for processing and installation, high cost and low popularity rate, missing knowledge and unknowns.

REFERENCES

[1] Sethi, G., N.S. Myers, and R.M. German. "An overview of dynamic compaction in powder metallurgy." *International Materials Reviews* 53.4 (2008): 219–234.

[2] Thirupathi, N., R. Kumar, and S.D. Kore. "Effect of electromagnetic force on the strength of electromagnetic impulse powder compaction." *Journal of Materials Engineering and Performance* 31 (2022): 1–14.

[3] Martin, C.L., D. Bouvard, and S. Shima. "Study of particle rearrangement during powder compaction by the discrete element method." *Journal of the Mechanics and Physics of Solids* 51.4 (2003): 667–693.
[4] R.S. Coates and K.T. Ramesh: in *'Shock wave and high strain rate phenomena in materials'* (ed.M.A. Meyers et al.), 1st edn, 203–212; 1992, New York, Marcel Dekker.
[5] Gourdin, W.H. "Dynamic consolidation of metal powders." *Progress in Materials Science* 30.1 (1986): 39–80.
[6] Prummer, R. "Explosive compaction of powders, principle and prospects." *Materialwissenschaft und Werkstofftechnik* 20.12 (1989): 410–415.
[7] E. Smugeresky and W.H. Gourdin: in *'Metallurgical applications of shock-wave and high-strain-rate phenomena'* (ed.L.E. Murr et al.), 1st edn, 107–128; 1986, New York, Marcel Dekker.
[8] Gourdin, W.H. "Energy deposition and microstructural modification in dynamically consolidated metal powders." *Journal of Applied Physics* 55.1 (1984): 172–181.
[9] Cui, J., et al. "Effect of discharge energy of magnetic pulse compaction on the powder compaction characteristics and spring back behavior of copper compacts." *Metals and Materials International* 27.9 (2021): 3385–3397.
[10] Moon, I.H. and K.H. Kim. "Relationship between compacting pressure, green density, and green strength of copper powder compacts." *Powder Metallurgy* 27.2 (1984): 80–84.
[11] Roman, O.V., A.P. Mirilenko, and I.M. Pikus. "Effect of high-speed loading conditions on the pressing mechanism." *Soviet Powder Metallurgy and Metal Ceramics* 28.11 (1989): 840–844.
[12] A.B. Sawaoka (ed.): *'Shock waves in materials science'*, 1st edn, 227; 1993, New York, Springer-Verlag.
[13] Meyers, M.A., L.E. Murr, and K.P. Staudhammer. *Shock-wave and high-strain-rate phenomena in materials*. 1992, Boca Raton, FL, USA, CRC.
[14] German, R.M. "Powder metallurgy science." *Metal Powder Industries Federation, 105 College Rd. E, Princeton, N. J. 08540, U. S. A, 1984. 279* (1984).
[15] C.Y. Chiem: in *'Shock wave and high strain rate phenomena in materials'* (ed.M.A. Meyers et al.), 1st edn, 69–86; 1992, New York, Marcel Dekker.
[16] B.O. Reinders and H.D. Kunze: in *'Shock wave and high strain rate phenomena in materials'* (ed.M.A. Meyers et al.), 1st edn, 127–136; 1992, New York, Marcel Dekker.
[17] J. Lankford, H. Couque, A. Bose, and C.E. Anderson: in *'Shock wave and high strain rate phenomena in materials'* (ed.M.A. Meyers et al.), 1st edn, 137–146; 1992, New York, Marcel Dekker.
[18] Kathrina, T. and R.D. Rawlings. *Br. Ceram. Trans.* 95 (1996): 233–237.
[19] Dawson, A.L., L. Piche, and A. Hamel. *Powder Metall.* 39 (1996): 275.
[20] A.L. Dawson and J.F. Bussiere: Proc. 1997 Int. Conf. on 'Powder metallurgy and particulate materials', Chicago, IL, USA, June– July 1997, Metal Powder Industries Federation, 1–539.
[21] Benson, D.J. "The calculation of the shock velocity—particle velocity relationship for a copper powder by direct numerical simulation." *Wave Motion* 21.1 (1995): 85–99.

[22] Skoglund P. "High density PM components by high velocity compaction." *Powder Metall* 44.3 (2002): 14–17.
[23] Bai, H. "A meso-simulation study on the effect of interfacial friction on iron powder compaction." *Hefei Univ Technol* (2009): 20–26.
[24] S. Elwakil and R. Davies: *Proceedings of 13th International Conference on 'Machine tool design research'*, 435–440; 1973, Oxford, Pergamon Press.
[25] Iyer, N.C., W.R. Lovic, and A.T. Male. "Dynamic compaction of P/M Aluminum alloy as related to powder form, microstructures and properties." *Industrial Heating* 53.5 (1986): 18–21.
[26] Stein, E.M., J.R. Van Orsdel, and P.V. Schneider. "High velocity compaction of iron powder." *Iron Powder Metallurgy*. Springer, Boston, MA, 1968. 154–160.
[27] Shaat, M., A. Fathy, and A. Wagih. "Correlation between grain boundary evolution and mechanical properties of ultrafine-grained metals." *Mechanics of Materials* 143 (2020): 103321.
[28] D. Raybould: Proc. 15th Int. Conf. on 'Machine tool design and research', Birmingham, UK, September 1974, 627–636; 1975, London, MacMillan.
[29] Rusnak, R.M. *The International Journal of Powder Metallurgy* 12 (1976): 91–99.
[30] Takashima, K., H. Tonda, I. Shimizu, N. Tomonoh, and M. Miyamoto. *The Japan Institute of Metals and Materials* 54 (1990): 581–588.
[31] S. Elwakil and R. Davies: Proc. 13th Int. Conf. on 'Machine tool design research', 435–440; 1973, Oxford, Pergamon Press.
[32] ElWakil, S.D. and R. Davies. "Lubrication effects in the compaction of sponge-iron powder at low and high speeds." *Powder Metallurgy* 16.31 (1973): 72–87.
[33] Bai, Y., et al. "A review on high velocity compaction mechanism of powder metallurgy." *Science Progress* 104.2 (2021): 00368504211016945.
[34] Kawakita, K. and K.H. Lüdde. "Some considerations on powder compression equations." *Powder Technology* 4.2 (1971): 61–68.
[35] Kostornov, A.G. "Mikhail Yulievich Bal'shin: Development of the scientific bases of fiber metallurgy." *Powder Metallurgy and Metal Ceramics* 42.11 (2003): 563–567.
[36] Heckel, R.W. "Density-pressure relationship in powder compaction." *Transactions of the Metallurgical Society of AIME* 221 (1961): 671–675.
[37] Huang P. *Principles of powder metallurgy*. Beijing: Metallurgical Industry Press, 1982, pp. 182–184.
[38] Johnson, G.R. "A constitutive model and data for materials subjected to large strains, high strain rates, and high temperatures." *Proc. 7th Inf. Sympo. Ballistics* (1983): 541–547.
[39] Guo, S., Y. Chi, F. Meng, et al. "Compaction equation of powder metallurgy high velocity compaction." *Powder Metallurgy Material Science Engineering* 11.1 (2006): 24–27.
[40] Richard, F. "HVC punches PM to new mass production limits." *Metal Powder Report* 57.9 (2002): 26–30.
[41] Wang, L.L. *Stress wave foundation*. Beijing: National Defense Industry Press, 1985, pp. 1–2.
[42] Ju, Y., Y. Yang, Y. Mao, et al. "Experimental study on stress wave propagation mechanism in porous media." *Chinese Science* 39.5 (2009): 904–917.

[43] Zhenhua, C. "Rheological problems of powder compression (II)-rheological model of powder compression." *Rare Met Mater Eng* 21.2 (1992): 1–11.

[44] Zhenhua, C. "Rheology of powder compression (I)-rheology of powder." *Rare Metal Materials and Engineering* 21.1 (1992): 3–7.

[45] Eriksson, M., et al. "New semi-isostatic high velocity compaction method to prepare titanium dental copings." *Powder Metallurgy* 47.4 (2004): 335–342.

[46] Zhou, J., et al. "Comparisons Between 2D anD 3D mpFem simulations in moDeling uniaxial HigH VeloCity CompaCtion BeHaViors oF ti-6al-4V powder." *Archives of Metallurgy and Materials* 67 (2022).

[47] Thirupathi, N., R. Kumar, and S.D. Kore. "Experimental and numerical investigations on electromagnetic powder compaction of Aluminium 6061 alloy powder." *Powder Technology* (2022): 117579.

[48] Wei, Z., et al. "Effect of particle size distribution on force chain evolution mechanism in iron powder compaction by discrete element method." 力学学报 54.9 (2022): 1–12.

[49] Ni, P., et al. "Laser shock dynamic compaction of aluminum powder." *Journal of Manufacturing Processes* 77 (2022): 694–707.

[50] Prasad, K.S., et al. "A rapid throughput system for shock and impact characterization: design and examples in compaction, spallation, and impact welding." *Journal of Manufacturing and Materials Processing* 4.4 (2020): 116.

Chapter 9

Effect of Different Modelling Condition on Finite Element Modelling of Electromagnetic Impact Welding

Alok Singh[1] *and Ramesh Kumar*[2]

[1]Indian Institute of Technology Guwahati, Assam, India

[2]Department of Mechanical Engineering, Saharsa College of Engineering (under Department of Science & Technology, Govt. of Bihar) Saharsa, Bihar, India

CONTENTS

DOI: 10.1201/9781003436072-9

9.1 INTRODUCTION

Electromagnetic impact welding is similar to that explosive welding. It has been differentiated based on the energy input provided in different explosive processes. In this method, for the energy input in place of explosive material and detonation high magnitude of current or voltage is applied. Miranda et al. have postulated a brief idea related to the joining process by the explosion welding process. The study revealed that the maximum velocity obtained for the impact is 400 m/s. This study has used steel, stainless steel (Z91), and aluminium alloys AA1050, AA6063, and AA6061 materials. This work revealed that a high current frequency is required for the lower conductive material whereas, for higher conductive material, the low-frequency current needed for the bond formation in between the material, stand-off distance should be optimum [1].

Mousavi et al. [2–4] have studied the modelling of electromagnetic impulse welding using the Williamsburg equation of state. Results were obtained in the simulation using AUTODYN and Johnson-Cook constitutive equation. The experiment got jetting and interfacial waves, revealing successful weld conditions. First, shear stress in the two plates must be of opposite sign and secondly, peak plastic strains should exceed a specific minimum value. The welding requirements include jetting, wavy interface, and relative velocity to create a bond between the two parts.

Zhidan et al. revealed that the joint strength depends upon the impact velocity. There should be optimum velocity to obtain maximum joint strength, such as joining an aluminium plate with steel if the impact velocity is 200 m/s. The joint strength is more than the strength of the aluminium itself [5–7]. Results revealed that as the velocity increases from 200 m/s to 350 m/s joint strength decreases because of a traditional jig-jag shape form observed in Ansys and peak and valley obtained. The excellent bonding weld region should deform plastically and have a wavy shape, and a good weld is found to be malleable, stated by Raoelison et al. [8–10]. It is also noted that wave height increases as discharge voltage increases, and with the gap, further, it also said that with the addition of discharge voltage and gap weld defect come into the picture.

Vivek et al. experimented on the copper by vaporizing foil actuator to define the morphology of the joint and came out with specific results [11,12]. The experiments revealed that for low velocity (around 460 m/s) the wavy interface was obtained whereas for high velocity (approximately 800 m/s) smooth interface was obtained. It was also revealed that due to the strain hardening and shock hardening, there is an increase in the hardness value near the interface. There is also a change in the grain size of the base metal from 10 μm to less than 500 nm. Nassiri et al. [13] have used the second-order method in the Abacus to study the impact of the two plates for welding or joining. The finite sliding formulation and penalty method describe the surface-to-surface contact between the two plates. Coulomb friction co-efficient (0.15) was used in the penalty method to determine the friction between the two plates. This study revealed that the flyer plate and the base plate material near the interface must move with the same velocity for welding.

Many papers showed metal joining had been done on the tube and observed different parameters. Kumar and Kore [14,15] concluded that first velocity and strain would increase if we increase the frequency and then decrease. Strain rate hardening has a profound impact on EM compression. In this simulation, the material model used for Al 1050 is the Johnson-Cook material model. Lee et al. [16–18] experiments have been conducted to join HT9 ferritic-martensitic steel tubes with end-plugs made of Gr91 oxide dispersion-strengthened (ODS) steel. This work has concluded that for joining the steel tube with the end-plugs, end-plug geometry and its taper angle play a vital role. The wavy nature of the joining interface depends on the taper angle and length of the end-plug. The metallurgical study also reveals the same regarding the wavy interface. Nassiri et al. modelled the electromagnetic impulse welding process for joining the Al2024-T3 and studied the workpiece's velocity and magnetic pressure distribution [13].

Xu and Sun [19] performed a weakly coupled simulation using an Eulerian-based thermo-mechanical collision model using various launch velocities to investigate the formation of a wavy interface in the bond region. But in this work, many thermo-physical variables were assumed to be constant. Still, in reality, they vary as the temperature changes. Properties like thermal expansion and thermal conductivity depend on temperature variation.

9.2 MATERIALS AND METHOD

The parts were modelled as a solid model and then meshed. Solid elements were meshed by eight-node solid hexahedral elements and assembly of the components performed in Ls-Dyna. The dimensions of the geometrical case used for simulation are in the literature. The materials for the field shaper and driver are copper, for tube and end-plug are STS 410 (HT9) and Grade91, respectively.

9.2.1 Field Shaper and Driver

The internal diameter of the field shaper is 8.7 mm, and external diameter is 106 mm, and the length of the field shaper is 79.3 mm. The material used for the field shaper is copper. Figure 9.1(a) shows a three-dimensional case of the field shaper in isometric view and Figure 9.1(b) front view. Figure 9.2 shows the cross-section view of the field shaper.

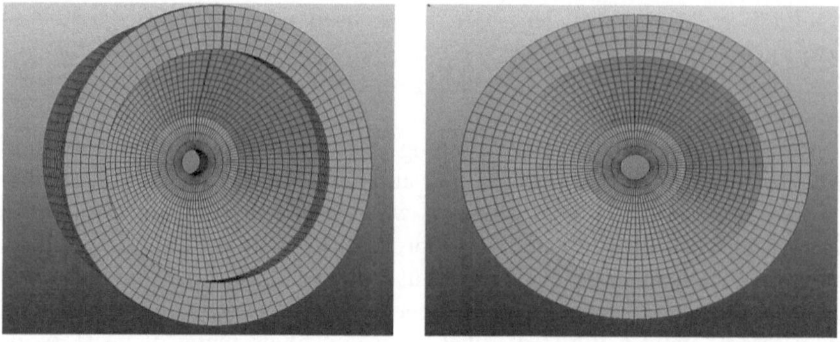

(a) (b)

Figure 9.1 Three-dimensional models of the field shaper: (a) isometric view and (b) front view.

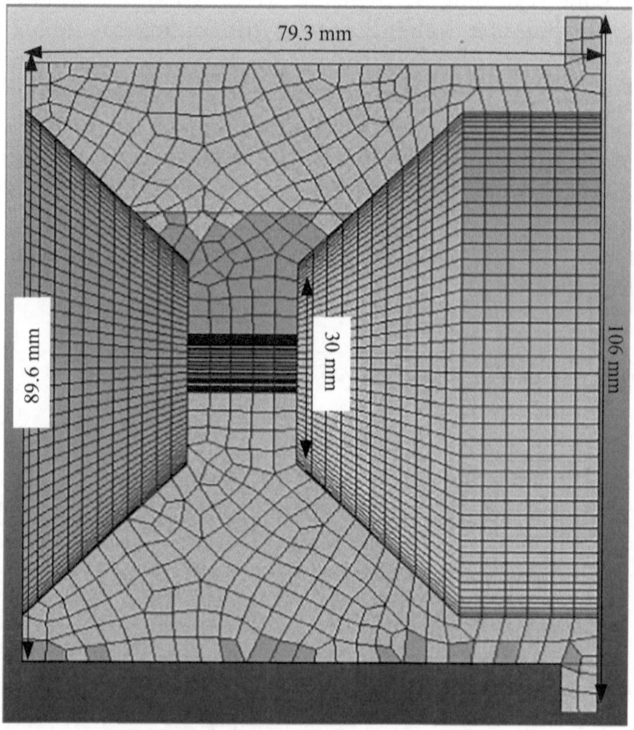

Figure 9.2 Dimensions of the field shaper used in the simulation.

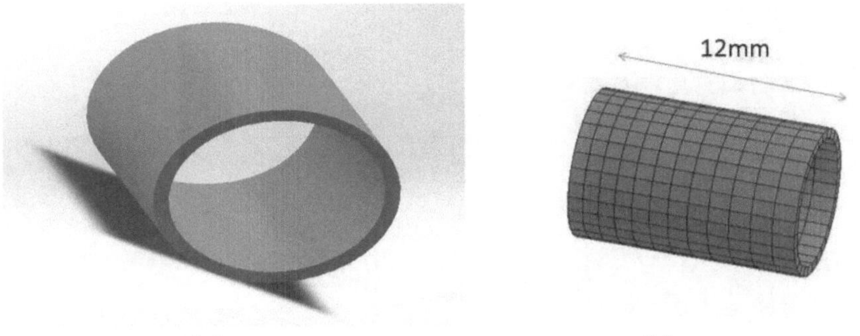

(a) (b)

Figure 9.3 Copper driver: (a) solid model and (b) meshed model.

The copper driver sleeve used has a length of 12 mm and a thickness of 0.5 mm. The driver is fixed over the end of the steel tubes. The material used for the driver is copper. Figure 9.3 shows the solid model and meshed model of the driver.

9.2.2 Flyer Tube and End-Plug

The steel tube used has a diameter of 7 mm, a thickness of 0.6 mm, and a length of 25 mm. The properties of various steels like 4340, D9, and STS 410 are similar to HT9 martensitic steel assigned to the non-conductive hollow tube for the numerical simulation. Figure 9.4 shows the solid and meshed models of the steel tube.

This study used an end-plug as a base rod for welding purposes. Figure 9.5 shows the dimensions of the end-plug. Figure 9.6(a) shows the complete assembly of the flyer tube, driver, end-plug, and the field shaper, and Figure 9.6(b) shows their cross-section view.

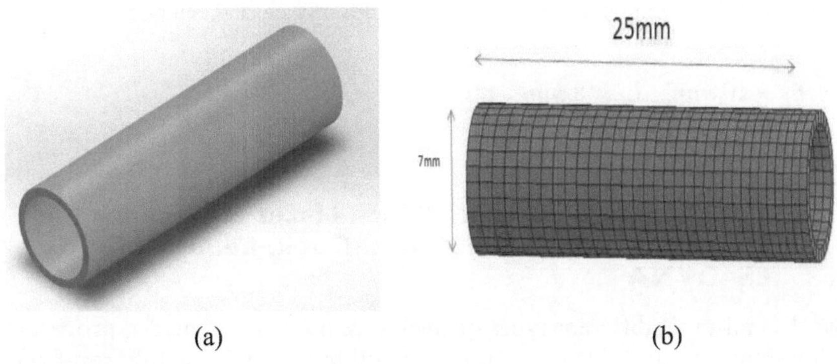

(a) (b)

Figure 9.4 Steel tube: (a) solid model and (b) meshed model.

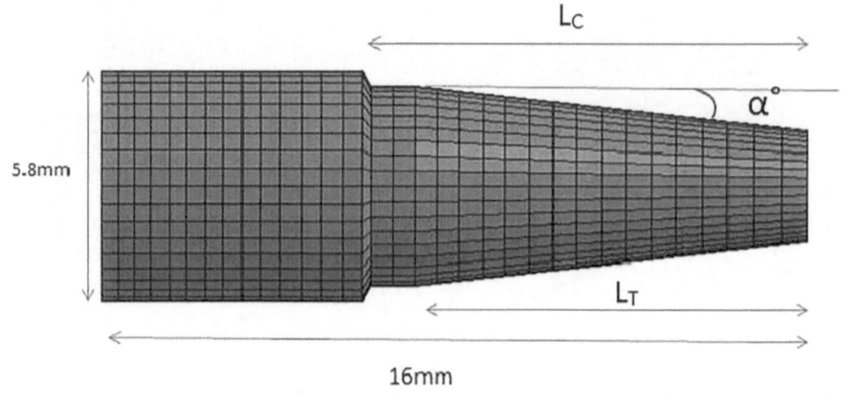

Figure 9.5 Meshed model of end-plug with different dimensions.

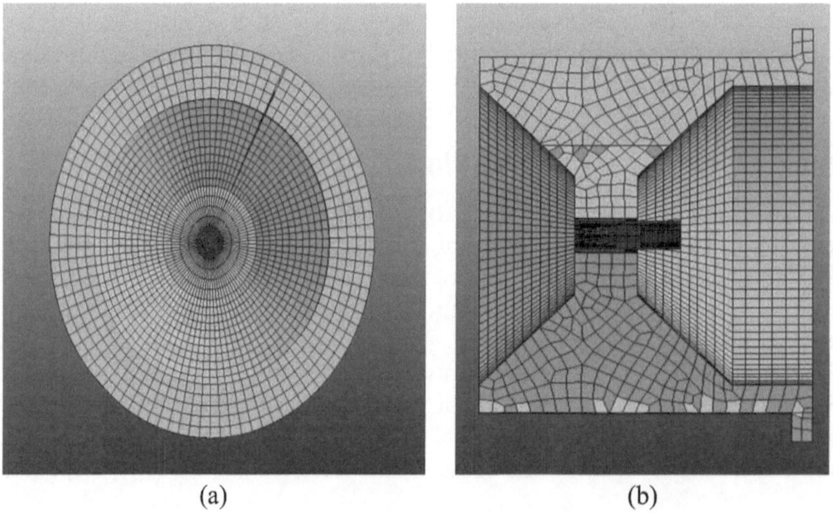

Figure 9.6 Assembled meshed model of the coil, driver, tube, and end-plug.

$$L_c = 10\ mm \quad L_t = 8\ mm \quad \alpha = 30$$

9.2.3 Case I: Input Current 704 kA Material Case for Field Shaper Is Simplified Johnson-Cook, i.e. mat098 in LS–DYNA

For the end-plug, different types of steel were used. The material properties of the types of steel used are shown in Table 9.1 and the Johnson-Cook constant of the materials is shown in Table 9.2.

Table 9.1 Material Properties of Case

Material	*Model*	*Mass density*	*Young's modulus*	*Poison's ratio*
Copper	Field shaper	8960	128 GPa	0.340
Copper	Driver	8960	128 GPa	0.340
STS410	Tube	7850	210 GPa	0.285
Grade91	End plug	7830	209 GPa	0.290

Table 9.2 Values of the Johnson-Cook Constants for Different Materials

Materials	*A(MPa)*	*B(MPa)*	*C*	*n*	*m*
Copper	92	292	0.025	0.31	1.09
STS 4340	450	738	0.02	0.388	0.8
Grade 91	113	211	0.218	0.073	0.881

Material cases are the constitutive relations that describe the response of materials under different mechanical and environmental conditions. Johnson and Cook developed a constitutive case for metal subjected to large strains, high strain rate, and high temperature. The Johnson-Cook plasticity case used to simulate the EM process involves high-strain rate deformation processes like explosive welding, electromagnetic welding, etc. The Johnson cook plasticity model incorporates material responses like strain hardening, strain-rate effect, and thermal-softening. It is empirical relation for flow stress (σ_y), is given in equation (9.1). In equation (9.1), $\dot{\epsilon}_p^*$ is the ratio of plastic strain rate and effective plastic strain rate, given by equation (9.2).

$$\sigma_y = (A + B(\epsilon_p)^n)\,(1 + Cln\dot{\epsilon}_p^*)\,(1 - T*^m) \tag{9.1}$$

$$\dot{\epsilon}_p^* = \frac{\dot{\epsilon}_p}{\dot{\epsilon}_{po}} \tag{9.2}$$

where A , B, C, n, m are material constants, ϵ_p is equivalent plastic strain, $\dot{\epsilon}_p$ is the plastic strain rate, and $\dot{\epsilon}_{po}$ represents the effective plastic strain rate of the quasi-static test used to determine the yield and hardening parameters A, B, and n.

9.2.4 Input Discharge Current

Electromagnetic forces were used as an energy source to accelerate the driver and the tube onto the end-plug. Figure 9.7 shows a current curve used for the simulation. Maximum current supplied for simulation 702 *kA*

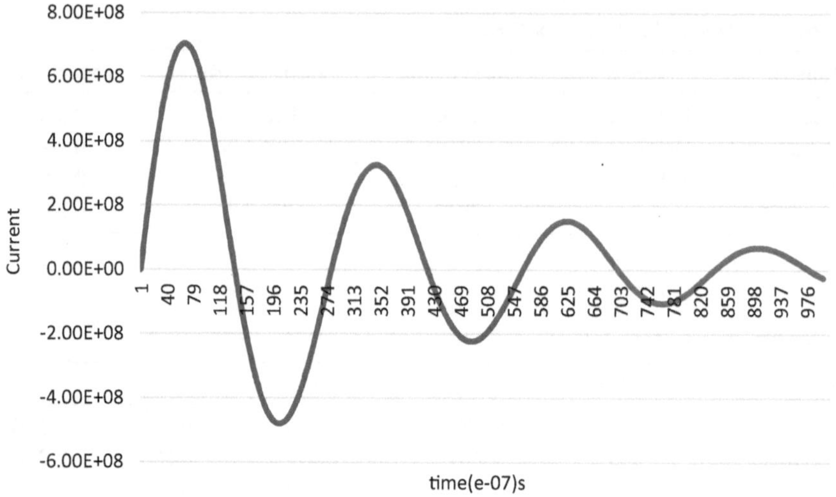

Figure 9.7 Current waveform used for simulation.

for proper plastic deformation in tube material. The termination time for the simulation is 15 μs since we use a positive cycle only, and we provide time step 0.1 μs. It is necessary to take a positive value only; otherwise, fluctuation in the material will start as the current will change to the negative side, the direction of magnetic force will change, and deformation will be opposite side; hence, the resultant will be zero.

9.2.5 Case 2: Input Current 704 kA Field Shaper Material Piecewise Linear Plastic Material, i.e. mat024 Ls-Dyna

The basis of the material type mat024 model is on the elastoplastic material model. This material model defines a strain rate dependency and stress vs strain curve. Failure is determined based on either plastic strain rate or minimum step size in this model.

Strain rate is calculated by using the Cowper and Symonds case, which scales the yield stress with the factor given by equation (9.3), where $\dot{\varepsilon}$ is the strain rate.

$$1 + \left(\frac{\dot{\varepsilon}}{C}\right)^{1/P} \tag{9.3}$$

All properties are similar to Case 1, except the case of field shaper—the values of the constants for copper material are given in Table 9.3. Where the strain rate is represented by $\dot{\varepsilon}$ and model constants are C, P for the

Table 9.3 Model Constants for Copper Material

Material	*Model*	*Mass density*	*Young's modulus*	*Poison's ratio*	*Yield stress*	*Tangent modulus*	*C*	*P*
copper	Field shaper	8960	200GPa	0.29	7.00e+07	0.590	40.0	5.0

Cowper and Symonds formulation, this case is susceptible towards strain rate-sensitive material.

9.2.6 Case 3: Input Current 704kA Material Case for Field Shaper Power-Law Plasticity Case, i.e. mat018 LS-DYNA

The power-law hardening rule is there in material type 18 (mat018). It is an example of the isotropic plasticity model that dramatically influences the rates. In this model, the yield stress (σ_Y) is a function of plastic strain, and equation (9.4) calculates it. In equation (9.4), elastic strain to yield point is represented by ε_{yp}; effective plastic strain is represented by ε^p. The current curve used in this case is same as it used in the above two cases.

$$\sigma_Y = K\varepsilon^n + K\left(\varepsilon_{yp} + \varepsilon^p\right) \tag{9.4}$$

The power-law plasticity model is there as the field shaper's materials model. Table 9.4 shows the values of the material's properties of it.

9.2.7 Case 4: A Material Model for Field Shaper Kinematic Plastic, i.e. Mat003 in Ls-Dyna

Material type 3 model, i.e., Mat003 material model, is suitable for kinematic hardening plasticity and isotropic model. This model has a speciality that rate effects can include with it. It is also cost-effective and used in different applications such as solid elements, shells, and beams. An additional criterion is there with this model: the viscoelastic formulation with the Cowper and Symonds formulations. The strain rate was calculated

Table 9.4 Model Constant for Mat018

Material	*Case*	*Mass density*	*Young's modulus*	*Poison's ratio*	*Strength coefficient*	*Hardening index*	*Yield stress*
Copper	Field shaper	8960	200GPa	0.29	1.438e+05	0.0713	7.0e+07

using the Cowper and Symonds model, which scales the yield stress. Using the factor given by equation (9.5):

$$1 + \left(\frac{\dot{\varepsilon}}{C}\right)^{1/P} \tag{9.5}$$

Where the strain rate is represented by $\dot{\varepsilon}$ and model constants are c,P for the Cowper and Symonds formulation, this case is susceptible towards strain rate sensitive material.

9.2.8 Case 5: Input Current 1100*kA* and Field Shaper of the Elastic Material Case of Copper, e.g., Mat.0001 in Ls-Dyna

Material type 1, i.e., Mat0001 material model, is a hypo-elastic isotropic material model used to analyze solid, shell, and beam. The speciality of this model is that one can model the fluid also by using this model. In this case, the current having a peak value of current is $1125kA$, and the current frequency used. The plot of the current curve used in this model is shown in Figure 9.8. The hypo-elastic material model may be unstable for the large magnitude of strains. Hyper-elastic material model, e.g., *MAT_002, will be mode suitable for large strain analysis. Bending and axial damping factors can dump down the numerical noise. The moment resultants (M_i) and force resultants (F_i) were incorporated in the calculation of the damping factors given by equation (9.6) and equation (9.7).

In equation (9.6), DA is the axial damping factor, and in equation (9.7), DB is the bending damping factor.

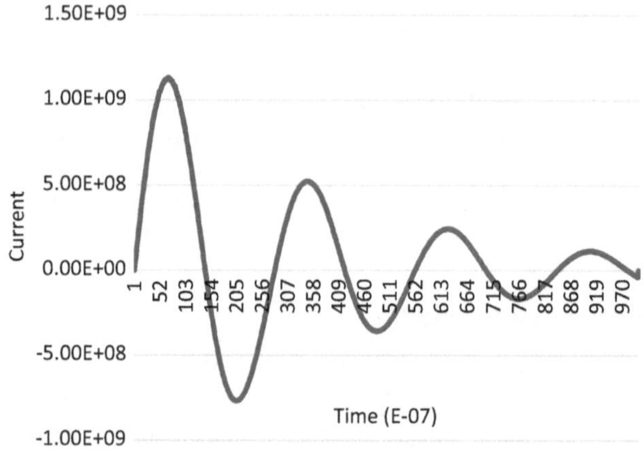

Figure 9.8 Discharge current curve with 1,100 kA peak current.

$$F_i^{n+1} = F_i^n + \left(1 + \frac{DA}{dt}\right)\Delta F_i^{n+1/2} \tag{9.6}$$

$$M_i^{n+1} = M_i^n + \left(1 + \frac{DB}{dt}\right)\Delta M_i^{n+1/2} \tag{9.7}$$

9.3 RESULTS AND DISCUSSIONS

The present work focuses on developing welding criteria based on simulated results. After solving the input deck, the post-processing part has performed in Ls-Dyna. It is important to study the impact behaviour of the tubes with end-plug colliding due to the electromagnetic (EM) force to understand the weld zone formation. Different output results obtained in the simulation are as follows.

9.3.1 Current Density

Current density depends upon various factors, such as the area and conductivity of the material. In these cases, the material case had changed although physical properties of remain are constant, their behaviour against external stimuli varies from case to case. The fringe pattern of current density obtained from the Ls-Dyna EM module did not observe any marginal change in the current density because current remains constant, but if current changes, it increases. The fringe pattern of the current density obtained is shown in Figure 9.9.

9.3.2 Lorentz Force

At $t = 0$, Lorentz's force is zero because the current is zero at the beginning. As the amplitude of the current increases, Lorentz force also increases. It also varies with the deformation of the driver. Maximum Lorentz force obtained at $t = 6.2\,\mu s$. Here at this instant, the maximum value of the Lorentz force is 5.243e+09 as per the Ls-Dyna module. The variation in the Lorentz force due to the variation of the material case is negligible. Still, in the fifth case, when the magnitude of current increases to 1125 kA, its magnitude changes abruptly. The fringe pattern of the Lorentz force for five different cases is shown in Figure 9.10.

Since the magnitude of current in Case 5 is very high (1100 kA), hence the magnitude of Lorentz's force is very high too. Its maximum magnitude is 1.32E+10.0 N, while its magnitude is less in the other four cases. The maximum Lorentz force in Case 1, Case 2, Case 3, and Case 4 are 5.5e+03 N, 4.95e+03 N, 4.95e+03 N, and 5.26e+03 N, respectively. The Lorentz force generated for five different cases is shown in Figure 9.11.

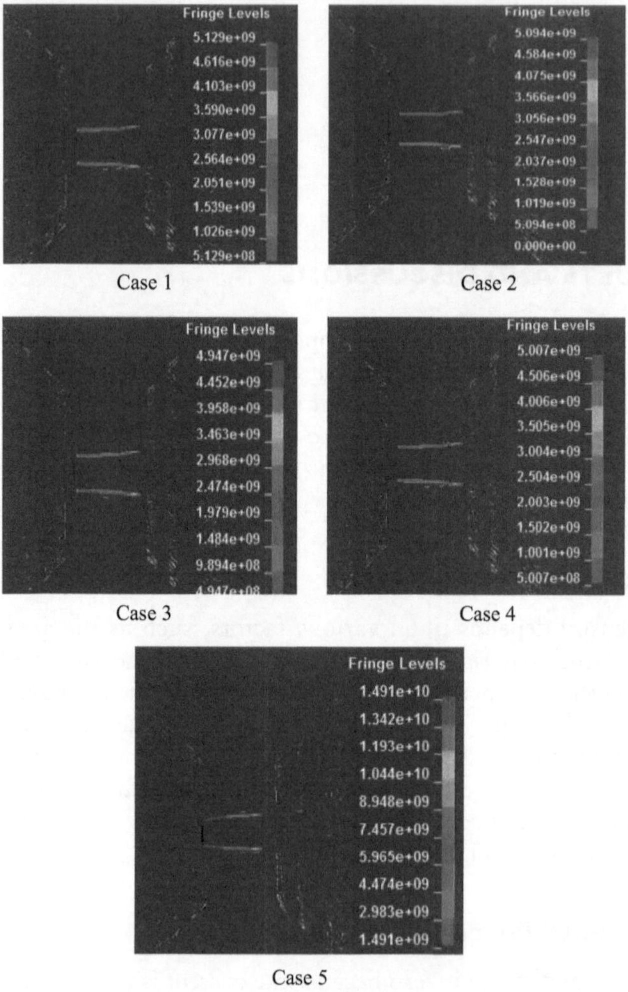

Figure 9.9 Current density for five different cases.

9.3.3 Electromagnetic Field

The electromagnetic field was generated by current and concentrated by field shaper. Fringe pattern and vector plot of surface electromagnetic field for Case 1 and Case 5 are given in Figure 9.12. As the current varies with time, the magnetic field also varies. Initially, the magnitude of the magnetic field is very low, but after certain time steps, it increases as current increases and further decreases. The fringe pattern of surface magnetic field and magnetic field BEM at different time intervals are taken from the output of Ls-Dyna, shown in Figure 9.12. The magnetic field direction is shown in Figure 9.12 by the right-hand rule; it can be defined that the direction of

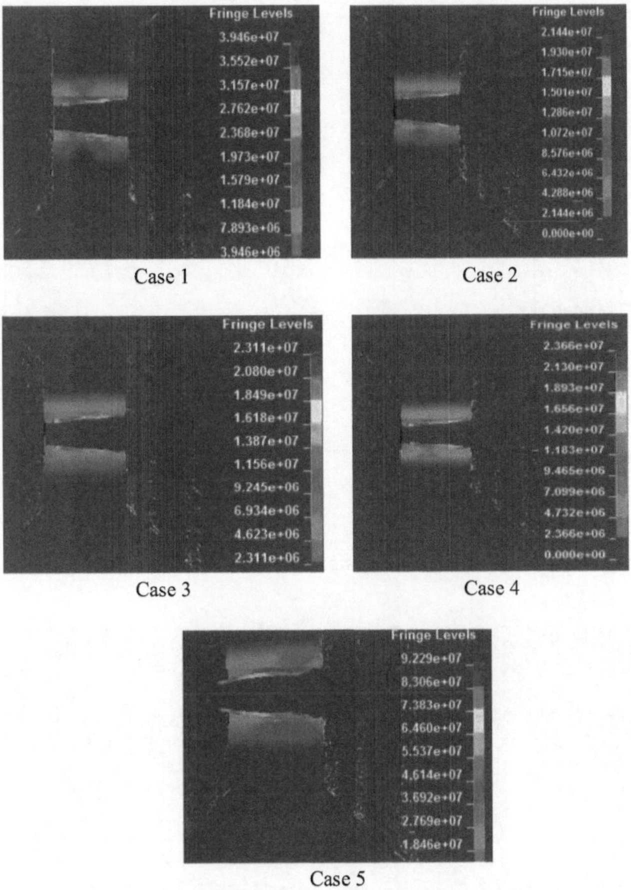

Figure 9.10 Lorentz force pattern for five different cases.

Lorentz force will be perpendicular to the driver since the magnitude of current in Case 5 is much higher than that of other cases. Here in Figure 9.12, it can be concluded that the magnetic field toward the end of the end-plug is more; hence, deformation at that end will also be more. The histogram plot for the magnetic field generated in five different cases is shown in Figure 9.13.

9.3.4 Pressure Distribution in the Tube

The first driver strike on the tube with high velocity makes the surface-to-surface contact; later, the tube strikes on the end-plug and makes surface contact. The fringe pattern of Case 1 and Case 5 are shown in Figure 9.14. It is obvious from the figure that pressure distribution is not uniform since

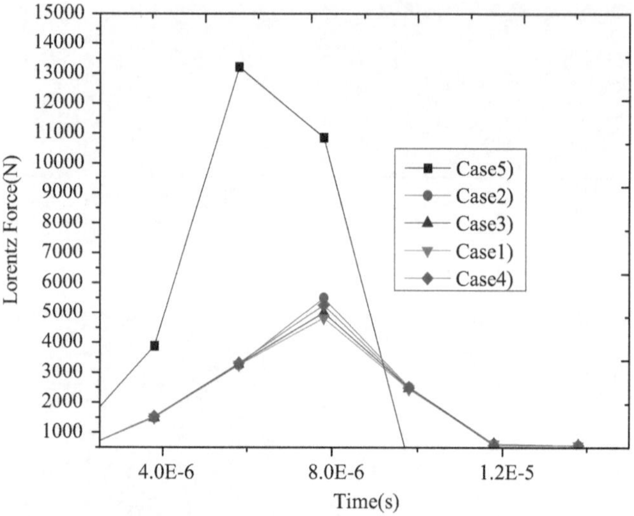

Figure 9.11 Lorentz force plot for five different cases.

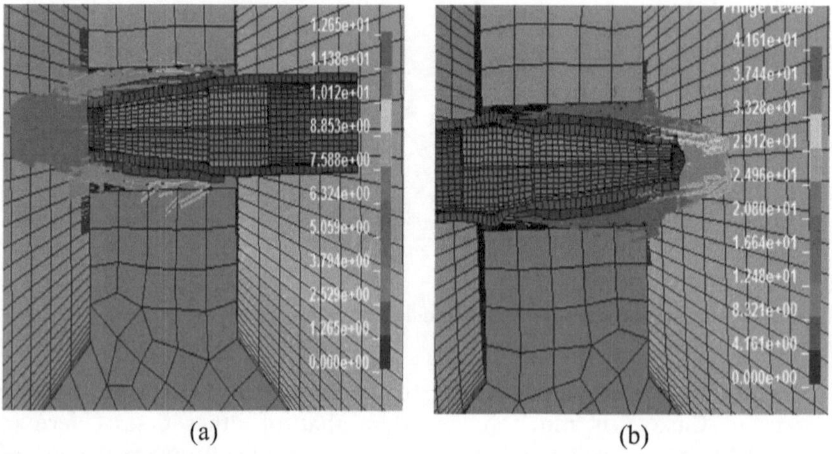

Figure 9.12 Vector plot of the electromagnetic field of (a) Case 5 and (b) Case 1.

the input current is the same from Case 1 to Case 4; hence, pressure variation is almost similar.

Here in this figure, magnitude of pressure increases from 1.245e+10 to 2.015e+10 when current input increases from Case 1 to Case 5. Initially, the magnitude of pressure in the tube was zero because the driver took time to strike on the tube. The deformation that resulted in the different parts of the assembly is shown in Figure 9.15.

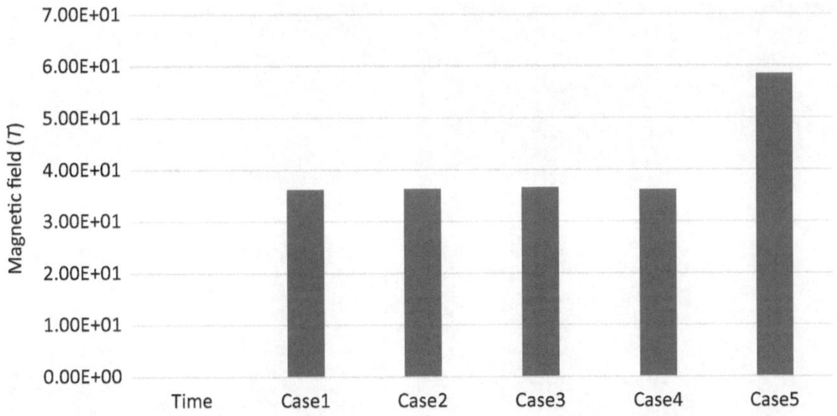

Figure 9.13 Histogram of peak electromagnetic field for five different cases.

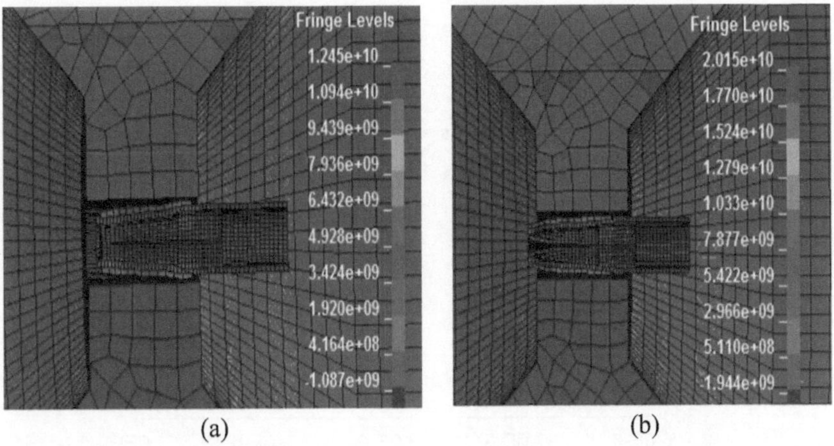

Figure 9.14 Pressure distribution of Case 1 and Case 5.

9.3.5 Impact Velocity

The results of the impact velocity obtained for the various cases are shown in Figure 9.16. The minimum velocity of impact required for weld occurrence is given by equation (9.8). In this study, the minimum impact velocity for 4340 steel is around 435 m/s.

$$v_{min} = \sqrt{\frac{UTS}{S}} = \sqrt{\frac{1110000000}{5850}} = 435 \text{ m/s} \qquad (9.8)$$

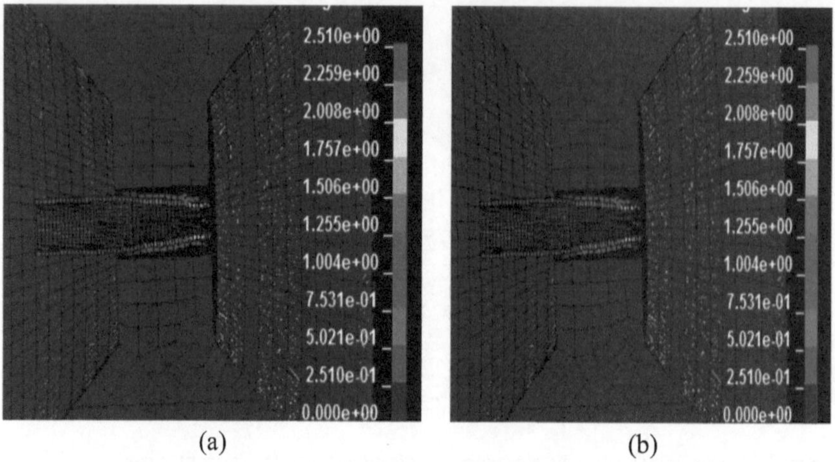

Figure 9.15 Deformation in Assembly: (a) Case 5 and (b) Case 3.

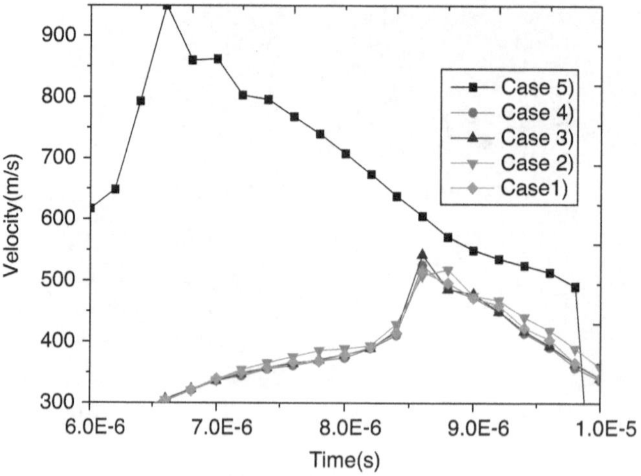

Figure 9.16 Velocity plot for five different cases.

Maximum velocity observed in Case 5 as elastic material has been taken for field shaper, and current input has a peak value of 1125 *kA*. Velocity curves of Cases 1, 2, 3, and 4 are almost similar; that's why the curve had coincided—the maximum magnitude of velocity found in Case 5. Among the four remaining cases, the velocity of Case 2 is maximum; its magnitude is more than 554.67 ms, while the maximum velocity of Case 1 is 962.64 *m/s*, which has a very high margin of magnitude than other cases.

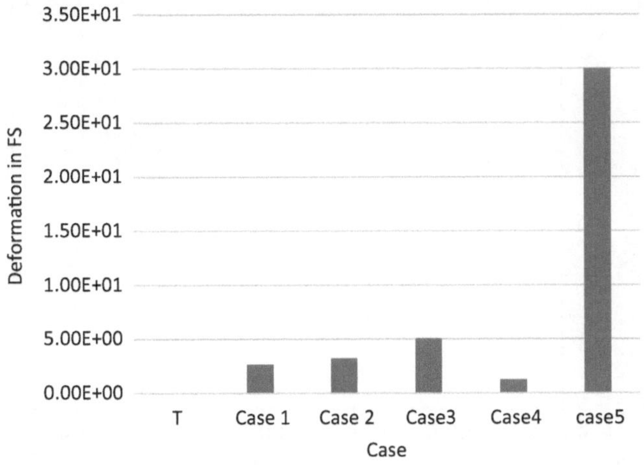

Figure 9.17 Deformation in FS for different cases.

9.3.6 Deformation in Field Shaper

The induced current in the driver generates a magnetic field in the opposite direction. It applies force on the field shaper, and deformation in the field shaper occurs. The deformation in the field shaper in five different cases is shown in Figure 9.17. In these cases, it has been observed that in Case 3, when the power-law plasticity case has used field shaper deform plastically a little although the current input is 702 kA. In Case 5, it deforms plastically up to 0.030 mm as the current input is very high, as shown in Table 9.5.

9.3.7 Effective Plastic Strain in Tube

The effects of the effective plastic strain were analyzed for the steel tube during the impact welding process along the taper length of the end-plug. The values of the effective plastic strain for different cases of the end-plug are shown in Figure 9.18.

Table 9.5 Results of All Five Cases

Variable parameter	*Case 1 Mat098*	*Case2 Mat024*	*Case 3 Mat018*	*Case 4 Mat003*	*Case 5 Mat0001*
Max. Lorentz force (kN)	4.95	5.01	4.95	5.02	13.2
Max. plastic strain in tube	0.859	0.865	0.863	0.863	2.161
Deformation in F.S. (*mm*)	0.00264	0.00324	0.00504	0.00123	0.03076
Plastic strain in FS.	0	0	0.00142	0	0.4

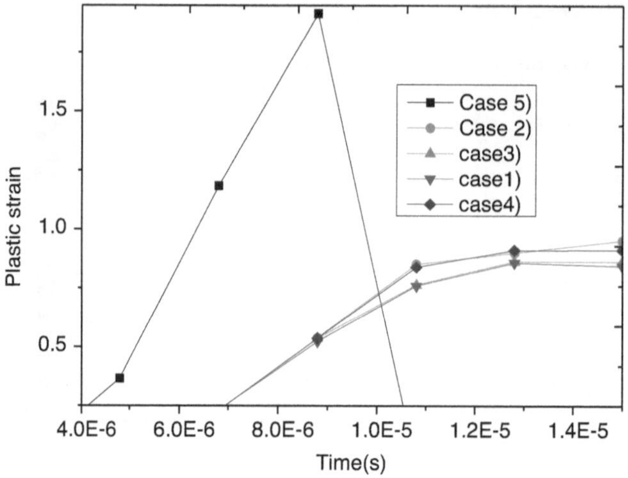

Figure 9.18 Results of the plastic strain obtained from simulation in various cases.

9.3.8 Deformation in Tube

As the driver strikes with the tube, it deforms the tube plastically, and deformation depends upon the impact velocity. Here in this study, deformation in the tube diameter is measured at the small tapered end of the end-plug. Maximum deformation obtained in the tube is 3.50491 *mm* in Case 5, and rebound magnitude is 0.199871 *mm*. The rebound occurred as the velocity was too high and the material had sufficient hardness. Figure 9.19 shows the deformation and rebound in tube diameter.

Here we found that the high current plastic strain is very high in the first case and in the remaining cases it almost the same. In the first case, the

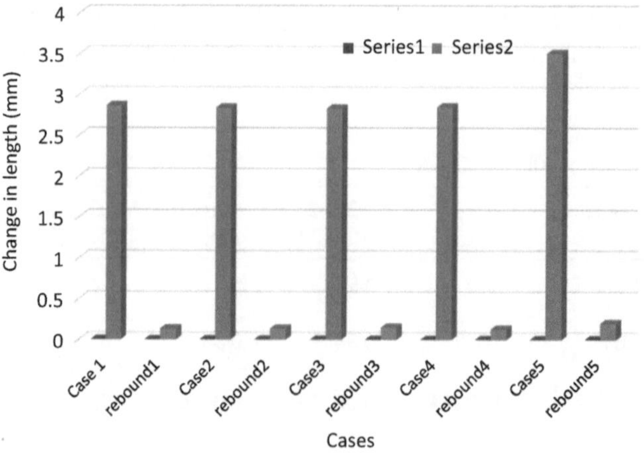

Figure 9.19 Deformation and rebound in tube diameter.

Table 9.6 Magnetic Field and Velocity of Tube in Different Cases

Parameters	*Case 1*	*Case 2*	*Case 3*	*Case 4*	*Case 5*
Maximum velocity (*m/s*).	519.41	554.67	529.07	515.6	962.64
Maximum magnetic field (*T*)	36.04	36.20	36.19	36.13	58.0

high value of the current plastic strain is very much higher than that of the other four cases. The highest value of plastic strain in the first case is 2.161, as shown in Table 9.5, while the remaining four cases where the peak current value is 704 kA maximum plastic strain is 0.865 of Case 2. The magnitude of the magnetic field and velocity are shown in Table 9.6.

9.4 CONCLUSIONS

This study aimed to model the deformation in the field shaper after constraining one end of the field shaper. From this study, it can be concluded that any material can be deformed with an increase in the current amplitude, such as in Case 1, although the material of the field shaper was elastic, it deformed plastically with high magnitude. When we decreased the amplitude of current up to 704 *kA*, no plastic deformation was observed in the remaining four cases except Case 3. A piecewise linear plasticity model obtained maximum velocity, and maximum magnetic field was obtained in Case 3. In Case 3, i.e., Mat018 power-law plasticity model, sufficient plastic deformation (1.416e-03) had been observed. The 77 steps in LS-DYNA with EM module have a time step 1.0e-007 and termination time 1.5e-05 seconds. Large deformation occurred in the field shaper after changing the material model.

REFERENCES

[1] R. M. Miranda, B. Tomas, T. G. Santos and N. Fernandes, "Magnetic Pulse Welding on the Cutting Edge of Industrial Applications," *Soldagem & Inspecao*, vol. 19, no. 01, pp. 69–81, 2014.
[2] S. A. A. A. Mousavi, "Numerical Studies of Explosive Welding of Three-Layer Cylinder Composites-part 2," *Mater Sci Forum*, Vols. 580-582, pp. 327–330, 2008.
[3] S. A. A. A. Mousavi, S. J. Burley and S. T. S. Al-Hassani, "Simulation of explosive welding using the williamsburg equation of state to model low detonation velocity explosives," *Int J Impact Eng*, vol. 31, pp. 719–734, 2005.
[4] S. A. A. A. Mousavi and P. F. Sartangi, "Experimental investigation of explosive welding of cp-titanium/AISI 304 stainless steel," *Materials & Design*, vol. 30, no. 3, pp. 459–468, 2009.

[5] X. Zhidan, Y. Haiping, L. Chunfeng and H. Yujie, "Interface Microstructure of Al-Fe Tubes Joint by Magnetic Pulse Welding," *Journal of Iron and Steel Research International*, vol. 19, pp. 442–445, 2012.
[6] P. Zhang, M. Kimchi, H. Shao, J. E. Gould and G. S. Daehn, "Analysis of the Electromagnetic Impulse Joining Process with a Field Concentrator," *In AIP Conference Proceedings*, vol. 712, no. 1, pp. 1253–1258, 2004.
[7] H. Zhang, M. Murata and H. Suzuki, "Effects of Various Working Conditions on Tube Bulging by Electromagnetic Forming," *Journal of Materials Processing Technology*, vol. 48, no. 1-4, pp. 113–121, 1995.
[8] R. N. Raoelison, D. Racine, Z. Zhang, N. Buiron, D. Marceau and M. Rachik, "Magnetic pulse welding: Interface of Al/Cu joint and investigation of intermetallic formation effect on the weld features," *Journal of Manufacturing Processes*, vol. 16, pp. 427–434, 2014.
[9] R. N. Raoelison, T. Sapanathan, N. Buiron and M. Rachik, "Magnetic pulse welding of Al/Al and Al/Cu metal pairs - Part 1: consequences of the dissimilar combination on the interface behaviour during the welding," *Journal of Manufacturing Processes*, vol. 20, pp. 112–127, 2015.
[10] R. Raoelison, M. Rachik, N. Buiron, D. Haye, M. Morel, B. D. Sanstos, D. Jouaffre and G. Frantz, "Assessment of Gap and Charging Voltage Influence on Mechanical Behaviour of Joints Obtained by Magnetic Pulse Welding," *5th International Conference on High Speed Forming*, pp. 207–216, 2012.
[11] A. Vivek, B. C. Liu and G. S. Daehn, "Collision Welding of Tungsten Alloy 17D and Copper Using Vaporizing Foil Actuator Welding," *6th International Conference on High Speed Forming*, pp. 181–188, 2014.
[12] A. Vivek, G. Taber, J. R. Johnson and G. S. Daehn, "Rapidly Vaporizing Conductors Used for Impulse Metalworking," *5th International Conference on High Speed Forming*, pp. 115–124, 2012.
[13] A. Nassiri and B. Kinsey, "Numerical studies on high-velocity impact welding: smoothed particle hydrodynamics (SPH) and arbitrary Lagrangian–Eulerian (ALE)," *Journal of Manufacturing Processes*, vol. 24, pp. 376–381, 2016.
[14] R. Kumar and S. D. Kore, "Effects of Surface Profiles on the Joint Formation during Magnetic Pulse Crimping in Tube-to-Rod Configuration," *International Journal of Precision Engineering and Manufacturing*, vol. 18, no. 8, pp. 1181–1188, 2017.
[15] R. Kumar and S. D. Kore, "Numerical Study on Pulsed Magnetic Cladding of Tubes," *International Journal of Materials and Product Technology*, 2018.
[16] H. Y. Lee, W. G. Kim and N. H. Kim, "Behavior of Grade 91 material specimens with and without defect at elevated temperature," *International Journal of Pressure Vessels and Piping*, vol. 125, pp. 3–12, 2015.
[17] S. H. Lee and D. N. Lee, "A Finite Element Analysis of Electromagnetic Forming for Tube Expansion," *Journal of Engineering Materials and Technology*, vol. 116, pp. 250–254, 1994.
[18] S. H. Lee and D. N. Lee, "Estimation of the magnetic pressure in tube expansion by electromagnetic forming," *Journal of Materials Processing Technology*, vol. 57, pp. 311–315, 1996.
[19] H. Yu, Z. Xu, Z. Fan, Z. Zhao and C. Li, "Mechanical property and microstructure of aluminum alloy-steel tubes joint by magnetic pulse welding," *Materials Science & Engineering A*, vol. 561, pp. 259–265, 2013.

Chapter 10

Friction Stir Welding and Its Advancement

Rituraj Bhattacharjee[1], *Susmita Datta*[2], *Tanmoy Medhi*[2], *and Pankaj Biswas*[2]

[1]Department of Mechanical Engineering, Indian Institute of Technology Guwahati, Assam, India

[2]Indian Institute of Technology Guwahati, Assam, India

CONTENTS

DOI: 10.1201/9781003436072-10

10.1 INTRODUCTION

10.1.1 Historical Background

The friction stir welding (FSW) process was invented in late 1991 at The Welding Institute (TWI). It is a relatively recent solid-state welding technology that many industry sectors (e.g., shipbuilding, defense, aerospace, etc.) have extensively endorsed. This technique is used for welding various metallic and non-metallic alloys that are strenuous to join, adopting traditional fusion welding procedures. Increasing demand for welding lightweight aircraft panels and vehicle body shells has come up with the help of this process in recent times. It eliminates defects during the fusion welding process, such as solidification cracking, porosity, crater crack, etc. FSW is remarkably a simple process that creates welds by combining mechanical deformation and friction heating. The highest temperature reached is over 0.8 over its melting point temperature. Because the base material is not heated, it is a solid-state welding procedure that produces a high-quality weld with minimal worker involvement. FSW was originally solely used to weld aluminum and its alloys, but it has since been utilized to weld magnesium, copper, titanium, nickel, steel, and stainless steel. The appropriateness of the material to be extruded determines the weldability or ease of welding of these materials. As a result, this technique can easily weld Al and Mg and their alloys compared to steel and stainless steel.

To generate heat at the interface, a non-consumable, specifically designed rotating tool plunges into the faying surfaces of the clamped workpieces in the first phase, plunging which is schematically depicted in Figure 10.1. Heat is generated by the tool's interaction with the workpiece, which softens it. Frictional heat is produced by the tool shoulder's contact with the surfaces of the workpiece material to be connected, which softens the materials. The tool pin surface also constantly stirs the localized material to be welded, albeit contributing to a lower percentage of heat dissipation. The tool rotates in almost the same position for a time during the dwelling phase to increase the temperature between the workpiece and the tool even more. The tool passes over the abutting edges to be joined in the third step, doing welding while constantly mixing the materials. The angular velocity of the tool (TRS), TS, tool dimension, and the axial downward load of the welding tool are all important aspects of in-process physics. The heat generated during welding is mainly due to rubbing of tool feces on the workpiece, i.e., due to friction between tool face and workpiece, and by visco-plastic dissipation of mechanical energy at high strain rates developed through interactions with the tool. As the tool moves forward, the weld is formed by material stirring. Due to its greater surface area, frictional heat generation occurs predominantly under the shoulder. The contact condition between a tool and the workpiece can be sliding and sticking.

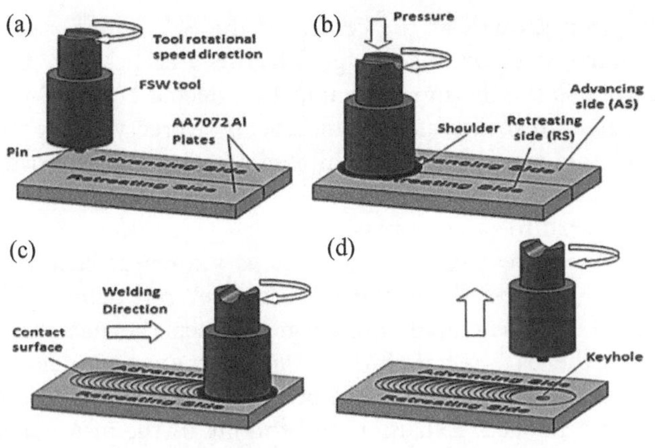

Figure 10.1 Schematic diagram of the FSW process [1].

As the tool passes along the longitudinal welding direction, severe plastic deformation and flow of this plasticized metal occur, leading to material transportation from the front edge of the tool to the rear edge, which is forged into an FS welded joint. The plate along which the direction of the rotating tool is in the same direction as that of the traverse is designated as the advancing side (AS). On the contrary, the plate alongside which the direction of the rotating tool is in the opposite sense to the travel direction, then it is designated as the retreating side (RS). Asymmetry in heat transfer, material flow, and the characteristics of the two surfaces of the weld can result from these discrepancies.

Four distinct regions in the FSW microstructure were identified [2], which are shown in Figure 10.2. The microstructural changes in various zones have a substantial impact on the material's post-weld properties: dynamically recrystallized region (DRZ) or weld nugget zone (WNZ) or stir zone (SZ) in which material recrystallization takes place from high-temperature deformation. It is evident from the microstructural studies that peak temperature during welding reached the austenite phase, allowing for

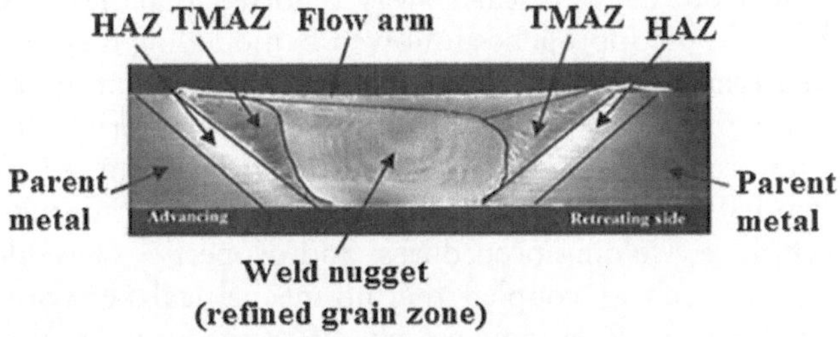

Figure 10.2 Various zones of the FSW process [2].

appreciable grain growth at the centre of the SZ, which resulted in the elongated mixed structure of polygon ferrite and pearlite. Outside SZ, the second zone is the thermo-mechanically affected zone (TMAZ), where the temperature and strain rates are insufficient to recrystallize the material and is severely deformed. TMAZ exhibited a grain-refined ferrite structure with minor amounts of pearlite in AS (DH36 steel), resulting in maximum hardness compared to other zones.

The third zone, just outside the TMAZ, is known as heat-affected zone (HAZ), in which the underlying material is not severely deformed. However, the high temperature generated in this region affects the microstructure. The fourth zone is the base metal (BM) or un-affected zone, where the effect of temperature and strain rates is not present, so the microstructure of this zone remains unchanged. Extrusion and forging of the metal at high strain rates are used in FSW to weld the pieces. A typical deformation strain rate of 10 s^{-1} was determined by evaluating particle size distribution and using a relationship between particle size distribution and the Zener–Holloman parameter characteristics, incorporating a temperature-adjusted strain rate [3]. Some of the researchers calculated effective strain rates in the 2–3 s^{-1} range [4]. Any theory for the process must include the plastic flow, the behavior of the metal at high strain rates, its dynamic recrystallization behavior, and the effects of heating and cooling.

The geometry of the tools and joints and other process parameters influence the mass and heat transfer behavior of the FSW joints. Due to some complex 3D FS welded geometries, it is sometimes difficult to predict its behavior even after employing modified governing equations. Furthermore, the non-linearity of the material due to its plasticity is challenging to integrate because the mathematical formulation gets incredibly complicated. Experiments might take a long time and be very expensive. Numerical analysis has been frequently used to solve these problems since the 1920s. The overall understanding of the FSW technique and the qualities of the weldment has dramatically improved. Numerical analysis was seen to find its widespread application in simulating the FSW process.

Not much comprehensive review of different numerical methodologies has been reported. A few research was reported that have extensively reviewed the different approaches followed in modelling thermal analysis, predicting material flow behavior, residual stress distribution, and defect detection during the FSW process of dissimilar materials [6]. Before this study, few researchers had extensively reviewed the techniques for modelling the FSW process. This study examines recent improvements in the FSW technique, welding procedures, and properties of welded joints. Important issues such as coupled thermo-mechanical behavior and the effect of varying process parameters on different output variables were also briefly discussed.

10.1.2 Advantages and Disadvantages of the FSW Process

FSW was invented in December 1991, and its industrial application started in mid-1992. It got the attention of researchers and industries because of its various advantages over the conventional arc welding process. The authors have described these advantages [2], which are:

- It can weld thick sections in a single pass.
- It has excellent mechanical properties.
- It provides low distortion, shrinkage, and residual stress.
- It doesn't require edge preparations, filler material, or shielding gas.
- Dissimilar alloys can be joined easily.
- It doesn't generate fumes and electromagnetic radiation.
- It is a solid-state process without any problem of solidification cracking.

Despite these advantages, there are a few disadvantages to this process which are as follows:

- Fillet welds and other complex geometrical shapes can't be welded.
- A hole is left at the end of this process.
- The workpiece must be rigidly clamped to react against the relatively high forces imposed by the process.
- Less flexible than manual and arc processes.

10.1.3 Scope and Applications

Although FSW is a relatively new technique, its benefits have led to it being adopted by numerous firms worldwide for various applications. FSW is currently used on many trains, vehicles, ships, and safety-critical equipment, such as fuel tanks on spacecraft, including the space shuttle. Welding grades of aluminum that cannot be welded with traditional fusion-gas or electric arc processes are no problem for this solid-state welding approach. In recent years, the method has also been utilized to weld magnesium and copper.

- *Railway Industry*

It has been established that commercial production of high-speed trains made from aluminum extrusions that FSW can assemble has begun. Applications include:

- High-speed trains
- Rolling stock of railways, trams, underground carriages
- Railway tankers and goods wagons

- ***Land Transportation***

The FSW technology is currently in use commercially and is also being evaluated for commercial use by various automobiles, businesses, and suppliers to this industry. Existing and potential applications include:

- Engines and chassis cradles
- Wheel rims
- Attachments to hydroformed tubes
- Space frames
- Truck bodies

- ***Aerospace Industries***

FSW is now used in aerospace to join prototype and production parts. Welding skin, spars, ribs, and stringers for military and civilian aircraft is possible with this approach. Compared to riveting and machining from solid, this approach has substantial advantages, such as lower production costs and weight reductions. Longitudinal butt welds in Al alloy fuel tanks have been FSWed and successfully utilized for space vehicles. The FSW process can, therefore, be considered for the following:

- Wings, fuselages, empennages
- Aviation fuel tanks
- External throw-away tanks of the military
- Some parts of the chassis of military and aircraft, hovercrafts

- ***Shipbuilding and Marine Construction***

The shipbuilding and marine industries are two of the first sectors that have adopted commercial applications.

The process is suitable for the following applications:

- Panels for decks, sides, floors
- Hulls and superstructure
- Helicopters landing platform
- Marine and transport structures
- Mast and booms, e.g., for sailing boats
- Refrigeration plant

Figure 10.3 depicts the usability of DH36 steel welded with mild steel and dissimilar aluminum-steel in various industries, leading from shipbuilding to aerospace, due to its lightweight, durability, and hardness.

Figure 10.3 Images of shipbuilding parts welded by FSW [5].

10.2 FSW PROCESS

Changing the welding process parameters can usually vary the FSW desired process output. As a result, welding process variables should always be selected to achieve the best possible microstructure. Furthermore, a thorough understanding of welding is essential for precise welded sample size determination, crystallographic morphology during the microstructural analysis, and mechanical testing of the FSWed joints. Because of the intricacy of the FSW process, knowing about welded joints during the forming process can be difficult. Numerical simulation aids in solving this problem by allowing for a thorough examination of the creation of FSW joints.

10.2.1 Thermo-Mechanical Analysis

To comprehend the FSW process's operating principle, it is vital to explore the mechanics associated with the heat produced during the in-process. The mechanical friction between the revolving tool and the materials to be welded produces heat in FSW. The revolving tool strains the underlying material, causing permanent plastic deformation, producing heat, and joining the two surfaces in contact. Heat is generated sporadically or in phases throughout the welding process. WELDSIM is a finite element analysis program designed specifically for a three-dimensional nonlinear thermal and mechanical simulation system during the FSW of 304 L stainless steel [7]. The TRS was changed from 300 to 500 rpm during the experimental examination. An inverse analysis method was provided to conduct the heat transfer analysis based on the expected simulation analysis based on the experimental transient temperature measurement records at multiple locations on the welded plate during the FSW of 304 L stainless steel. The influence of the creation of residual stresses on the weldment was analyzed using a three-dimensional thermal-mechanical analysis considering the material's elastic-plastic behavior. The residual stress data produced using a non-destructive approach (NDT), such as the neutron diffraction technique, was well matched with the mathematical modelling results. For tube-sheet seal connections, FSW was applied and numerically simulated with the help of a developed 3D thermo-mechanical FE model [8]. The tool-workpiece interaction's heat effect and axial load were included in the model, but the material flow around the tool was not taken. The tool-workpiece interaction was taken into consideration in the model.

The model aimed to assess the workpiece temperature profile and residual stress distribution caused primarily by the heating cycle and axial load throughout welding. It was done for various process parameters and to investigate how residual stresses in adjoining enlarged roller pipes were affected during the FSW process. An experimental setup was developed and fabricated to demonstrate the method's applicability in restricted-size connections and to confirm the theoretical conclusions. Thermo-mechanical behavior and microstructural development of different AA6061-T6 and AA5086-O materials during FSW were successfully reported [9]. Abaqus 3D FE software was used to estimate the thermal-mechanical response of materials under comparable and differing FSW operations. The mechanical characteristics and microstructural morphology developed on the welded specimens were also studied using experimental data and model predictions. Different strengthening processes in AA5086 and AA6061, as per the research, results in intricate behavior in hardness value when the cross-section is taken along with the welded specimen. Recrystallization and the production of fine grains in weld nuggets are the major sources of hardness variation in equivalent AA5086-O joints.

Later ageing events influence the hardness variations inside the welded joint of pairs such as AA6061 and AA6061, AA6061, and AA5086 seam welds. During the FSW process, a significant quantity of residual stresses is typically created, resulting in critical component structural integrity and performance degradation. The FSW joints are subjected to both tensile and compressive residual stresses. The HAZ developed the maximum amount of residual stresses. The AS beyond the weldment just next to the BM, on the other hand, has the lowest compressive residual stresses [10]. The ability of various established numerical simulations to properly forecast residual stress distributions, strains developed, and deformities in the workpiece is a crucial outcome. A single-block technique was employed to model the friction stir welding process with different butt joint configurations using a 3D FE model with general validity [11]. Thanks to a new time-saving strategy, the model can predict residual stresses only based on thermal processes. The estimated and experimentally measured values were found to be in good agreement. The efficiency of the proposed computational approach was evaluated by comparing the calculation times of the two numerical procedures. Figures (10.4 & 10.5) show the welded joint's design and residual stress distribution modelling in detail.

The plates are heated, plastically deformed, and welded locally during the FSW process by moving the probe and shoulder down the weld line. One of

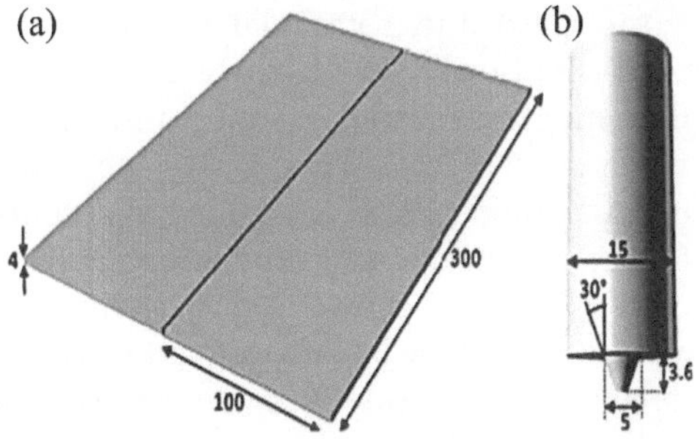

Figure 10.4 Schematic diagram of the **(a)** welded and **(b)** tool geometry [11].

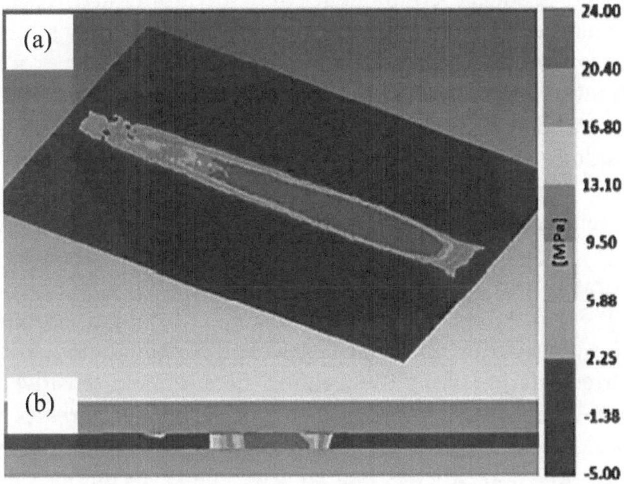

Figure 10.5 Residual stress distribution along with the longitudinal direction **(a)** considering the upper surface of the plate and **(b)** along the transverse section of the welded joint — taking 500 rpm and 225 mm/min as for the case study [11].

the significant downsides of this approach is residual stress, which is induced by the metal's thermal-mechanical properties and the fixture's limits on the weldment. Estimating the clamp pressure acting on the plates throughout FSW should facilitate easier maintaining stress concentration and joint performance. Furthermore, projecting force history in the FSW process will assist in understanding the physics of the procedure and giving appropriate modelling methods for controlling the operation, particularly during robotic FSW.

A 3D FE model may be used to measure the temperature and stress evolved during the FSW process, which can then be utilized to calculate mechanical stresses in the longitudinal and transverse directions [12–14]. The connection between the estimated residual stress of the weld and process parameters, such as tool traversal speed and rotational speeds, and fixture removal has been investigated intensively. In all three dimensions, FSW thermo-mechanical modelling may predict the overall transient temperature distribution, residual stresses developed, and forces, and it can also estimate active stresses generated in the model. For examining the contact conductance value, a uniform FE thermo-mechanical simulation, including machine tool load, was developed [15]. The model was used to predict the amount of stress developed on the workpiece and the backplate-workpiece interface. The non-uniform adaptive pressure distribution was defined using pressure distribution contours. The thermal model uses contact conductance to forecast the workpiece's thermal history. The thermo-mechanical model was then used to forecast how stresses in FSW would evolve. As a result of the FSW processes, the final produced components may have undesired residual stress and distortions and localized mechanical property loss at weld joints. Various studies were performed to identify and describe the primary process effects of FSW assembly methods on stiffened panel static strength [16]. Using experimentally validated FE modelling approaches, it was established that induced residual stresses during welding significantly impacted failure behavior. The panel's preliminary buckling behavior is less vulnerable to effective system influences and magnitudes. According to the simulation, authors [17] constructed a three-dimensional thermo-mechanical model that incorporated heat transmission and elastoplasticity evaluations to trace residual stress concentrations during the FSW of copper containers. The residual stress distribution inside a thick-walled copper container is found to be asymmetric in nature along the welding direction and sensitive along the circumferential direction. Compressive and tensile residual stresses emerge all along the welded joint as well as its vicinity. The maximum tensile stress cannot exceed 50 MPa, either estimated using the FE technique or evaluated using the hole-drilling methodology, or in particular using the X-ray diffraction (XRD) procedure. Overall stresses generated all throughout thermal cycles during the FSW of metals were computed using a coupled FE model [18]. This simulation is done in two phases. The thermal history is calculated from a thermal model based on input torque in the first phase. The temperature distribution obtained by the thermal model is fed as an input to the mechanical FE model that successively predicts residual thermal stresses in a sequential manner. Plastically deformed weld nugget zone (WNZ) and TMAZ are not considered by the developed simulated model to simplify it. This could result in a difference between the residual stress predicted by the model and the residual stress measured experimentally. The model incorporates temperature-dependent material thermo-mechanical properties. Three different materials were studied, such as AA2024, AA6061, and SS304. The FSW process was studied using a steady-state computational model developed by [19]. In the simulation, there were two main steps. The first uses an Eulerian representation of the thermo-mechanical issue: In the first stage, a three-dimensional FE Model is developed based on the

Computational Fluid Dynamics (CFD) approach, which was used to determine the necessary flow of material and temperature distribution during the in-process. In the second stage, the residual stress induced by the process was calculated using a steady-state method based on an elastic viscoplastic constitutive law. The longitudinal residual stress field was discovered to have two peaks, with the two peaks located in the zone with the highest soluble precipitating gradient. As per the parametric analysis, the welding speed and lower rotating speed will result in substantially lower temperature and residual distortions. The influence of alloy composition and welding processing parameters on the magnitude and distribution of residual stress concentration during the FSW process was investigated by [20]. The magnitude of the generated strains is much smaller than the elastic modulus of the underlying material at ambient temperature. An "M-shape" distribution was always discovered (albeit more noticeable in AA6082-T6 than in AA2024-T3) on age-hardenable alloys. But a "plateau" shape was seen developing in the strain hardenable AA5754 H111. Microstructural variations in the weld centre are primarily responsible for the low magnitude and longitudinal residual stress distribution variations. A three-dimensional explicit computational domain of the FSW process was modeled to investigate the effect of welding speed on the distribution of heat and developed residual stress [21]. The suggested model incorporated material properties as temperature-dependent parameters. The heat transfer rate varied and was asymmetrical throughout the thickness, as evidenced by the generated residual stresses. Only the heat effect was considered when estimating the resultant residual stress. This was identified as the primary factor driving modest discrepancies in the simulated results compared to experimental data.

10.2.1.1 Frictional Heat Generation

Mechanical friction work is initiated whenever rotational tool edges make contact with the workpiece's adjacent static surface under normal load conditions. The rotating tool creates a velocity differential between the dynamically revolving tool and the stationary workpiece surfaces while sliding, culminating in frictional work and heat produced. This mechanical friction work is described by Amonton's laws, which first demonstrate how friction among two individual materials is directly related to the normal applied load to the materials. In this law, the static friction coefficient is a constant variable that depends upon temperature; nevertheless, kinetic friction would only be examined whenever the interaction conditions are non-sticking or sliding. Second, the apparent contact area has no bearing on the friction force (Popov 2010). Generally, frictional heat generation is governed by both Columb's and Amonton's laws. The release of high local heat energy generated due to the agitation, deformation, and breaking down of softer material results in a rise in temperature [22]. The heat is transmitted and retained between the revolving tool and the workpiece chunk material. This thermal heat transfer to the workpiece material leads to softening, reducing the strength of the workpiece, breaking and deforming into a softer workpiece forged between the tool and the workpiece.

10.2.1.2 Plastic Heat Generation

This phenomenon occurs at elevated temperatures as a result of frictional heat dissipation, although it is only noticeable during the dwelling and actual welding phases. The frictional heat mechanism elevated the temperature of the workpiece material under the revolving tool to the point where the material layer just at the moving tool interface started to lose rigidity, yield, adhere, and combine with the tool. This phenomenon creates shearing between the workpiece layer interfaces, increasing the thermal softening impact from friction heat. It decreases frictional heat but simultaneously introduces a high strain rate of plastic deformation. The highly concentrated heat is produced inwardly inside the workpiece material, away from the revolving tool, due to the dynamic velocities differential and boundary sliding condition [23–25]. During the tool traverse stage, the temperature rises due to plastic dissipation, and the shearing of the workpiece at the trailing edge is also more visible. As a result, the operating temperature increases even more since the tool is exposed to novel contact pressure somewhere at the leading edge for a while. Therefore, the mechanical frictional heat generation process is reactivated [15].

10.2.1.3 Heat Transfer

During the FSW process, the total heat generated caused by friction and plastic thermal dissipation is split between the workpiece material, the revolving tool, the clamp attachment, and the back plate, with the remainder going to the environment. Heat generation is susceptible to 3D heat movement away from the heat source under boundary conditions. The heat generated due to the constant rubbing of the tool with the workpiece material leads to its dissipation between the tool and the workpiece surface utilizing conduction and convection modes of heat transfer. For heat exchange at the top surface of the workpiece, past the shoulder, peripheral, convective, and radiative heat transfers are evaluated, while only convective heat transfer conditions are studied for the bounding surface of the workpieces [26,27]. In the case of the base plate, the relevant variable gap conductance is taken into account depending on the specific heat transfer resistance condition, contact pressure, or temperature dependence between the workpiece and the backing plate [15,18,26]. In the end, all of the energy created by mechanical work is transformed into heat and physical deformation, with roughly 88% of the heat being conducted and spread globally between the workpiece chunk, backing plate, and rotating tool [28]. The heat generated is vital not only for the joint's physical performance and the process's temperature profile. It is also significant for heat transfer, internal strain, and residual stress distribution in the workpiece material. It significantly also affects the final weldment's microstructural morphology and residual stress characteristics, which are closely connected to welding process factors and temperature-dependent material attributes [29].

10.2.2 Material Flow Analysis

Material flow is critical in the FSW process since it dictates how effective the joints are. Most of the material movement happens on the trailing edge, formed by the plasticized material transported to the tool's backside. Depending on the tool design, processing parameters, and forged material, the flow of materials around the tool is exceedingly intricate. As a result, to achieve the best design approach and process variables combinations, it's critical to understand the material flow characteristics. Previous literature has suggested several models to simulate the effect of heat transfer and material flow during the FSW process. The purpose of the study was only to create a three-dimensional Lagrange iterative FE approach for simulating the friction stir processing (FSP) process using Deform-3D software [30]. This particular simulation methodology can also be used to evaluate FSW material flow. Point tracking was used to acquire 3D findings for the flow of material along with the domain of AS, RS, and centre of the welded joint. The formation of a hollow tunnel and a groove underneath the pin is depicted in Figure 10.6.

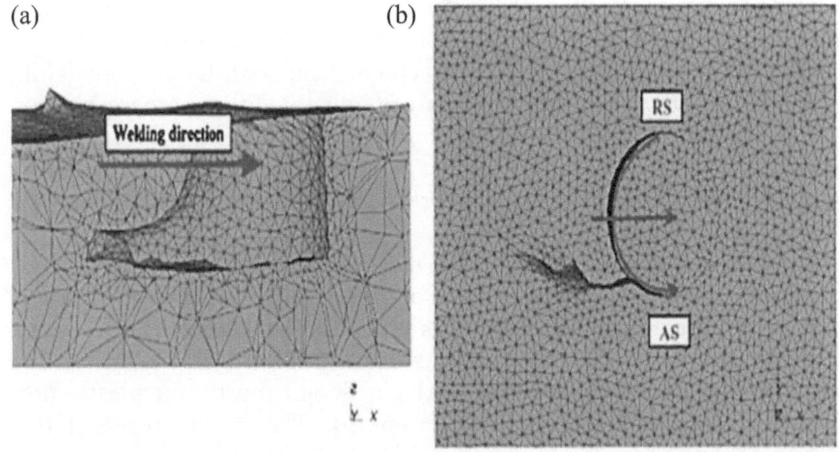

Figure 10.6 (a) Formation of tunnel cavity behind the region of the pin and (b) formation of slot around AS behind the pin [30].

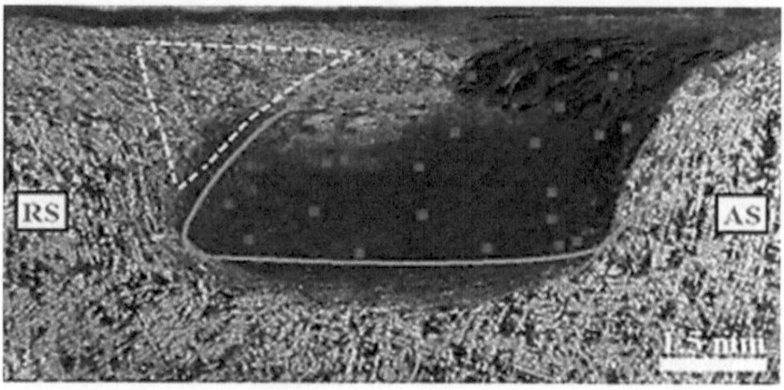

Figure 10.7 Comparison of simulated and experimental SZ shape [30].

According to the data, most material movement occurs on the stirring tool's upper surface and the AS. The top layer material was strained to the AS, resulting in the non-symmetrical shape of the SZ. Figure 10.7 compares the formation of simulated and obtained experimental stir zones. The material flow behavior investigation provides valuable information about the mixed material and improves optimum tool design and process parameter choices.

Some publications described the use of the Fluent CFD code to predict 2D or 3D material flow in FSW operation [31–33]. The study used a standard threaded tool profile, including tool rake angle, temperature rise, and heat flow. His research's primary goal has been to understand better materials mixing around a complex FSW tool and explain how tool rake angle, welding, and rotational speed affect the material flow behavior. A few researchers used computational methods to study material flow due to continuous stirring and mixing during the FSW process [34]. The workpiece material was considered an Eulerian component, whereas the FSW tool was considered a Lagrangian component. The FSW process was studied using a two-way thermo-mechanical coupled Eulerian-Lagrangian (CEL) computational analysis. The temperature influences the mechanical components of the model via material properties, which vary depending on temperature change. The workpiece was modeled using the modified Johnson-Cook plasticity model, whereas the tool was taken as isotropic linear-elastic material.

The data collected from material flow during FSW were matched with experimental results. A finite volume model (FVM) of FSW was built and used to investigate the rotating tool's practical shape and the connection between material parameters and temperature using Ansys fluent software [35,36]. The impact of shoulder and pin design mostly on plastic flow plasticity characteristics was also examined. The results revealed that decreasing the cone angle of the pin or decreasing the breadth of the screw groove enhanced the flow velocity of material inside the weld. When a rotational tool with a left screw pin rotates clockwise, material near the pin flows downward, while material near the TMAZ affected zone flows upward, which is the opposite of what happens with a right screw pin. The flow patterns and residual stresses developed in the FSW process have been studied using FE models based on computational solid mechanics (CSM)–based approaches [37]. The flow of material during FSW was examined using tracer particle technology. The material flow on the AS and RS was asymmetrical. The material rotating around the tool pin sloughs off in the wake of the pin after several rotations, primarily on the AS. A three-dimensional FE simulation was published to investigate material flow during the FSW process [38]. According to earlier results, the material in front of the pin moves upwards while rotating due to its extrusion action. The material moves behind the rotating tool and settles in the wake. The majority of the flow of material in the FSW process is tangential. On the AS, there is a swirl, and the reverse flow of material accelerates as the welding speed increases. The shoulder increased material velocity distribution close to the top surface for both radial and tangent directions. The FSW process was modeled using a general-purpose FE-based program to create a material thermal mapping and corresponding mass flow [39]. Computational temperature results were compared to experimental thermal imaging maps with close agreement. And computational material flow data were consistent with material flow optimization techniques with reasonable understanding. A coupled nonlinear thermal-mechanical simulation of the FSW process was developed [40,41]. According to the author's findings, increasing the tool shoulder's rotation improves the corresponding flow of material at the welded plate's upper surface. The evolution of the welded specimens at the microstructural level is caused by material deformation and temperature distribution history. The texture and appearance of the FSWed welded joints can be closely related to the equivalent plastic strain (PEEQ) distributions on the top surface of the weld profile joint. When the same material was used, the temperature field was found to be fairly symmetric all around the weld joint. Asymmetrical material flow is more visible with varying thickness and physical aspects of the workpiece. Author [42] developed an FSW method with various material thicknesses while investigating FSW AA2024-T3. According to the computed results, the movement of materials on the trailing edge of the workpiece material and the front portion of the welding tool is higher. As a result, slipping rates on the RS and the front sides are substantially lower than that on the trailing and AS. As a result, heat fluxes are also higher, leading to increased temperatures for thick and thin sheets in this region. During FSW of AA2024-T3, the welding plate's energy accounts for more than half of the total energy, with frictional heat accounting for nearly 85% of the total energy. Mechanical effects are responsible for maintaining the thermal balance. The FE method was used to predict Al 6061-T6 alloy material flow in the FSW process and develop a relationship between material movement and tool pin angular velocity [37]. As the angular velocity rises, the material flow improves. The material at the front of the pin flows around the rotational direction and upward because of the extrusion effect of the pin. The material behind the pin is dragged downwards during the continued FSW process.

10.3 EFFECT OF INPUT PROCESS PARAMETER VARIABLES

The key independent factors employed to control the FSW process are welding speed, tool rotating speed, vertical pressure on the tool, tool tilt angle, and tool design. These variables influence the heat generation rate, temperature field, cooling rate, x-direction force, torque, and power.

10.3.1 Tool Design

The effect of tool design plays a vital role in optimizing the weld quality and predicting the material flow plastic behavior through numerical simulation around the rotating tool. Tool design influences heat generation, plastic flow, the amount of power required, and the welded joint's regularity. The shoulder generates a significant portion of the heat, preventing the plasticized material from escaping from the workpiece. At the same time, the material flow is influenced by both the shoulder and the tool pin. In recent years, some significant innovations have been introduced into tool design. Figure 10.8 lists several TWI-developed tools. It was found that the pin volumes of the Whorl and MX-Triflute-shaped tool are smaller than those of cylindrical pin one [43].

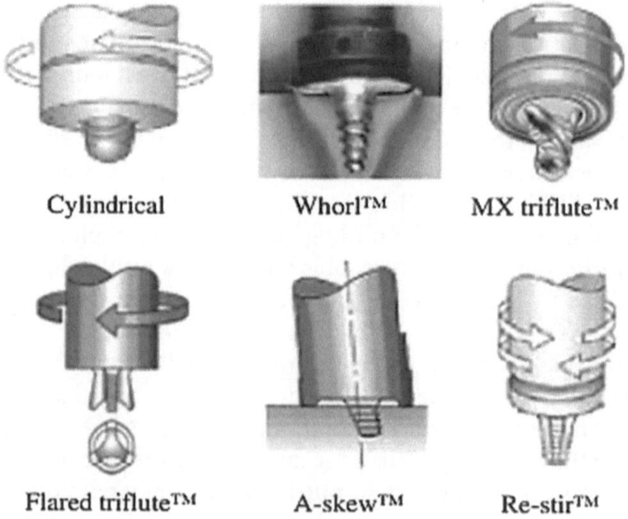

Figure 10.8 Various tool designs at TWI [44].

The whorl style of tool geometry utilizes tapered threads to generate a vertical gradient of velocity that enhances plastic flow. The flutes in the MX-triflute increase the tool-to-workpiece interface area, resulting in rapid thermal radiation, softening, and material flow. As a result, higher stirring reduces the forward tool motion traverse force and the welding torque [43]. During lap welding, due to the excessive thinning of the upper base plate and oxides being adhered between the overlapping surfaces, cylindrical, whorl, and triflute tool designs can't produce good welded joints. So, to overcome this problem, TWI came up with some newer tool designs, such as flared-triflute and A-Skew tools, to achieve interfacial oxide layer fragmentation and a larger weld formation compared to conventional weld nugget size. Tool motion induced by its rotation and translation causes an imbalance in the flow of material and temperature control throughout the tool pin. Material flow on the RS of the welded joint is more evident during the FSW process [45–48]. TWI designed a new tool, Re-stir, to solve this challenge. In this configuration, the tool rotates in the other direction regularly. This rotational reversal eliminates most issues with the conventional FSW's inherent asymmetry.

10.3.2 Tool Material

Compared to the workpiece, the tool material for FSW must be rigid and robust. In addition, the tool's mechanical qualities, such as strength and thermal properties, should be on the higher end. These are the essential and notable factors for efficient and uniform heat dissipation [49]. Lightweight materials, including aluminum and magnesium alloys, could integrate with readily accessible commercial tool materials. Examples include high-speed steel (HSS), low-carbon steel, and H13 tool steel. These are also useful materials to weld in various process input parameter conditions [50]. A tool with both longevity and superior weld qualities is in great demand for structural applications, particularly in the military, aviation, petroleum and gas, and heavy industrial sectors. More research and development are needed for tool materials for high melting point metals like steel and titanium alloys. For welding, high-strength materials, including steel, copper, and titanium, and specialized tool materials, such as tungsten carbide and polycrystalline cubic boron nitride (pCBN) are commonly utilized. The tool must be cost-effective, long-lasting, and mechanically robust for commercial FSW operations [51]. Weld quality and cost, as well as material choice and tool performance monitoring, are all significant challenges.

10.3.3 Tool Rotational Speed and Welding Speed

TRS and WS are two essential input parameters that govern the welded joints' productivity and quality. The rotating tool pushes off the material, transforming it into a viscous substance to form a joint that determines the joint effectiveness and efficiency. Because they are interconnected, establishing a relationship between them is more complicated. With a higher rotating speed and a reduced welding speed, more frictional heat is produced between the workpiece and tool, leading the specimen to soften more rapidly.

On the other hand, traverse speed increases the peak temperature by increasing contact duration at the interface. An experimental investigation was carried out on AA6060-T6 aluminum sheets with a thickness of 2 mm. The FSSW approach has been applied to pairs of the overlapping sheet by altering the rotating tool speed while keeping the other process parameters constant, such as rate of axial loading, plunge depth, and dwelling time duration. Tensile testing and metallurgical investigations were also carried out preliminary to get an in-hand idea of the weld joint quality. The commercial FE code Deform 3D was used to build and implement the FSSW method. The model parameters were chosen based on the results of the experiments. A 3D transient thermal model was constructed and experimentally validated using the Ansys FE code to quantify the thermal history. Nine experiments were conducted using a full factorial design to systematically explore the influence of input parameters such as tool rotational and traverse speed [44].

The impact of changing input parameters on the joints' thermal history and mechanical properties was investigated using variance analysis (ANOVA). The rotational tool speed more substantially influenced the temperature under the tool than the traverse speed. According to the results, travel speed impacted the mechanical characteristics of the weldment significantly. Using a three-dimensional numerical model of the FSW process, the influence of tool travel speed upon thermal transfer and residual stresses was explored [21]. Temperature-dependent material properties parameters in the suggested model. The obtained stress concentration revealed that the heat transfer rate fluctuates and is discontinuous across the plate thickness. Only the heat effect was considered when evaluating the induced residual stresses. This was identified as the primary factor driving modest discrepancies in simulation findings compared to experimental data. The relative motion between both the tool as well as the workpiece is strongly influenced by the rotational speed. As a result, the welding speed does not affect the rate of heat formation. High traverse speeds have the effect of reducing heat input and lowering the in-process temperature. Since material flow becomes much more challenging at relatively low temperatures, increased welding speed increases the torque. The power required can be calculated using the torque on the tool as $p = \omega M$, where M is the total torque on the tool. In short, peak temperature rises as rotational speed rises and falls slightly as welding speed increases. With an increase in axial pressure, the peak temperature increases as well. Figure 10.9 demonstrates a significant increase in peak temperature as the rotating speed increases.

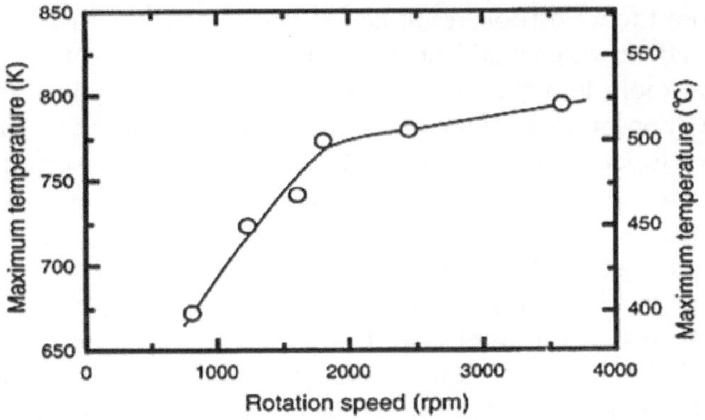

Figure 10.9 Relation among rotational speed and peak temperature obtained during FSW of AA6063 [52].

10.3.4 Tool Tilt Angle and Plunge Depth

Researchers [53] investigated the effect of process variables such as plunging depth feed to the tool, TRS, and WS on the process variables such as normal forces, etc., by developing some empirical models maintaining a dynamic relationship. Establishing relatively constant correlations between the process parameters and the process variables was used to examine the relative influence of each processing parameter upon every process variable. The results showed that a nonlinear power relationship could describe the steady-state relationship between the process parameters and the variables. In contrast, low-order linear equations may well describe dynamic responses. The torque generated by FSW is affected by factors, including the applied axial load, tool design, tilt angle, local shear stress at the tool-material interface, friction coefficient, and the amount of slip between the tool and the material. For specific welding scenarios, recorded torque readings provide insight into the mean flow stress around the tool and the amount of slip between the tool and workpiece when all other parameters are held constant. The plunging depth of the stirring tool is also crucial in obtaining a strong welded joint. This is done to provide a proper forging load and surface contact area, which aids in even better heat generation. Disproportionate plunge depth can result in more flash generation and lower weld quality due to certain flaws [54]. Another essential component affecting the forging process during FSW, which reduces flash production, and enhances the weld's mechanical qualities, is the tool tilt angle [55]. The plasticizing material can flow more freely around the tool when the angle is not zero. Enhanced downward force and frictional heat arose from increasing the tilt angle during welding, leading to increased plastic deformation and softening of the material [56]. Weld quality is governed by adjusting the optimum tool tilt angle [57]. A higher tilt angle may destroy the pin from the weld root, resulting in erroneous or fractured welds. The material is efficiently carried from the front side to the rear of the pin due to the optimal tool tilt angle, resulting in a smooth welded joint. As a result, finding an appropriate tool inclination angle and a proper combined effect of all other input parameters like TRS and WS are necessary. Increasing the tool inclination angle in the workpiece increases stress concentration more toward the tool's front edge. As material flow along the back AS is recorded, raising the tilt angle induces a significant rise in temperature and a decrease in von Mises stress value. Also, it increases the amount of material stirred from behind the tool edge toward the AS [58]. During FSW lap joints of AA5083 and SS400, the voids created all along the weldment rise with an increase in tool pin profile diameter and tool inclination angle [59]. Pin profile diameter (> 5 mm) generates much heat to produce intermetallic compounds (IMC's) such as $FeAl_3$ and Fe_2Al_5. FeAl, on the other hand, is produced at lower temperatures. Aluminum containing $FeAl_3$ and Fe_2Al_5 are harder and more brittle than FeAl, with low shear joint properties. Increased tool tilt angle (> 1°) leads to an increase in heat generation, the formation of precipitates rich in aluminum compounds, and a reduction in the strength of the welded joint [59]. The influence of tool tilt angle on defect generation was investigated by [60]. They experimented with tilt angles ranging from 0 to 2°. A faulty weld resulted from a 0° tilt angle. A 1° tilt angle caused a tunnelling defect along the welding direction. The inclination angle of 2° ultimately results in a smooth, defect-free weld due to the enhanced axial force value. Weld effectiveness between 0° and 1° were 35% and 50%, correspondingly, due to defects. A 90% % weld efficacy was recorded for a tilt angle of 2°. The influence of tool inclination angle during the friction stir welding of copper with aluminum alloy was investigated by [61]. At a constant rotational speed of 1,300 rpm and a 40 mm/min welding speed, tilt angles ranged from 0° to 4° with a 1° interval; 0° and 1° tilt angles resulted in a poor weld, whereas 2°, 3°, and 4° tilt angles resulted in a defect-free weld. The greatest weld strength at a 4° tilt angle was 117 MPa. For a 4° tilt angle, the highest micro-hardness value recorded was about 186 HV, while the minimum macro-hardness was about 58 HV.

10.4 EFFECT ON OUTPUT FIELDS OF FSW PROCESS

10.4.1 Temperature Distribution

Friction between the tool and the workpiece generates heat in the FSW process. Due to the strong plastically deformed material caused due to the tool's translational and rotational effect, temperature measurements within the stirring zone are highly challenging. Due to the complexity, the numerical analysis provides information on the temperature distribution during the FSW process. Thermo-mechanical simulation of the welding process can forecast the thermal history varying with time, peak temperatures obtained, active stress forces, and residual stress can all be calculated [62–65]. Ulysse computed the whole FSW process using the 3D viscoplastic FE modelling technique [66].

Parameterized experiments were conducted to establish the effect of varying tool rotational and translational speeds on surface temperature. The results obtained with the help of the simulated model are in turn compared with the existing experimental data. It was found that the model can aid in developing welding tools that can produce effective thermal differential. Figure 10.10 shows the FE mesh and temperature distribution. The temperature distribution during the FS welded aluminum plate was determined using transient thermal FE studies [67]. A moving heat source with a heat distribution computing the heat created by frictional heat generation due to rubbing between the tool shoulder and the workpiece material was utilized to mimic the heat transfer study. The commercial Ansys and HyperXtrude software were used for the 3D modelling. APDL

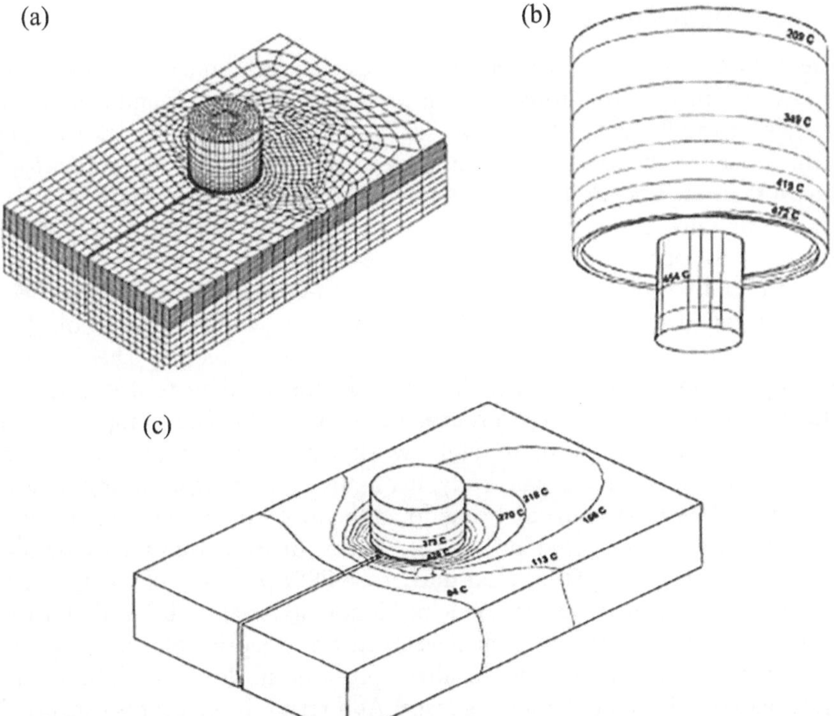

Figure 10.10 (a) Finite element modelling of the FSW process, (b) predicted isotherms (in °C) on a magnified view of the FSW tool, (c) temperature isotherms (in °C) on the tool and workpiece surfaces [66].

(ANSYS parametric design language) code was developed to model the moving heat source and as well as to define the boundary conditions. For the FSW of steels, a 3D Lagrangian implicit, coupled, and resilient visco-plastic continuum-based FE model was proposed [68]. The model can predict temperature, strain, strain rate distribution, and thermo-mechanical stresses on the welding tool at various significant processing parameters. The computational results of AA6061-T6 revealed an area with less plastically strained material found at the underside and along the back of the welding tool. Due to the material flow pathlines, these regions are produced [40]. The plastically strained top surface of the workpieces resembles the onion textures seen in appearances on the weld seam. Researchers [69,70] used FE analysis to conduct numerical examinations of the temperature distribution in the FSW plates. Friction between the workpiece and the stationary element is included in their model, resulting from the heat dissipation caused by plastically deformed material. The model helped determine important welding properties such as the HAZ's properties and susceptibility to different welding circumstances.

10.4.2 Axial Force

The authors [71] used an adaptive re-meshing technique to create a FE model to depict the tool forces during welding. It has been found from their study that the forces in the tool attain peak values during the start of the welding stage. The axial forces acting perpendicularly on the welding tool have a more significant impact than all the 3D forces on the tool. This axial force acting just above the weld seam joint should be minimum so that the tool does not penetrate the workpieces. Results depict the fact that with an increase in TWS and a decrease in the tool's angular velocity. Researchers [72] developed a finite element model incorporating the Johnson-Cook (JC) plastic material constitutive law to visualize the effect of shear stress developed and vertical axial load acting on the welding tool during the plunging stage of the FSW process of AISI 1045. The computational domain incorporates a deformable workpiece, donor material, and a welding tool considered a rigid body. It was also evident from the numerical study that donor material can bring down the tool wear to a certain extent. This idea was implemented in the experimental investigation for validation purposes. To reduce the tool wear during the FSW process's plunging stage, the authors [73] proposed introducing donor material to bring down the tool wear. The model has been developed on the principle of using the softer material as a donor and localized pre-heating the area of the workpiece material during the plunging stage. As a result, more heat is generated along the softer 'donor' material. The heat is then transferred by conduction to the harder material in contact. This research includes multiple computational models consisting of the donor material. The basic idea is to use several donor materials and plain carbon steel as the workpiece in the experimental investigation. A substantial decrease in the contact pressure and axial force was seen when the 'donor' soft material was used in the plunging area. This decrease in the output parameter responses contributes to lessening the tool wear during the welding process.

10.4.3 Defects

Due to improper selection of process parameter variables, different welding defects are formed. The inappropriate design of the tool used, its dimension, rotational tool speed (TRS), welding speed (WS), plunging depth, and tilt angle contribute to developing any defect in the weldment. Minimal temperature rise and less welded material softening sometimes lead to defect formation. High temperatures are undesirable during FSW because they alter the welded specimen's microstructural and mechanical properties. Frictional heat loss on the moving tool's leading and trailing sides results in inefficient material movement during the welding process [58,74].

Researchers [75] analyzed the literature on FSW modelling to better understand the link between process variables by simulating different parameter conditions. Apart from material flow, the distributions of a thermal gradient, residual stress variations, strain, and strain rates developed were examined throughout by taking various sections across the weldment section to evaluate the efficiency of the process and weld quality. Researchers in this discipline rely on the distribution of processing parameters to estimate the likelihood of welding failures and pinpoint problematic locations. Modelling the FSW process revealed that in conditions such as maximum strain rate and elevated temperatures, a lack of comprehensive material constitutive characteristics and other structural and thermal parameters appears to be a significant bottleneck. To simulate the deformation and thermal history of the material in the FSW process, a thermo-mechanical model was constructed [41]. The particles on the top surface of the material accumulate around the wake's edge on the RS. However, the material doesn't enter the trail. According to the numerical results, this leads to the production of weld flash during the FSW process. Both were raising the rotational speed of the tool and decreasing the welding speed of the welding tool enhanced the stirring effect of the pin, which improved the weld quality as well. To eliminate probable welding imperfections like voids, the traverse speed must be increased in combination with the rotational speed. Increasing the TRS and WS of the tool simultaneously increases the residual stress being developed on the plates. Square and tapered tool pin configurations were used to develop the FSBW of 6063-T4 aluminum alloy [76,77]. Tunnel imperfections were visible during the tensile testing of transverse weld specimens welded by tapered pin profiles at higher welding speeds. Mechanical instability emerged from these flaws far before macroscopic necking due to abrupt reductions in load-displacement curves. Some significant defects arising from improper selection of process variables are tunnel defects, microcracks, void, root flaws, kissing bonds, ribbon flashes, surface defects, and oxide entrapment.

10.5 MECHANICAL PROPERTY

10.5.1 Tensile Property and Hardness Variation

When AA2219 and AA6061 aluminum alloys were friction stir processed independently, it was found that the square pin designed tool yielded defect-free welded specimen. This tool design exhibited maximum tensile strength compared to other tool designs [78]. A cylindrical pin, threaded cylindrical pin, and tapered pin were used to perform FSW on AZ31B magnesium alloy. A negative defect area was detected using a tapered and threaded cylindrical pin. The tapered pin exhibited the highest tensile strength, hardness, and elongation of all three pins [79]. Defect-free areas were found when taper and threaded cylindrical tools were used to weld the AA6061–AA5086 aluminum alloy. The threaded cylindrical tool records maximum tensile strength, according to [80]. Authors [81] measured maximum tensile strengths of 21.44 and 26 MPa during welding HDPE sheets with flat and concave shoulders, respectively. It was found that the joining strength of HDPE was improved using the concave shoulder at lower rotational and welding speeds. FS welded joint strength by flat shoulder was reported at increased TRS and WS. The investigation of the study done by researcher [82] depicts that the FSW plate forged at 900 rpm TRS and 100 mm/min WS with TCT tool had better mechanical qualities like hardness and tensile strength when compared with other boundary conditions. When the fracture occurs in the body of the BM or the HAZ, which has a lower hardness value than the SZ, the welded specimen depicts higher hardness values. Because of the lower hardness value, the joints fractured in the HAZ of the AA6061 side. The lower value is due to a lack of fusion between the two plates. As the distance from the central weld line decreases, the slope of the variation of the hardness curve decreases. The microhardness curve becomes smoother after a certain point and gets closer to the weld centre as the welding speed drops, and the HAZ's size also decreases. After a certain point, the microhardness curve tends to deviate less and becomes smooth in nature as it is closer to the central weld line. It was found from their study that, the dimension of the HAZ zone decreases when there is an increase in welding speed value. The HAZ zone endured a thermal cycle without plastic deformation. This tends to decrease the hardness value of the HAZ due to over-ageing and coarsening of strengthening precipitates. Precipitate over-ageing in the HAZ is influenced by heat vulnerability, time and temperature, as well as rotational and welding rates, which also play an important part as the deciding factors for the hardness distribution in the HAZ.

10.5.2 Metallurgical Aspect (Microstructure and Grain Size)

Ti–6Al–4 V is by far the most common titanium alloy, with a heterogeneous microstructure consisting of hexagonal-close packed α and body-centered cubic β-phases. These phases are highly stable at high-temperature values. Cooling from the α-phase field to form Widmanstätten β produces the usual microstructure, with approximately half of the persisting α-phase field in the microstructure. Ti-6Al-4V alloy is popular due to its ultimate strength of about 1100 MPa, resistance to creep at around 300 °C, resistance to fatigue behavior, and formability. This titanium alloy covers around half of all the titanium produced and processed for industrial use. The heat and deformation produced during the FSW process have a crucial impact on the underlying microstructure. But the effects on performance during manufacturing and service need to be investigated. According to general fatigue research, based on the dimension of the specimen, crystalline morphology, and amount of residual stresses generated, the fracture propagation rate in the HAZ is found to be maximum or minimum than the underlying BM used [83]. It is noteworthy that due to mechanical twinning, which causes distortional issues, pure titanium tries to remain in its close-packed hexagonal α-form [84]. The SZ of an FSW joint has a high density of dislocations and mechanical twins, with transmission microscopy revealing a fine elongated structure. At the same time, the overall grain shape seems to be equiaxed on the optical microscope scale. Recrystallization is thought to have happened during welding; however, also accompanied by minimum plastic deformation. The HAZ merely indicated the growth of grain structures, minimum hardness value, and thus the fracture region in cross-weld testing.

As is characteristic of aluminum FS welds, the TMAZ area on the welded specimens can't be clearly defined. A novel approach known as the cellular automata finite element (CAFE) technique is adopted to predict microstructural grain size morphology of any material. This technique is also used to indicate the formation of the weld defects during the forming process. Abaqus 6.8 FE code was built coupled with the DFLUX subroutine to simulate the FSW process of the elements made up of FSW blanks [85]. The thermal history and strain-rate analytical models are being incorporated into this FE model. By coupling the temperature evolved and strain rate developed through transition rules during the process, one can predict the grain size morphology and yield strength of the material. The CAFE model's predictions of grain size morphology and yielding strength variation matched well with the experimental results for all FSW circumstances. The true stress-strain data of the experimental FSWed welded blank compared well with the computed stress-strain behavior from the CAFE model. This includes learning about the grain size morphological distribution and yield strength variation.

10.6 ADVANCEMENTS IN THE FSW PROCESS

Friction stir welding has spawned a slew of variations due to its popularity. For about a decade, these advancements in the FSW process made welding techniques, surface modification, surface composite manufacture, microstructural and mechanical modifications, and additive manufacturing. Some of the advanced variations of the FSW process are discussed below.

10.6.1 Underwater Friction Stir Welding (UFSW)

FSW is a solid-state welding technique that weld materials difficult to join using the fusion welding process. In addition, as compared to fusion welding, this procedure is more energy-efficient and environmentally beneficial. Despite the benefits of FSW over fusion welding, the temperature cycles generated during the FSW process soften the welded joints of heat-treatable aluminum alloys (AA's) due to the dissolution or coarsening of the strengthening precipitates, resulting in a reduction in mechanical behavior. Underwater friction stir welding (UFSW) is a viable option to overcome these constraints. This method is commonly used for heat-treatable AAs and is ideal for alloys sensitive to welding heat. Authors [86] investigated basic principles such as material flow, temperature generation, process parameters, microstructure, and mechanical characteristics generated during the UFSW process. According to the study results, UFSW is a better strategy for building joint strength than FSW. Figure 10.11 depicts the schematic representation of a UFSW process.

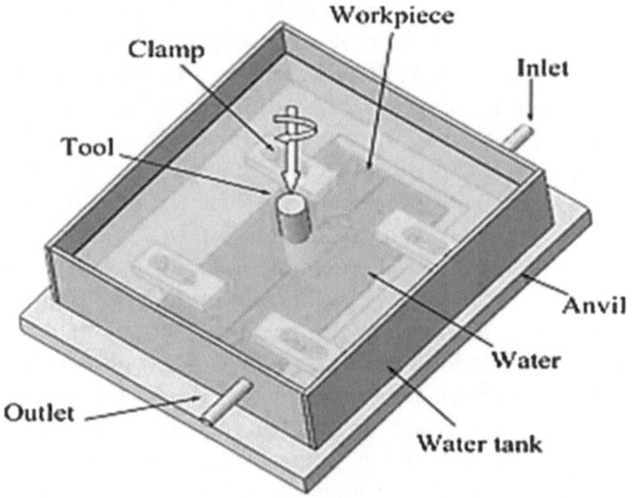

Figure 10.11 Schematic diagram of underwater FSW (UFSW) process [86].

Due to intense plastic deformation and dynamic recrystallization, FSW causes fine grain structure in the SZ, resulting in significant microstructural development [3,87–89]. FSW has superior mechanical qualities because of its ultrafine microstructure. Even though the heat input in FSW is lower than in friction welding (FW), it is high enough, in general, to produce softening in heat-treatable AA's. The softening is caused by the dissolving or coarsening of the strengthening precipitates, which reduces the welded joints' mechanical characteristics [90–92]. To address these concerns, a strategy that improves the cooling rate while lowering the peak temperature can improve the joint characteristics in FSW. External cooling has been used in numerous solid-state joining methods to improve joint performance in this area [93,94]. Because of its broad flow movement and exceptional heat absorption capacity, water cooling has also been examined to provide a cooling effect on the samples during FSW.

10.6.2 Friction Stir Additive Manufacturing (FSAM)

Friction stir additive methods are novel approaches that use the layer-by-layer additive manufacturing (AM) principle in combination with the solid-state FSW methodology. These are a subset of friction-based additive methods (FATs) that can be used in various ways [95]. This brings it into the category that this approach is considered a significant advancement in metal additive manufacturing (MAM). MAM method's ability to fabricate sophisticated parts has led to their being viewed as a viable choice for different industrial sectors like aviation, automotive, and marine industries. Fusion-based technologies, which rely on MAM approach MAM methods, have several drawbacks, most of which are linked to solidification. Shear strength issues are a problem. FATs, due to their solid-state behavior, primarily address the shortcomings of fusion-based MAM approaches. They produce high-quality, defect-free components and exhibit good mechanical and structural qualities.

Since FSW is the foundation of FSAM and AFS, these techniques are referred to as FSATs. Most of the MAM flaws are successfully addressed by FSATs. In these methods, metal layers are bonded in layered patterns in a solid-state, preventing undesirable phases and flaws. FSATs' parts have several advantages, including a finely equiaxed granular structure, significantly improved mechanical properties, and fewer flaws [96–98]. Figure 10.12 depicts the schematic representation of an FSAM process.

10.6.3 Friction Stir Spot Welding

In the automobile, shipbuilding, and aerospace sectors, friction stir spot welding (FSSW) is a version of linear FSW, which was designed to compete with resistance spot welding (RSW) and the riveting of lightweight alloys. The use of FSSW has been expanded to include a wide range of metals and

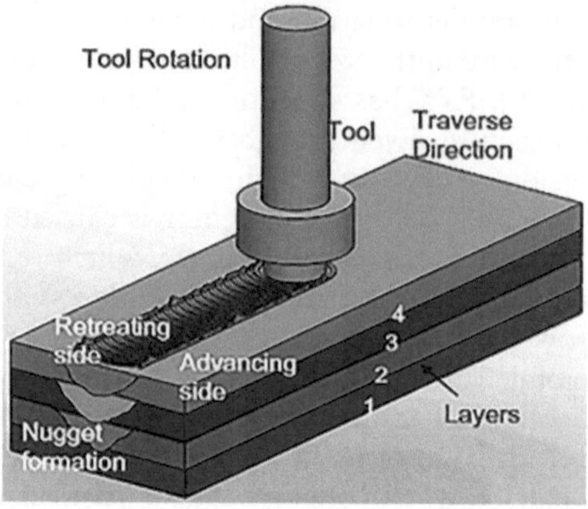

Figure 10.12 Schematic diagram of friction stir additive manufacturing (FSAM) process [95].

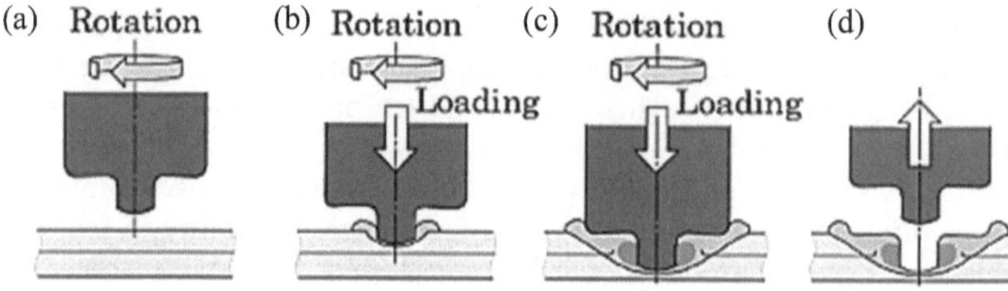

Figure 10.13 Schematic diagram of conventional friction stir spot welding (FSSW) process: **(a)** rotating, **(b)** plunging, **(c)** stirring (dwell), **(d)** drawing out [99].

nonmetals. Figure 10.13 depicts the schematic representation of an FSSW process. Conventional FSSW was initially utilized in Mazda RX-8 rear door panels in 2003 and then in the 2005 Mazda MX-5 sports car to attach the aluminum trunk lid to the steel bolt retainers.

It's also been utilized on the decklid and hood of Toyota's Prius gasoline/electric hybrids. When compared to the RSW of Al alloys, it is projected that using conventional FSSW saves 90% of energy and 40% of capital investment since it eliminates the need for numerous pieces of equipment, such as a large electric power source, a cooling unit, and an electrode dresser.

10.6.4 Friction Stir Processing

Friction stir processing (FSP) is an emerging metalworking technique capable of providing concentrated reconfiguration and regulation of crystal structure morphologies in near-surface layers of processed metallic components. It is based on the basic principles of FSW; a solid-state joining process originally developed for aluminum alloys. The FSP results in

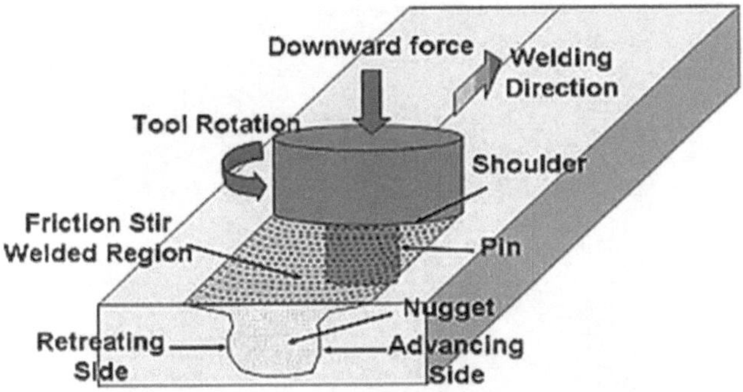

Figure 10.14 Schematic diagram of friction stir processing (FSP) [100].

considerable microstructural refinement, densification, and homogeneity of the processed zone due to solid plastic deformation, material mixing, and thermal exposure. Figure 10.14 depicts the schematic representation of an FSP process.

A rotating tool with a pin and shoulder is inserted into a single material. This results in the enhancement of localized microstructures for specific property enhancement. FSP, for example, was used to create a fine-grained microstructural morphology for high-strain-rate superplasticity in the commercial 7075Al alloy. The FSP technique has also been employed to homogenize powder metallurgy (PM) aluminum alloys, metal matrix composites, cast aluminum alloys, and create a surface composite on an aluminum substrate [101].

10.6.5 Micro Friction Stir Welding (μFSW)

Significant progress has been made in FSW, with many research outputs concentrating on materials, tools, and process optimization, among other things. Micro-friction stir welding (μFSW) is a kind of FSW used to join materials with a thickness of 1 mm or less. Materials, flaws, tools, and new applications are all considered. FSW has experimented with butt and lap welds, focusing on butt welds [102].

There was a lot of concern in the early stages of the FSW of very thin plates about the practicality of the process with such thin materials. As a result, the initial research concentrated on determining if micro-friction stir welding could be done. Scialpia et al. looked into the possibility of micro FSW, focusing on the mechanical properties of FSW joints between 2024-T3 and 6082-T6 alloys [103]. The alloys were successfully bonded, and the joint's total strength was similar to that of the weaker 6082-T6 alloy. Figure 10.15 depicts the schematic representation of an µFSW process.

μFSW, like FSW, was first used on aluminum and its alloys. Since then, applications have been expanded to include copper, brass, steels in

Figure 10.15 μFSW in progress through 300 μm thick aluminum alloy sheet [5].

similar and dissimilar configurations, and other materials such as polymers, composites, and even wood. When adopting FSW, the problem of intermetallics, which causes dissimilar joints between copper and aluminum when fusion welding, can be reduced. μFSW is a particularly appealing technology for creating aluminum and copper joints for electrical contacts because of this. Similarly, owing to the lower welding temperatures, using μFSW to weld thermoplastic polymers helps to alleviate the problem of thermal deterioration.

10.7 CONCLUSION

The use of the FSW process to link materials is a novel concept. Despite its origin in aluminum alloys, FSW has since been applied to several materials, from steels to polymers. Despite fewer advancements in the in-process physics of the technique, it has a wide range of uses in various sectors. This method can now effectively weld hard materials like steel and other essential high-melting point alloys. A far better comprehension of welded joints' morphology and characteristics has also advanced significantly. FSW is expected to become a more commonly used solid-state technique in the future, given the current rate of progress. The current discovery has helped reduce flaws and improve weld property consistency while extending the use of FSW to advanced technological alloys. Effective tools have been developed due to a better analytical understanding of heat transfer, material flow, tool-workpiece contact conditions, and the effects of various process parameters. While FSW has effectively combined difficult-to-weld materials, it is still in its infancy. The FSW process has been empirically developed for each new application so far. Numerical studies based on scientific knowledge are quite valuable for comprehending the FSW process. The mechanism of cross-bonding during the friction stir welding process has led to several FSW modifications, including FSP, FSSW, and FSAW, in addition to welding. These several joining techniques offer a lot more flexibility in component manufacture. Because it is a solid-state joining technique, it has several advantages, including eliminating solidification flaws, enhanced strength properties due to finer morphological crystallographic structures, minimum energy, and so on.

REFERENCES

[1] A. O. Mosleh, F. H. Mahmoud, T. S. Mahmoud, T. A. Khalifa, Microstructure and static immersion corrosion behavior of AA7020-O Al plates joined by friction stir welding, *Proc. Inst. Mech. Eng. Part L J. Mater. Des. Appl.* 230 (2016) 1030–1040. 10.1177/1464420715594484.

[2] P. L. Threadgill, R. Johnson, The potential for friction stir welding in oil and gas applications, *Proc. Int. Offshore Polar Eng. Conf.* 1 (2004) 1–7.

[3] K. V. Jata, S. L. Semiatin, Recrystallization during friction stir welding of high strength aluminium alloys, *Scr. Mater.* 43 (2000) 743–749.

[4] K. Masaki, Y. S. Sato, M. Maeda, H. Kokawa, Experimental simulation of recrystallized microstructure in friction stir welded Al alloy using a plane-strain compression test, *Scr. Mater.* 58 (2008) 355–360. 10.1016/j.scriptamat.2007.09.056.

[5] https://www.twi-global.com/technical-knowledge/published-papers/developments-in-micro-applications-of-friction-stir-welding

[6] R. Bhattacharjee, P. Biswas, Review on thermo-mechanical and material flow analysis of dissimilar friction stir welding, *Weld. Int.* 35 (2021) 295–332. 10.1080/09507116.2021.1992256.

[7] X. K. Zhu, Y. J. Chao, Numerical simulation of transient temperature and residual stresses in friction stir welding of 304L stainless steel, *J. Mater. Process. Technol.* 146 (2004) 263–272. 10.1016/j.jmatprotec.2003.10.025.

[8] F. Al-Badour, N. Merah, A. N. Shuaib, A. Bazoune, Experimental and finite element modeling of friction stir seal welding of tube-tubesheet joint, *Adv. Mater. Res.* 445 (2012) 771–776. 10.4028/www.scientific.net/AMR.445.771.

[9] H. Jamshidi Aval, S. Serajzadeh, A. H. Kokabi, Evolution of amicrostructures and mechanical properties in similar and dissimilar friction stir welding of AA5086 and AA6061, *Mater. Sci. Eng. A.* 528 (2011) 8071–8083. 10.1016/j.msea.2011.07.056.

[10] R. S. Mishra, Z. Y. Ma, Friction stir welding and processing, *Mater. Sci. Eng. R Reports.* 50 (2005) 1–78. 10.1016/j.mser.2005.07.001.

[11] G. Buffa, A. Ducato, L. Fratini, Numerical procedure for residual stresses prediction in friction stir welding, *Finite Elem. Anal. Des.* 47 (2011) 470–476. 10.1016/j.finel.2010.12.018.

[12] C. M. Chen, R. Kovacevic, Finite element modeling of friction stir welding - Thermal and thermomechanical analysis, *Int. J. Mach. Tools Manuf.* 43 (2003) 1319–1326. 10.1016/S0890-6955(03)00158-5.
[13] C. Chen, R. Kovacevic, Thermomechanical modelling and force analysis of friction stir welding by the finite element method, *Proc. Inst. Mech. Eng. Part C J. Mech. Eng. Sci.* 218 (2004) 509–520. 10.1243/095440604323052292.
[14] C. M. Chen, R. Kovacevic, Parametric finite element analysis of stress evolution during friction stir welding, *Proc. Inst. Mech. Eng. Part B J. Eng. Manuf.* 220 (2006) 1359–1371. 10.1243/09544054JEM324.
[15] V. Soundararajan, S. Zekovic, R. Kovacevic, Thermo-mechanical model with adaptive boundary conditions for friction stir welding of Al 6061, *Int. J. Mach. Tools Manuf.* 45 (2005) 1577–1587. 10.1016/j.ijmachtools.2005.02.008.
[16] A. Murphy, W. McCune, D. Quinn, M. Price, The characterisation of friction stir welding process effects on stiffened panel buckling performance, *Thin-Walled Struct.* 45 (2007) 339–351. 10.1016/j.tws.2007.02.007.
[17] L. Z. Jin, R. Sandström, Numerical simulation of residual stresses for friction stir welds in copper canisters, *J. Manuf. Process.* 14 (2012) 71–81. 10.1016/j.jmapro.2011.10.001.
[18] M. Z. H. Khandkar, J. A. Khan, A. P. Reynolds, M. A. Sutton, Predicting residual thermal stresses in friction stir welded metals, *J. Mater. Process. Technol.* 174 (2006) 195–203. 10.1016/j.jmatprotec.2005.12.013.
[19] A. Bastier, M. H. Maitournam, F. Roger, K. Dang Van, Modelling of the residual state of friction stir welded plates, *J. Mater. Process. Technol.* 200 (2008) 25–37. 10.1016/j.jmatprotec.2007.10.083.
[20] K. Deplus, A. Simar, W. Van Haver, B. De Meester, Residual stresses in aluminium alloy friction stir welds, *Int. J. Adv. Manuf. Technol.* 56 (2011) 493–504. 10.1007/s00170-011-3210-0.
[21] M. Riahi, H. Nazari, Analysis of transient temperature and residual thermal stresses in friction stir welding of aluminum alloy 6061-T6 via numerical simulation, *Int. J. Adv. Manuf. Technol.* 55 (2011) 143–152. 10.1007/s00170-010-3038-z.
[22] B. A. Sherwood, W. H. Bernard, Work and heat transfer in the presence of sliding friction, *Am. J. Phys.* 52 (1984) 1001–1007. 10.1119/1.13775.
[23] G. Ravichandran, On the Conversion of Plastic Work into Heat During High-Strain-Rate Deformation, (2003) 557–562. 10.1063/1.1483600.
[24] H. Schmidt, J. Hattel, J. Wert, An analytical model for the heat generation in friction stir welding, *Model. Simul. Mater. Sci. Eng.* 12 (2004) 143–157. 10.1088/0965-0393/12/1/013.
[25] R. Nandan, G. G. Roy, T. Debroy, Numerical simulation of three dimensional heat transfer and plastic flow during friction stir welding, *Metall. Mater. Trans. A Phys. Metall. Mater. Sci.* 37 (2006) 1247–1259. 10.1007/s11661-006-1076-9.
[26] M. Z. H. Khandkar, J. A. Khan, A. P. Reynolds, Prediction of temperature distribution and thermal history during friction stir welding: Input torque based model, *Sci. Technol. Weld. Join.* 8 (2003) 165–174. 10.1179/136217103225010943.

[27] M. Song, R. Kovacevic, Thermal modeling of friction stir welding in a moving coordinate system and its validation, *Int. J. Mach. Tools Manuf.* 43 (2003) 605–615. 10.1016/S0890-6955(03)00022-1.
[28] H. B. Schmidt, J. H. Hattel, Thermal modelling of friction stir welding, 58 (2008) 332–337. 10.1016/j.scriptamat.2007.10.008.
[29] R. Nandan, T. DebRoy, H. K. D. H. Bhadeshia, Recent advances in friction-stir welding - Process, weldment structure and properties, *Prog. Mater. Sci.* 53 (2008) 980–1023. 10.1016/j.pmatsci.2008.05.001.
[30] S. Tutunchilar, M. Haghpanahi, M. K. Besharati Givi, P. Asadi, P. Bahemmat, Simulation of material flow in friction stir processing of a cast Al-Si alloy, *Mater. Des.* 40 (2012) 415–426. 10.1016/j.matdes.2012.04.001.
[31] P. A. Colegrove, H. R. Shercliff, Two-dimensional CFD modelling of flow round profiled FSW tooling, *Sci. Technol. Weld. Join.* 9 (2004) 483–492. 10.1179/136217104225021832.
[32] P. A. Colegrove, H. R. Shercliff, Experimental and numerical analysis of aluminium alloy 7075-T7351 friction stir welds, *Sci. Technol. Weld. Join.* 8 (2003) 360–368. 10.1179/136217103225005534.
[33] P. A. Colegrove, H. R. Shercliff, 3-Dimensional CFD modelling of flow round a threaded friction stir welding tool profile, *J. Mater. Process. Technol.* 169 (2005) 320–327. 10.1016/j.jmatprotec.2005.03.015.
[34] M. Grujicic, G. Arakere, B. Pandurangan, J. M. Ochterbeck, C. F. Yen, B. A. Cheeseman, A. P. Reynolds, M. A. Sutton, Computational analysis of material flow during friction stir welding of AA5059 aluminum alloys, *J. Mater. Eng. Perform.* 21 (2012) 1824–1840. 10.1007/s11665-011-0069-z.
[35] S. D. Ji, Q. Y. Shi, L. G. Zhang, A. L. Zou, S. S. Gao, L. V. Zan, Numerical simulation of material flow behavior of friction stir welding influenced by rotational tool geometry, *Comput. Mater. Sci.* 63 (2012) 218–226. 10.1016/j.commatsci.2012.06.001.
[36] S. Ji, A. Zou, Y. Yue, G. Luan, Y. Jin, F. Li, Numerical simulation of effect of rotational tool with screw on material flow behavior of friction stir welding of Ti6Al4V alloy, *Acta Metall. Sin. (English Lett.* 25 (2012) 365–373.
[37] H. W. Zhang, Z. Zhang, J. T. Chen, The finite element simulation of the friction stir welding process, *Mater. Sci. Eng. A.* 403 (2005) 340–348. 10.1016/j.msea.2005.05.052.
[38] H. W. Zhang, Z. Zhang, J. T. Chen, 3D modeling of material flow in friction stir welding under different process parameters, *J. Mater. Process. Technol.* 183 (2007) 62–70. 10.1016/j.jmatprotec.2006.09.027.
[39] D. Santiago, S. Urquiza, G. Lombera, L. de Vedia, 3D modeling of material flow and temperature in Friction Stir Welding, *Soldag. Inspeção.* 14 (2009) 248–256. 10.1590/s0104-92242009000300008.
[40] Z. Zhang, H. W. Zhang, A fully coupled thermo-mechanical model of friction stir welding, *Int. J. Adv. Manuf. Technol.* 37 (2008) 279–293. 10.1007/s00170-007-0971-6.
[41] Z. Zhang, H. W. Zhang, Numerical studies on controlling of process parameters in friction stir welding, *J. Mater. Process. Technol.* 209 (2009) 241–270. 10.1016/j.jmatprotec.2008.01.044.
[42] Z. Zhang, J. T. Chen, Z. W. Zhang, H. W. Zhang, Coupled thermo-mechanical model based comparison of friction stir welding processes of

AA2024-T3 in different thicknesses, *J. Mater. Sci.* 46 (2011) 5815–5821. 10.1007/s10853-011-5537-1.

[43] W. M. Thomas, K. I. Johnson, C. S. Wiesner, Friction stir welding-recent developments in tool and process technologies, *Adv. Eng. Mater.* 5 (2003) 485–490. 10.1002/adem.200300355.

[44] X. He, F. Gu, A. Ball, A review of numerical analysis of friction stir welding, *Prog. Mater. Sci.* 65 (2014) 1–66. 10.1016/j.pmatsci.2014.03.003.

[45] T. U. Seidel, A. P. Reynolds, Visualization of the material flow in AA2195 friction-stir welds using a marker insert technique, *Metall. Mater. Trans. A Phys. Metall. Mater. Sci.* 32 (2001) 2879–2884. 10.1007/s11661-001-1038-1.

[46] R. Nandan, G. G. Roy, T. J. Lienert, T. Debroy, Numerical modelling of 3D plastic flow and heat transfer during friction stir welding of stainless steel, *Sci. Technol. Weld. Join.* 11 (2006) 526–537. 10.1179/174329306X107692.

[47] R. Nandan, G. G. Roy, T. J. Lienert, T. Debroy, Three-dimensional heat and material flow during friction stir welding of mild steel, *Acta Mater.* 55 (2007) 883–895. 10.1016/j.actamat.2006.09.009.

[48] S. Xu, X. Deng, A. P. Reynolds, T. U. Seidel, Finite element simulation of material flow in friction stir welding, *Sci. Technol. Weld. Join.* 6 (2001) 191–193. 10.1179/136217101101538640.

[49] F. C. Liu, Z. Y. Ma, Influence of tool dimension and welding parameters on microstructure and mechanical properties of friction-stir-welded 6061-T651 aluminum alloy, *Metall. Mater. Trans. A Phys. Metall. Mater. Sci.* 39 (2008) 2378–2388. 10.1007/s11661-008-9586-2.

[50] K. Elangovan, V. Balasubramanian, Influences of tool pin profile and tool shoulder diameter on the formation of friction stir processing zone in AA6061 aluminium alloy, *Mater. Des.* 29 (2008) 362–373. 10.1016/j.matdes.2007.01.030.

[51] R. Rai, A. De, H. K. D. H. Bhadeshia, T. DebRoy, Review: Friction stir welding tools, *Sci. Technol. Weld. Join.* 16 (2011) 325–342. 10.1179/1362171811Y.0000000023.

[52] Y. S. Sato, M. Urata, H. Kokawa, Parameters controlling microstructure and hardness during friction-stir welding of precipitation-hardenable aluminum alloy 6063, *Metall. Mater. Trans. A.* 33 (2002) 625–635. 10.1007/s11661-002-0124-3.

[53] X. Zhao, P. Kalya, R. G. Landers, K. Krishnamurthy, Empirical dynamic modeling of friction stir welding processes, *J. Manuf. Sci. Eng. Trans. ASME.* 131 (2009) 0210011–0210019. 10.1115/1.3075872.

[54] C. Devanathan, A. S. Babu, Effect of Plunge Depth on Friction Stir Welding of Al 6063, *2nd Int. Conf. Adv. Manuf. Autom.* (2013) 482–485. http://www.scopus.com/inward/record.url?eid=2-s2.0-84887219505&partnerID=40&md5=1b2b99a2ecd377f9829460d7f984f4a8.

[55] K. R. Seighalani, M. K. B. Givi, A. M. Nasiri, P. Bahemmat, Investigations on the effects of the tool material, geometry, and tilt angle on friction stir welding of pure titanium, *J. Mater. Eng. Perform.* 19 (2010) 955–962. 10.1007/s11665-009-9582-8.

[56] K. Kumar, S. V. Kailas, The role of friction stir welding tool on material flow and weld formation, *Mater. Sci. Eng. A.* 485 (2008) 367–374. 10.1016/j.msea.2007.08.013.

[57] S. Zhang, Q. Shi, Q. Liu, R. Xie, G. Zhang, G. Chen, Effects of tool tilt angle on the in-process heat transfer and mass transfer during friction stir welding, *Int. J. Heat Mass Transf.* 125 (2018) 32–42. 10.1016/j.ijheatmasstransfer.2018.04.067.

[58] N. Dialami, M. Cervera, M. Chiumenti, Effect of the tool tilt angle on the heat generation and the material flow in friction stir welding, *Metals (Basel).* 9 (2019). 10.3390/met9010028.

[59] K. Kimapong, T. Watanabe, Effect of welding process parameters on mechanical property of FSW lap joint between aluminum alloy and steel, *Mater. Trans.* 46 (2005) 2211–2217. 10.2320/matertrans.46.2211.

[60] P. Chauhan, R. Jain, S. K. Pal, S. B. Singh, Modeling of defects in friction stir welding using coupled Eulerian and Lagrangian method, *J. Manuf. Process.* 34 (2018) 158–166. 10.1016/j.jmapro.2018.05.022.

[61] K. P. Mehta, V. J. Badheka, Effects of tilt angle on the properties of dissimilar friction stir welding copper to aluminum, *Mater. Manuf. Process.* 31 (2016) 255–263. 10.1080/10426914.2014.994754.

[62] Frigaard, Grong, O. T. Midling, A process model for friction stir welding of age hardening aluminum alloys, *Metall. Mater. Trans. A Phys. Metall. Mater. Sci.* 32 (2001) 1189–1200. 10.1007/s11661-001-0128-4.

[63] H. N. B. Schmidt, J. Hattel, Heat source models in simulation of heat flow in Friction Stir Welding, *Proc. Int. Offshore Polar Eng. Conf.* 14 (2004) 41–49.

[64] H. Schmidt, J. Hattel, Modelling heat flow around tool probe in friction stir welding, *Sci. Technol. Weld. Join.* 10 (2005) 176–186. 10.1179/174329305X36070.

[65] T. Long, W. Tang, A. P. Reynolds, Process response parameter relationships in aluminium alloy friction stir welds, *Sci. Technol. Weld. Join.* 12 (2007) 311–317. 10.1179/174329307X197566.

[66] P. Ulysse, Three-dimensional modeling of the friction stir-welding process, *Int. J. Mach. Tools Manuf.* 42 (2002) 1549–1557. 10.1016/S0890-6955(02)00114-1.

[67] B. G. Kiral, M. Tabanoglu, H. T. Serindag, Finite element modeling of friction stir welding in aluminum alloys joint, *Math. Comput. Appl.* 18 (2013) 122–131. 10.3390/mca18020122.

[68] G. Buffa, L. Fratini, Friction stir welding of steels: Process design through continuum based FEM model, *Sci. Technol. Weld. Join.* 14 (2009) 239–246. 10.1179/136217109X421328.

[69] M. Rangel Pacheco, J. Paul Kabche, I. Thesi, F. Nunes Diesel, Temperature distribution analysis in plates joined by Friction Stir Welding, *ASME Int. Mech. Eng. Congr. Expo. Proc.* 11 (2010) 163–169. 10.1115/IMECE2009-12892.

[70] M. Rangel Pacheco, P. M. Calas Lopes Pacheco, Analysis of the temperature distribution in friction stir welding using the finite element method, *Mater. Sci. Forum.* 758 (2013) 11–19. 10.4028/www.scientific.net/MSF.758.11.

[71] Z. Zhang, Z. Y. Wan, Predictions of tool forces in friction stir welding of AZ91 magnesium alloy, *Sci. Technol. Weld. Join.* 17 (2012) 495–500. 10.1179/1362171812Y.0000000039.

[72] J. M. Rice, S. Mandal, A. A. Elmustafa, Microstructural investigation of donor material experiments in friction stir welding, *Int. J. Mater. Form.* 7 (2014) 127–137. 10.1007/s12289-012-1110-y.

[73] S. Mandal, J. Rice, G. Hou, K. M. Williamson, A. A. Elmustafa, Modeling and simulation of a donor material concept to reduce tool wear in friction stir welding of high-strength materials, *J. Mater. Eng. Perform.* 22 (2013) 1558–1564. 10.1007/s11665-012-0452-4.

[74] M. Grujicic, S. Ramaswami, J. S. Snipes, V. Avuthu, R. Galgalikar, Z. Zhang, Prediction of the Grain-Microstructure Evolution Within a Friction Stir Welding (FSW) Joint via the Use of the Monte Carlo Simulation Method, *J. Mater. Eng. Perform.* 24 (2015) 3471–3486. 10.1007/s11665-015-1635-6.

[75] R. K. Uyyuru, S. V. Kallas, Numerical analysis of friction stir welding process, *J. Mater. Eng. Perform.* 15 (2006) 505–518. 10.1361/105994906X136070.

[76] M. Imam, K. Biswas, V. Racherla, On use of weld zone temperatures for online monitoring of weld quality in friction stir welding of naturally aged aluminium alloys, *Mater. Des.* 52 (2013) 730–739. 10.1016/j.matdes.2013.06.014.

[77] M. Imam, K. Biswas, V. Racherla, Effect of weld morphology on mechanical response and failure of friction stir welds in a naturally aged aluminium alloy, *Mater. Des.* 44 (2013) 23–34. 10.1016/j.matdes.2012.07.046.

[78] K. Elangovan, V. Balasubramanian, Influences of tool pin profile and welding speed on the formation of friction stir processing zone in AA2219 aluminium alloy, *J. Mater. Process. Technol.* 200 (2008) 163–175. 10.1016/j.jmatprotec.2007.09.019.

[79] P. Motalleb-Nejad, T. Saeid, A. Heidarzadeh, K. Darzi, M. Ashjari, Effect of tool pin profile on microstructure and mechanical properties of friction stir welded AZ31B magnesium alloy, *Mater. Des.* 59 (2014) 221–226. 10.1016/j.matdes.2014.02.068.

[80] M. Ilangovan, S. Rajendra Boopathy, V. Balasubramanian, Effect of tool pin profile on microstructure and tensile properties of friction stir welded dissimilar AA 6061–AA 5086 aluminium alloy joints, *Def. Technol.* 11 (2015) 174–184. 10.1016/j.dt.2015.01.004.

[81] K. Mustapha, B. Abdessamad, B. Azzeddine, Z. Mokhtar, Experimental Investigation of Friction Stir Welding Process on High-Density Polyethylene, *J. Fail. Anal. Prev.* 20 (2020) 590–596. 10.1007/s11668-020-00867-0.

[82] S. Ravikumar, V. Seshagiri Rao, R. V. Pranesh, Effect of Process Parameters on Mechanical Properties of Friction Stir Welded Dissimilar Materials between AA6061-T651 and AA7075-T651 Alloys, *Int. J. Adv. Mech. Eng.* 4 (2014) 101–114.

[83] R. John, K. V. Jata, K. Sadananda, Residual stress effects on near-threshold fatigue crack growth in friction stir welds in aerospace alloys, *Int. J. Fatigue.* 25 (2003) 939–948. 10.1016/j.ijfatigue.2003.08.002.

[84] W. B. Lee, C. Y. Lee, W. S. Chang, Y. M. Yeon, S. B. Jung, Microstructural investigation of friction stir welded pure titanium, *Mater. Lett.* 59 (2005) 3315–3318. 10.1016/j.matlet.2005.05.064.

[85] R. S. Saluja, R. Ganesh Narayanan, S. Das, Cellular automata finite element (CAFE) model to predict the forming of friction stir welded blanks, *Comput. Mater. Sci.* 58 (2012) 87–100. 10.1016/j.commatsci.2012.01.036.

[86] M. A. Wahid, Z. A. Khan, A. N. Siddiquee, Review on underwater friction stir welding: A variant of friction stir welding with great potential of

improving joint properties, *Trans. Nonferrous Met. Soc. China* (English Ed. 28 (2018) 193–219. 10.1016/S1003-6326(18)64653-9.
[87] W. Woo, L. Balogh, T. Ungár, H. Choo, Z. Feng, Grain structure and dislocation density measurements in a friction-stir welded aluminum alloy using X-ray peak profile analysis, *Mater. Sci. Eng. A*. 498 (2008) 308–313. 10.1016/j.msea.2008.08.007.
[88] G. Liu, L. E. Murr, C. S. Niou, J. C. McClure, F. R. Vega, Microstructural aspects of the friction-stir welding of 6061-T6 aluminum, *Scr. Mater.* 37 (1997) 355–361. 10.1016/S1359-6462(97)00093-6.
[89] S. Benavides, Y. Li, L. E. Murr, D. Brown, J. C. McClure, Low-temperature friction-stir welding of 2024 aluminum, *Scr. Mater.* 41 (1999) 809–815. 10.1016/S1359-6462(99)00226-2.
[90] H. J. Liu, H. Fujii, M. Maeda, K. Nogi, Tensile properties and fracture locations of friction-stir-welded joints of 2017-T351 aluminum alloy, *J. Mater. Process. Technol.* 142 (2003) 692–696. 10.1016/S0924-0136(03)00806-9.
[91] W. A. Baeslack, K. V. Jata, T. J. Lienert, Structure, properties and fracture of friction stir welds in a high-temperature Al-8.5Fe-1.3V-1.7Si alloy (AA-8009), *J. Mater. Sci.* 41 (2006) 2939–2951. 10.1007/s10853-006-5089-y.
[92] M. Cabibbo, H. J. McQueen, E. Evangelista, S. Spigarelli, M. Di Paola, A. Falchero, Microstructure and mechanical property studies of AA6056 friction stir welded plate, *Mater. Sci. Eng. A*. 460–461 (2007) 86–94. 10.1016/j.msea.2007.01.022.
[93] L. Fratini, G. Buffa, R. Shivpuri, In-process heat treatments to improve FS-welded butt joints, *Int. J. Adv. Manuf. Technol.* 43 (2009) 664–670. 10.1007/s00170-008-1750-8.
[94] T. D. Clark, <An Analysis of Microstructure and Corrosion Resistance in Underwa.pdf>, (2005).
[95] M. Srivastava, S. Rathee, S. Maheshwari, A. Noor Siddiquee, T. K. Kundra, A Review on Recent Progress in Solid State Friction Based Metal Additive Manufacturing: Friction Stir Additive Techniques, *Crit. Rev. Solid State Mater. Sci.* 44 (2019) 345–377. 10.1080/10408436.2018.1490250.
[96] M. Yuqing, K. Liming, H. Chunping, L. Fencheng, L. Qiang, Formation characteristic, microstructure, and mechanical performances of aluminum-based components by friction stir additive manufacturing, *Int. J. Adv. Manuf. Technol.* 83 (2016) 1637–1647. 10.1007/s00170-015-7695-9.
[97] C. M. Lewandowski, N. Co-Investigator, C. M. Lewandowski, Microstructure and mechanical properties of WE43 alloy poduced via additive friction stir technology, *Eff. Br. Mindfulness Interv. Acute Pain Exp. An Exam. Individ. Differ.* 1 (2015) 1689–1699.
[98] J. Rodelas, J. Lippold, Characterization of Engineered Nickel-Base Alloy Surface Layers Produced by Additive Friction Stir Processing, *Metallogr. Microstruct. Anal.* 2 (2013) 1–12. 10.1007/s13632-012-0056-2.
[99] Z. Shen, Y. Ding, A. P. Gerlich, Advances in friction stir spot welding, *Crit. Rev. Solid State Mater. Sci.* 45 (2020) 457–534. 10.1080/10408436.2019.1671799.
[100] Z. Y. Ma, Friction stir processing technology: A review, *Metall. Mater. Trans. A Phys. Metall. Mater. Sci.* 39 A (2008) 642–658. 10.1007/s11661-007-9459-0.

[101] R. Butola, M. S. Ranganath, Q. Murtaza, Fabrication and optimization of AA7075 matrix surface composites using Taguchi technique via friction stir processing (FSP), *Eng. Res. Express*. 1 (2019). 10.1088/2631-8695/ab4b00.

[102] S. Ahmed, A. Shubhrant, A. Deep, P. Saha, Development and Analysis of Butt and Lap Welds in Micro-friction Stir Welding (μFSW), *Top. Mining, Metall. Mater. Eng.* (2015) 295–306. 10.1007/978-81-322-2355-9_15.

[103] A. Scialpi, L. A. C. De Filippis, P. Cuomo, P. Di Summa, Micro friction stir welding of 2024-6082 aluminium alloys, *Weld. Int.* 22 (2008) 16–22. 10.1080/09507110801936069.

Chapter 11

Studies on Stellite 6 GTA (Gas Tungsten Arc) Weld Claddings' Microstructural and High Temperature Wear

Adarsh Prakash[1] *and Amandeep Singh Shahi*[2]

[1]School of Mechanical Engineering, Indian Institute of Technology GOA, India
[2]Department of Mechanical Engineering, Sant Longowal Institute of Engineering and Technology Punjab, India

CONTENTS

11.1 INTRODUCTION

11.1.1 Hardfacing

Hardfacing can be defined in a way that it is a process of depositing a hard and wear-resistant material in the form of thick coatings over another material surface in order to ensure enhanced tribological properties at the surface. Commonly used processes for hardfacing layer deposition on substrate material are spray-fuse method, thermal spraying method and welding processes [1]. This process is widely used either for repairment of worn-out parts or to harden the surface of a new component. This process is a surface modifying process that is generally used for repairing of earth moving and agricultural machineries,

DOI: 10.1201/9781003436072-11

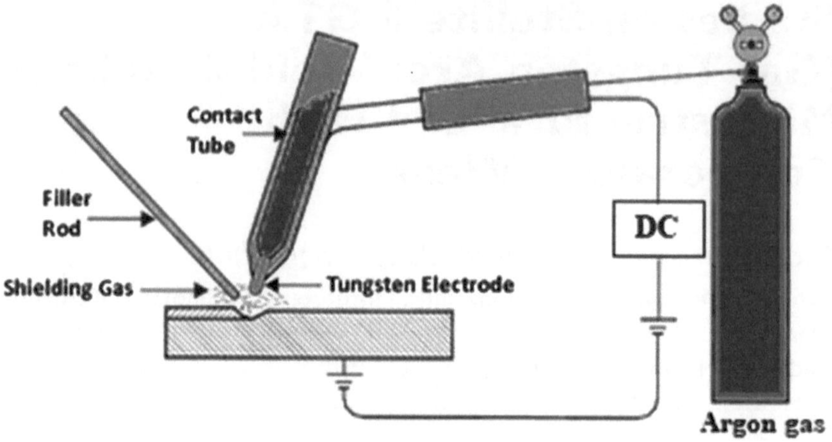

Figure 11.1 Schematic diagram of hardfacing operation.

repairment of the railway rolling stock, large and medium-sized gear wheels, turbine parts, conveyor belts and many other components. The schematic diagram of hardfacing is shown in Figure 11.1.

11.1.2 Need of Hardfacing in Industries

Either hardfacing or hardfaced components are used in almost all the heavy industries that are employed in the manufacturing and production sector. Hardfacing is used in industries to:

- Reduce costs of component – Hard surfacing or hardfacing of a worn-out component converted the component to like-new condition, which usually saves the cost of replacement that is approximately 30–70% of the cost of replacement of components.
- Enhance equipment life – Hardfacing enhances the life of component from 40 to 300 times, as compared to that of a non-hardened component, depending upon their applications.
- Reduce downtime – Hardfacing reduces the industry downtime because components last longer, fewer shutdowns are required on regular basis to replace the defected component and this directly reduces the productivity of the industry.

11.1.3 Objectives of Hardfacing

It is desired to enhance or increase the service life of equipment as well as assemblies and also to save expensive material used. Apart from increasing the service life of equipment, it is also equally important to know the technique used, which the deposition of hardfacing layer can be laid on these critical equipments as well as assemblies made up of metals or alloys that need enhancement in their hardness as well as wear resistance. These depositions provide resistance to:

a. Abrasion
b. Impact
c. Erosion
d. Cavitation or pitting wear
e. Oxidation
f. Corrosion

11.1.4 Application of Hardfacing

Hardfacing is applicable in the manufacturing of new components as well as reclamation of worn components. In the above-mentioned cases, hardfacing is commonly used in small to large industries for extending the service life of their products, which directly reduces the consumption of their expensive materials, as discussed earlier. A large number of engineering components, engineering equipment and engineering products in industries are surfaced on a regular schedule to keep them working; this makes the industry economically stable also. This surface modifying method is used in the manufacturing or reclaiming of the following equipment in industries.

- Earth-moving equipment like bulldozer trunnion, bulldozer loaders, cultivator sweeps, dragline buckets, digging arms, excavator buckets, plough shares, etc.
- Steel and foundry industry, which contains coke wagons, feed rollers, forging moulds and punches, gaskets, ore and ash handling equipment, speed reducer, sheet metals conveyor guide, smooth-faced rollers, ventilators and blower parts.
- Coal and cement crushing equipment like asphalt mixer paddles, blast furnace cones, coal recovery augers, conveyor screws, cruiser rolls and hammers, crusher jaws, etc.
- In textile industries, it is mainly used in cloth pullers, diagonal cutter, filament guides, heating plates and rollers.
- In special applications like biscuit industry, babbitt and bronze bearings, calipers, compressors and engines, crank and transmission shafts for pumps, cylinder sleeves, cylinder sleeve inserts, electric insulators, electric motor shafts and bodies, shafts, gas and water piston pumps, gate valve seats, heat treatment and paper-mill equipment, pump and blower shafts, pistons, piston segments, rollers and cylinders for sheet metal and tobacco industry.

11.2 STELLITES

Stellites are a group of a special kind Co-based hardfacing superalloys. The components that are to be subjected to high-temperature applications use this kind of cobalt- and nickel-based hardfacing alloys on a large scale. Because of high temperature, the components require better resistance to erosion and corrosion, which mainly involves abrasive wear, impact wear and metal-metal wear. Chromium with other alloying elements like molybdenum, tungsten, carbon etc is added in large amounts. The first cobalt-based alloy was in 1900 by Elwood Haynes, who developed the first co-based superalloy with composition 52Co-28Cr-4W-l.lC (wt %) that is now globally known as Stellite 6 [2]. With the progress of time and demand of best superalloys in the growing world, approximately more than 20 co-based superalloys have been developed and are being extensively used in small to heavy industries all around the world [3].

When the composition of Co-based alloy is varied, it always resultes in varied microstructure. The microstructure of this alloy is found either as primary dendrites of cobalt in the form of hypoeutectic structure and eutectic carbides are present surrounding these Co solid solutions or in the form of large idiomorphic primary chromium-rich carbides of the hyper eutectic type [4]. The alloying elements of Stellite 6 (Figure 11.2) are discussed later, in which carbon is considered to be the main alloying element that has the highest influence on the microstructure. The above-discussed hypereutectic and hypoeutectic structures are differentiated on the basis of carbon and this carbon is the real cause of changing from hypoeutectic Stellite to a hypereutectic Stellite. Tungsten (in W6C form) act as a carbide forming element and because of its presence in high amounts, it reduces carbon required for achieving the eutectic composition. Chromium forms $^{M}7^{C}3$ and $^{M}23^{C}6$ type carbides that enhance strength and wear resistance [5]. Molybdenum and tungsten are known as refractory metals that contribute to solid solution hardening in the alloys through precipitation hardening due to which MC and $^{M}6^{C}$ type carbides formed along with

Figure 11.2 Stellite 6 hardfacing TIG filler wire.

(W, MO) Co3 type intermetallic phase that consequently increases the strength as well as wear resistance of this superalloys [6]. Other alloying elements like Fe and Ni affect the stacking fault energy (SFE) and hence influence the properties of alloy. On the basis of chemical composition, stress generated during work and temperature of working directly determine the crystal structure alloy either f.c.c. or h.c.p., which further determines the wear characteristics on the basis of alloying elements as a particular phase stabilizer element provides better wear resistance with that crystal structure alloy. The broad discussion related to microstructure and wear characteristics of Stellite 6 will be discussed later in the Results and Discussion.

11.3 WEAR TESTING

Pin-on-disc wear testing (Figures 11.4 and 11.5) is a wear testing method used to determine the wear characteristics of a material on the basis of the wear loss calculated and coefficient of friction achieved for various testing conditions. These results for various materials were compared quantitatively to identify the better wear-resistant material. The wear testing is carried out following the ASTM G-99 testing procedure that is accepted globally. Due to formation of hard intermetallic carbides around the Co eutectic islands, the Stellite 6 alloy exhibits higher hardness, which further enhances the wear characteristics of this alloy [7]. In this case, dilution also affects the wear characteristics of Stellite 6 at high temperatures, where, due to diffusion of particles from a harder surface to a softer surface, resulted in high wear loss in a corrosive environment. Dilution has always had a significant impact on a material's mechanical properties. But because of the presence of chromium in Stellite 6, an oxide layer forms at the surface when employed in high-temperature applications where this layer acts as a protective layer and lubricating layer that further reduces wear. Wear mechanism are also identified with the help of some metallurgical tests like SEM of worn-out parts [8] (Figure 11.3).

11.4 LITERATURE REVIEW

Jing Q. et al., 2013 [9], made a comparison for microstructure, hardness, wear resistance and mass loss between the substrate (45 carbon steel) and Stellite 6 coated material. Coating was done using an electro spark deposition (ESD) method with 0.5 mm thick deposition and it was determined that the coating's microhardness had greatly improved, being about 3.6 times greater in comparison to the substrate, whereas the mass loss of the substrate being about 4.2–4.5 times greater than that of the coating, demonstrating superior wear resistance.

Figure 11.3 High temperature pin-on-disc tribometer.

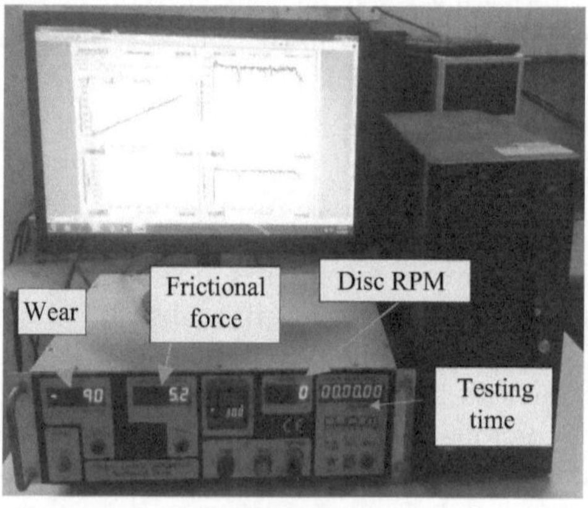

Figure 11.4 Tribometer controller attached with computer system.

Sreedhar B.K. et al., 2015 [10], utilised two distinct hard-faced coatings, Ni-based Colmonoy5 and Co-based Stellite 6 alloy for hardfacing, and the outcomes were matched with those of 316 L austenitic stainless steel. In this investigation, it was discovered that the cavitation resistance of Colmonoy5 and Stellite 6 deposits was much higher than that of 316 L austenitic stainless steels in sodium. This proved that any of these alloys can significantly limit the level of damage expected in components in fast breeder reactors (FBRs) that are likely to experience cavitation in service by hardfacing the surfaces.

Pradeep G.R.C. et al., 2012 [11], studied for the effect of sliding velocities on wear results when hardfacing was done on AISI 1020 steel substrate. The hardfacing was done using three welding processes at same heat input: TIG welding, SMA welding and gas welding. The wear test was done on pins and the results were further compared for different sliding velocities and different method of hardfacing. From the results, it was concluded that the pins prepared from the TIG samples had shown better wear resistance when tested up to a sliding velocity of 1.256 m/s as compared to pins prepared from SMA welding and gas welding. But as the sliding velocity increases above 1.571 m/s, the TIG welding samples showed the worst wear characteristics among all.

Kapoor S., 2012 [12], studied about a group of Stellite alloys for hardness, wear resistance and their variation with temperature. He found that the wear resistance of Stellite alloys, in general, decreased with temperature. However, for low-carbon Stellite alloys at high temperatures, the wear resistance was increased. This abnormal behavior was attributed to the surface hardening, which resulted in strain hardening (aggravated dislocations), the fcc to hcp transformation and the "glazing" effect.

Altan T. et al., 2011 [13], compared the hardness of commercially available die materials, and then presented a method for comparing commercially available hot work tool steels based on the hardness data available in the material datasheets, did work on the selection of die raw material and surface treatments for enhancing die life in hot as well as warm forging.

Ferozhkhana M.M. et al., 2016 [14], studied the microstructure, hardness and wear resistance of cladded specimen that is prepared by first depositing austenitic stainless steel on Cr-Mo steel (ASME Grade 91) using the FCAW process and next to this layer, Stellite 6 (Co-based alloy) was coated using the PTAW process for improving wear resistance when subjected to high-temperature applications. It was concluded that PTA hardfacing process of Stellite 6 had good wear-resistant characteristics in high-temperature applications and also the hardness was uniform throughout the coating, decreased when it crossed the fusion line and at HAZ the hardness was found at a minimum.

Zhu Z. et al., 2019 [15], investigated the metallurgical as well as wear characteristics of Stellite 6 coating on Q235 steel substrate. The coating was done using a plasma arc surfacing method. The wear test results showed less wear when pins were tested at high temperature due to formation of oxide layer in between the pin and disc. The microstructure and phase composition was studied with the help of optical microscopy and XRD.

Kusmoko A. et al., 2013 [16], investigated the effect of heat input on the mechanical as well as metallurgical characteristics when a coating of Stellite 6 was laid on a mild carbon steel substrate. The laser cladding method was used for laying the weld overlay at two heat inputs of 1 kW and 1.8 kW. The metallurgical analysis was done on the basis of results obtained from optical microscopy: SEM and XRD. The results reported showed less pore development and less cracking in the case of less heat input. More hardness of coating was also found for low heat input coating i.e., for 1 kW. The wear test for MS 1.8 showed higher wear in comparison to MS 1 due to lower hardness.

Apay S. et al., 2014 [17], coated a low-carbon steel AISI 1015 surface by Stellite 6 powder that is a cobalt-based superalloy using micro laser welding. The cross-section of the coated sample was examined for microstructural analysis. The microstructural and mechanical properties were investigated using optical microscopy, SEM, XRD, hardness testing and wear testing. Line analysis method and elements mapping analysis were used to investigate the dilution of chromium and cobalt in a substrate material. The results thus obtained in the above-stated tests were compared with the results obtained by other researchers in their paper with the same experimentation details like the same substrate, filler, heat input, coating method, etc.

On the basis of the literature survey, it can be concluded that there has been little research on Stellite 6's high-temperature wear resistance. The features of cladded alloy at room temperature have been the subject of a lot of research. Ambiguities concerning the function and impact of different phases and compounds contained in the cladding on the mechanical properties of the cladded material were evident in the literature survey done up to this point. Also, a threshold thickness of the cladding layer can be found after which the extra cladded material is not going to enhance wear characteristics in any way when tested at a high temperature. In this way, a database related to this alloy can also be strengthened by working out a threshold Stellite 6 clad thickness, beyond which these claddings would exhibit uniform wear behavior.

11.5 EXPERIMENTATION

The base material for this investigation was AISI 1020 mild steel, which is frequently utilised in the production of valves. For wear testing, single-layer and double-layer claddings were chosen. In Tables 11.1 and 11.2, respectively, the nominal chemical compositions of AISI 1020 and the chosen cladding material are tabulated.

Table 11.1 Substrate Composition AISI 1020 (wt %)

Element	*C*	*Mn*	*Si*	*Cr*	*Ni*	*Mo*	*P*	*S*	*Cu*	*Fe*
%age	0.227	0.346	0.077	0.188	0.0803	0.003	0.039	0.0445	0.119	Bal.

Table 11.2 Stellite 6 – AWS A5.21 ERCoCr-A (wt %)

Element	*C*	*Mn*	*Si*	*Cr*	*Ni*	*Fe*	*Mo*	*W*	*Co*
%age	0.70–1.4	2.0 max	2.0 max	25–32	3.0 max	5.0 max	1.0 max	3.0–6.0	Bal.

Stellite alloy was deposited using a tungsten inert gas welding (TIG) process. A FRONIUS make TIG welding machine was used for weld overlay of Stellite 6 on 50 mm thick substrate material, as shown in Figure 11.5. The cladding parameters are tabulated in Table 11.3. The weld overlay was laid onto the substrate in four different conditions.

I. Single-layer weld overlay
II. Single-layer weld overlay with remelting of the cladded surface when it gets cool.

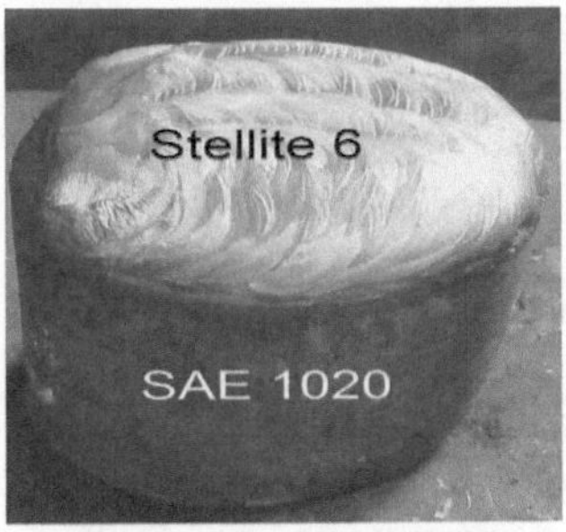

Figure 11.5 Stellite 6 cladded SAE1020 workpiece.

Table 11.3 Deposition Parameters Used during Cladding in This Research Work

Deposition arameters	*Values*
Deposition current	109 Amp
Deposition voltage	11.3
Deposition speed	1.32 mm/s
Gas flow rate (Argon)	10 lit/min
Deposition interpass temperature	150°C
Electrode diameter	2.4 mm
Filler rod diameter	3.2 mm

III. Double layer weld overlay taking 150°C as interpass temperature.
IV. Double layer with remelting considering both interpass temp and cooling of workpiece before remelting.

In order to evaluate the top layer, it was studied virtually undiluted (i.e., with substrate material). Visual inspection was used to assess the weld quality. Light optical microscopy was used to characterise the microstructures of all the claddings in their as-clad state. The specimen for optical microscopy was extracted from the cladded material containing both cladded as well as base metal using a hand cutter. The specimen size was kept more than 1 cm × 1 cm. The polishing machine had a rotating disc over which we fit the emery paper of grit size varying from 150 to 3,000 in increasing order and the specimen is polished. At last, the specimen is polished with velvet cloth and then etched with Aaqua regia (HCl:HNO3:: 3:1). The polished specimen was dipped into the etchant for 20–25 seconds and then cleaned under running water, clearly visualizing the dendrites, grain boundaries, etc.

The testing sample was in direct contact with a disc during sliding wear testing on a high-temperature pin-on-disk tribometer. The wear testing parameters are mentioned in Table 11.4. The pins for the wear testing were extracted in the form of cylindrical bar having a diameter of 8 mm and length of 53–57 mm depending upon the type and condition of weld overlay using wire cut EDM, as shown in Figure 11.6. After extracting the cylindrical piece, the cladded portion of the specimen is made as a hemispherical shape of 4 mm radius and the total size of pin was kept 32 mm, so that point contact of the specimen could be ensured with the rotating disc of tribometer [18]. The final pins prepared for wear testing are shown in Figure 11.7. The disk material was EN 31 hardened to 63 HRc and ground to 1.6 μm Ra surface roughness. The stationary pin specimen, 32 mm in height and 8 mm in diameter, was made from the cladded specimen. The pins prepared for the wear testing have 3 mm, 4 mm, 5 mm and 6 mm clad thicknesses so that a threshold clad thickness can be obtained. The wear tests were carried out in dry conditions at two temperatures, 3000 C and room temperature, with a standard load of 44.1 N (4.5 kgf) rotating at 350 rpm over a sliding distance of 3,000 m. The contacting surfaces were

Table 11.4 Wear Testing Parameters Used in This Research Work

Testing parameters	*Value*
Load	4.5 kgf
Disc RPM	350
Wear track diameter	90 mm
Pin sliding distance	3000 m
Test duration	30 minutes
Testing temperatures	Room temperature and 300°C

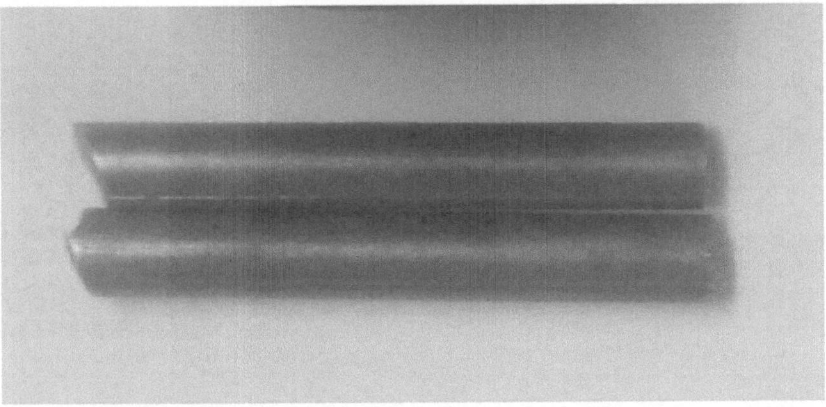

Figure 11.6 Pins extracted from cladded specimen using wire cut EDM.

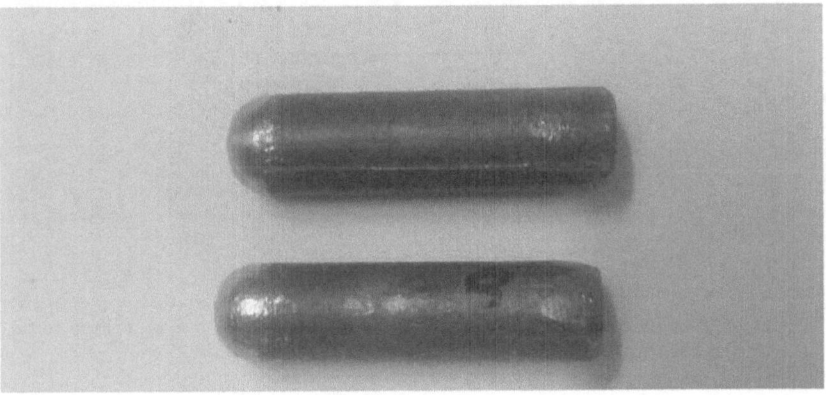

Figure 11.7 Pins prepared for wear testing.

polished with 1,000 mesh sandpaper before being cleaned with an ultrasonic cleaner and acetone, dried and weighed with a precision of 0.1 mg. For tests at high temperatures, the specimens were given adequate time to reach the test temperature.

11.6 RESULTS AND DISCUSSION

11.6.1 Metallurgical Studies of Stellite 6 Claddings

Optical micrographs of Stellite 6 claddings laid under different conditions are presented from Figures 11.8–11.11. Various observations made based on these photomicrographs are discussed next.

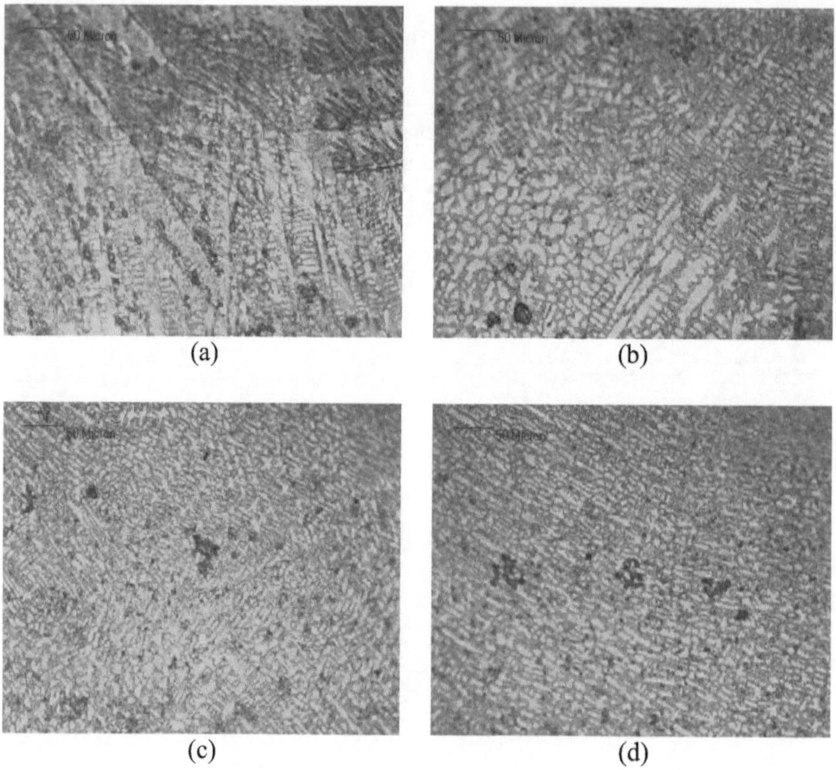

Figure 11.8 Microstructure of single layer cladded Stellite 6 on AISI 1020 (a) near fusion boundary layer; (b) transition zone; (c) weld zone of clad layer; (d) top surface of cladding.

The interface formed between the substrate metal and the Stellite 6 layer was continuous, indicating a strong fusion bond between the two and without any lack of fusion defect. As shown in Figure 11.8(a), the optical micrograph was taken near the fusion boundary zone, which shows epitaxial solidification that showed accelerated growth in the direction of the welding centre. It was made up of interdendrites produced by the eutectic of Co and Cr with carbides having h.c.p. structure and dendrites of Co solid solution with f.c.c. structure (bright area) (dark region). It was discovered that the carbides formed when the intermetallics (Co, Cr and W) reacted with carbon to produce M_7C_3 (Cr) and M_2C (W), which were the ones that increased the hardness, abrasion and ultimately wear resistance of Stellite clads. The top part of the cladding shown in Figure 11.8(d) shows finer microstructure in comparison to the bottom part of the cladding because of the reason that the top part of the cladding was directly exposed to the surrounding that resulted into a relatively higher cooling rate as compared to the lower part of the cladding.

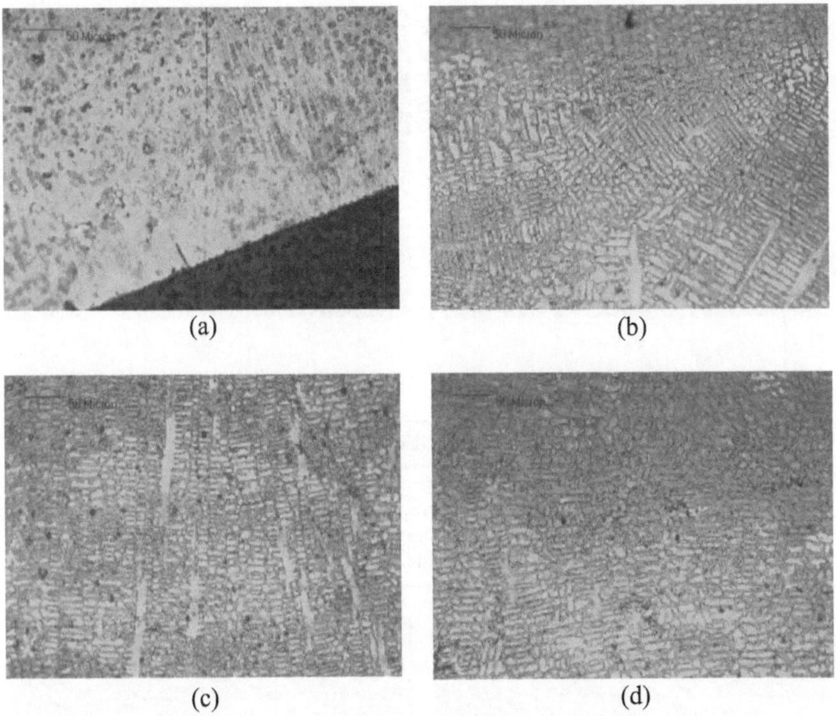

Figure 11.9 Microstructure of single layer cladded and remelted Stellite 6 on AISI 1020 (a) near fusion boundary layer; (b) transition zone; (c) weld zone of clad layer; (d) top remelted surface of cladding.

Figure 11.9 shows the optical micrograph of single layer Stellite 6 cladding with remelted pass. As shown in Figure 11.9(b), the Co eutectic islands and dendritic morphologies became finer than the transition zone, as shown in Figure 11.9(b) because of heat received from the remelted pass. The upper black region in Figure 11.9(d) shows finer grains because of surface remelting.

Figure 11.10 represented the optical micrographs of double layer Stellite 6 cladding. Figure 11.10(a) shows clearly that microstructural grain coarsening occurred within the weld zone. In Figure 11.10(b), the top and bottom layers of Stellite 6 cladding were clearly separated from one another by an intermediate HAZ, which caused dendrites to get coarsen as a result of a slow cooling rate brought on by heat from the second clad pass. Because the top layer of the clad was directly exposed to the environment and cooled more quickly than the bottom layer of the clad, the difference in grain size between the first layer and second layer (top layer) could be seen in Figure 11.10(c) and 11.10(d), respectively. Because of this, the cladding's bottom layer had relatively coarser cladding grains when compared with the top layer's Stellite 6 cladding grains.

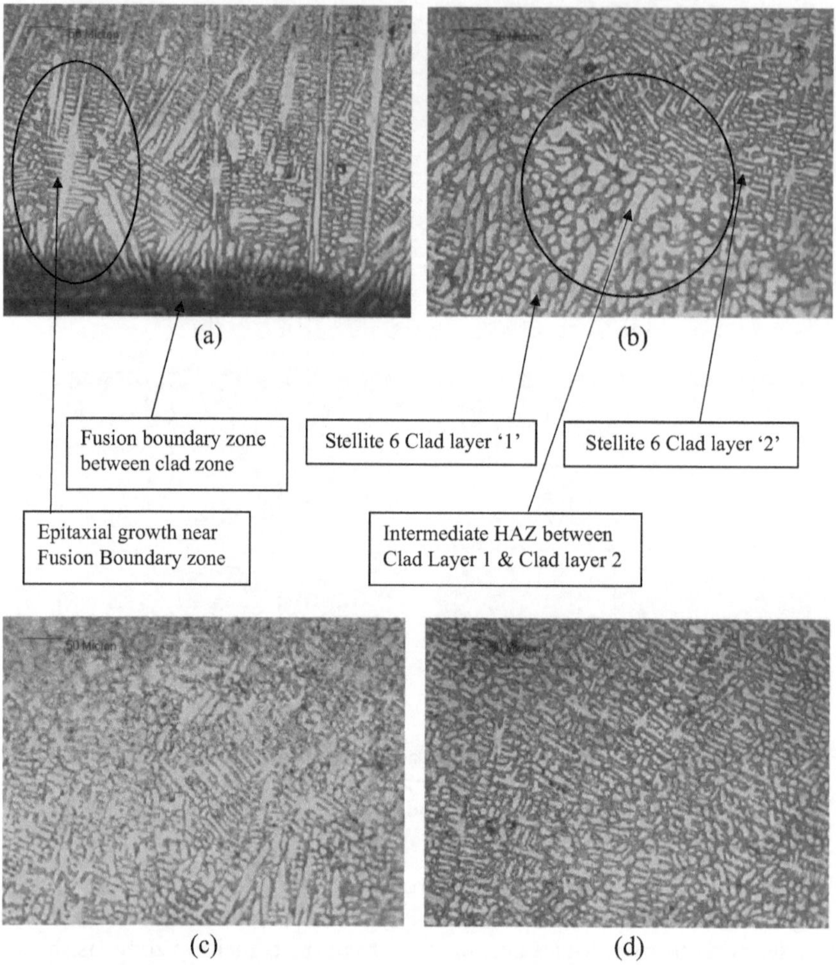

Figure 11.10 Microstructure of double layer cladded Stellite 6 on AISI 1020 (a) near fusion boundary layer; (b) composite zone between first and second layer; (c) weld zone of first layer; (d) weld zone of second or top layer.

Figure 11.11 represents the optical micrograph of double layer Stellite 6 cladding with remelted pass. In Figure 11.11(b), the HAZ region between top and bottom clad layer was finer in comparison to grain coarsened region of intermediate HAZ of Figure 11.11(b) due to heat received from the remelted pass. The extent of uniformity in terms of morphological changes were more in Figure 11.11(b) than in Figure 11.11(b). Figure 11.11(c) and 11.11(d) represent the grains of bottom clad layer and top clad layer, respectively. This grain refinement of the top layer occurred due to a higher cooling rate of the top layer, whereas lower heat dissipation was experienced by the bottom layer. From Figure 11.11(e),

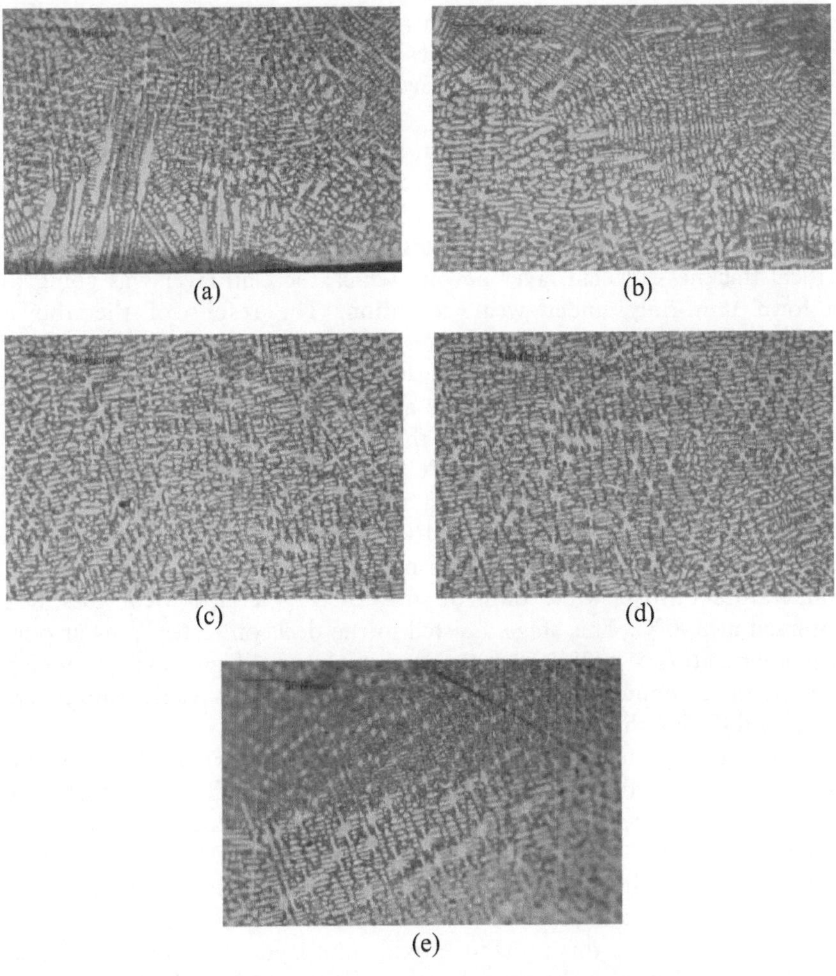

Figure 11.11 Microstructure of double layer cladded and remelted Stellite 6 on AISI 1020 (a) near fusion boundary layer; (b) composite zone between first and second layer; (c) weld zone of first layer; (d) weld zone of second layer; (e) top remelted region.

a black carbide region was seen with incredibly small Co eutectic islands in the uppermost layer's remelted region.

11.7 WEAR TEST FINDINGS OF STELLITE 6 CLADDINGS

Wear testing of different pins of Stellite 6 that were extracted from the Stellite claddings were subjected to pin-on-disc tribometer. The procedures of wear testing have been discussed earlier in Introduction and

Experimentation. The computer-generated data were acquired with the help of dedicated software and the results were reported accordingly. The method used for comparing the wear testing results was the weight loss measurement method.

This wear testing was carried out at room temperature as well as at high temperature (300°C). It was carried out in three phases; the first two phases of testing were completely dedicated to establish the wear testing parametric range. The third phase of wear testing was solely aimed for finding a critical thickness of clad layer beyond which the clad layer was going to perform uniformly under wear condition. The results of the above-mentioned three rounds of wear testing are tabulated sequentially in Table 11.5, 11.6, and 11.7, respectively.

A 3.5 kg load and 350 rpm were applied to the single layer Stellite 6 cladded pins during the first phase of the wear test, which was conducted at room temperature, 150°C, and 300°C (HT). Following the test, the wear loss was calculated as 70 mg, 62 mg and 34 mg for the circumstances of room temperature, 150°C, and 300°C, respectively. According to these findings, just 8 mg less wear loss was observed at 150°C compared to room-temperature testing, while only 36 mg, or 51.4%, less wear loss was observed at 300°C. This stage assisted in the decision to test pins at only two temperatures: 300°C and room temperature. Figure 11.13, which displays the computer-generated results, displays various results parameters calculated during the testings (Figure 11.12).

Because oxides such Cr_2O_2, Fe_2O_3 and CoO had developed in the contact surface area between the pin and rotating disc at high temperatures, the wear loss in the cladded sample tested at a high temperature was found to be lower than the sample tested at room temperature. The oxide coatings that formed over the contact surfaces got caught between the two mating surfaces and, as a result, created a shield over Stellite 6. While being tested at high temperatures, this protective oxide film layer atop Stellite 6 served as a lubricant between the two mating surfaces, and resulted in reduced wear. The term "glazing layer" has been used to describe this type of oxide layer development that leads to lubricating tendencies [19].

Figure 11.14 below, shows the actual view of the disc that was used for pin on disc experiment both at room temperature as well as at 300°C. It is clearly seen from this disc that the wear tracks differed from each other. For instance, at the room temperature, the dark rings were not seen, whereas corresponding to a 300°C testing condition, a dark layer formation occurred on the disc, indicating the formation of an oxide layer. This dark oxide layer on the disc also formed on the pin that got worn out during the test.

The cladded pins underwent similar testing in the second stage of wear testing at two different temperatures. Specifically, at ambient temperature and 300 °C with three distinct loading situations (3.5 kg, 4.5 kg and 7 kg) while maintaining a constant disc speed of 350 rpm. When wear testing was

Table 11.5 Stellite 6 Claddings' First Round of Wear Testing Parameters and Findings

S No.	*Specimen*	*Testing conditions*					*Coefficient of friction (μ)*	*Wear loss (mg)*	*Remarks*
		Te mp (°C)	*Lo ad (kg)*	*Rpm*	*Track diameter (mm)*	*Time (min)*			
1	P1	RT	3.5	350	90	30	0.365	70	Both, μ and wear was observed to be at its highest point because there was no oxidative layer between the pin and the disc surface that was being worn.
2	P2	150	3.5	350	90	30	0.314	62	Again, no effect of the oxide layer was seen at this temperature.
3	P3	300	3.5	350	90	30	0.299	34	At high temperatures, oxidative layer formation was seen, μ and wear was also reduced.

Table 11.6 Stellite 6 Claddings' Second Round of Wear Testing Parameters and Findings

S No.	*Specimen*	*Testing conditions*					*Coefficient of friction (μ)*	*Wear loss (mg)*	*Remarks*
		Temp (°C)	*Load (kg)*	*Rpm*	*Track diameter (mm)*	*Time (min)*			
1	S1	RT	3.5	350	90	30	0.365	70	At room temperature, μ was high. Moreover, wear was considerable due to the lack of an oxidative barrier between the pin and the disc.
2	S2	300	3.5	350	90	30	0.299	34	COF reductions and wear are caused by the formation of an oxide layer at high temperatures.
3	S3	RT	4.5	350	90	30	0.392	155	The coefficient of friction risen similarly as loading and wear both rose.
4	S4	300	4.5	350	90	30	0.338	103	An efficient oxide layer was observed
5	S5	RT	7	350	90	30	0.431	332	Absence of oxide layer
6	S6	300	7	350	90	30	0.436	323	Absence of oxide layer

Table 11.7 Stellite 6 Claddings' Third Round of Wear Testing Parameters and Findings

Pin No.	*Temperature*	*Clad thickness*	*Wear loss (mg)*
F1	Room temperature	6 mm	125.70
F2		5 mm	127.30
F3		4 mm	155.10
F4		3 mm	199.70
F5	300°C	6 mm	84.70
F6		5 mm	88.80
F7		4 mm	103.10
F8		3 mm	139.80

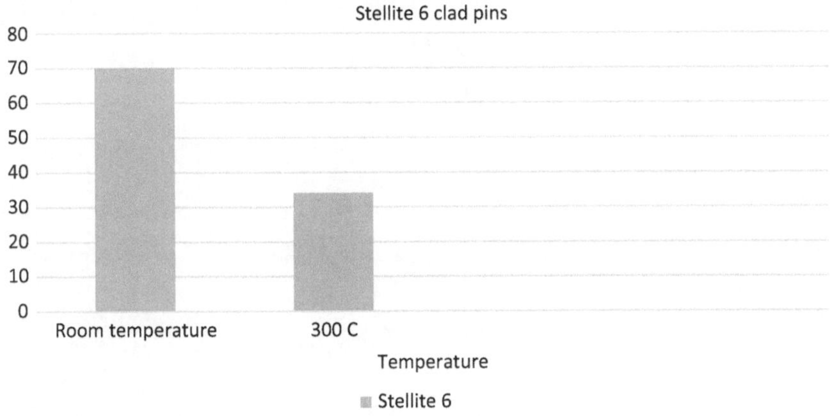

Figure 11.12 Graph depicting the Stellite 6 weld claddings' wear rate during the pin-on-disc test.

carried out at room temperature and 300°C, respectively, the wear loss measured for a 4.5 kg loading condition was 155 mg and 103 mg. When Stellite 6 cladded pins were tested for high-temperature wear less than 4.5 kg of loading, the influence of the oxide layer was seen to be evident.

The wear loss measured for the pins, however, was 332 mg and 323 mg, respectively, under heavy loading conditions of 7 kg at ambient temperature and 300°C. It was determined from this stage of the work that the oxide layers between the contacting surface could not support themselves during wear tests under such high-stress conditions. As a result, when tested for wear at room temperature, cladded pins saw a similar wear loss.

The double layer cladded Stellite 6 pins were created for the third stage of wear testing by maintaining clad thicknesses of 3 mm, 4 mm, 5 mm and 6 mm. Two temperatures, 300°C and room temperature, were used to measure the wear of these pins on a disc. A 4.5 kg of force was applied

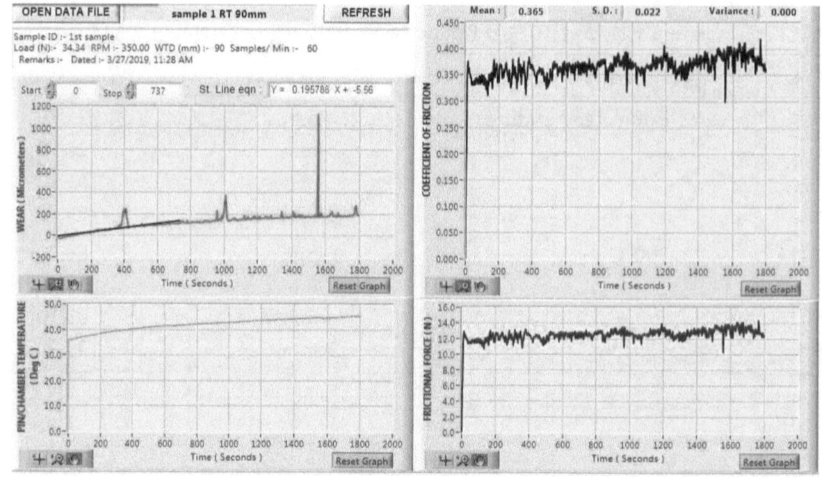

(a)

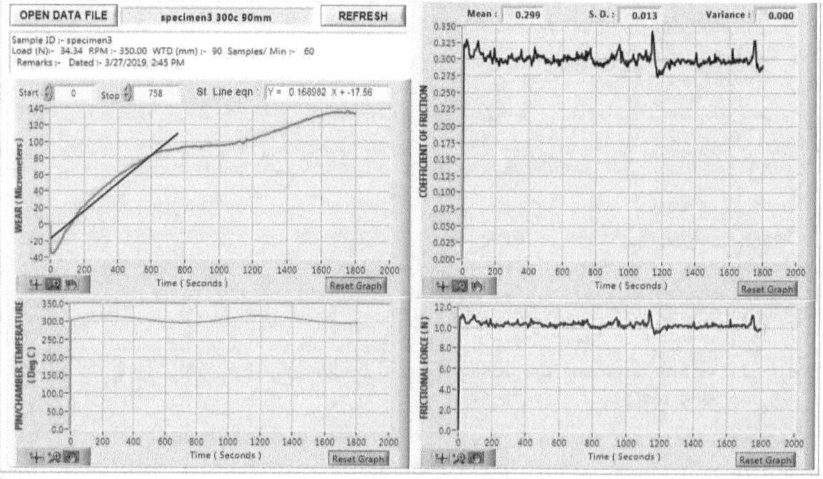

(b)

Figure 11.13 For pins tested at (a) RT and (b) HT (300°C), a graph showing wear over time, coefficient of friction over time and frictional force over time is shown.

during the test while the disc was spinning at 350 rpm for 30 minutes with a 90 mm worn track diameter. Following is the tabulated results obtained from the wear testings.

For a clad thickness of 3 mm, the computed wear loss for testing at ambient temperature was 199.80 mg in comparison to 139.00 mg for testing at 300°C. When pins with a 4 mm clad thickness were examined, the wear loss at ambient temperature was 155 mg, while the wear loss

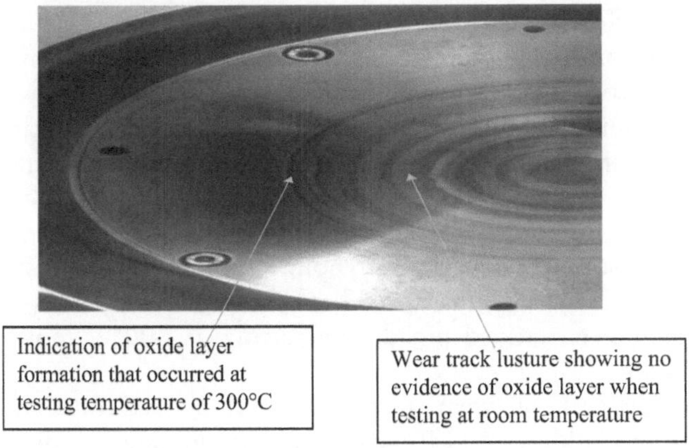

Figure 11.14 Image depicting wear tracks on the disc.

calculated at 300°C was 103 mg. Pins with a 5 mm clad thickness tested at ambient temperature measured a wear loss of 127.4 mg, while testing at 300°C showed a wear loss of 88.90 mg. Identical findings were made for 6 mm clad thickness, where wear loss was calculated to be 125.60 mg at ambient temperature and 84.80 mg at 300°C, respectively.

The wear rate of Stellite 6 was significantly impacted by the dilution of several layers. According to wear studies done on various cladded layer thicknesses, it was found that the wear was 29% lower for a 4 mm clad thickness than for a 3 mm clad thickness. Furthermore, it was discovered that the wear was 20% lower for 5 mm of cladding than for 4 mm. It was discovered that the wear loss was nearly identical for clad thicknesses of 5 and 6 mm. According to the wear testing procedures and conditions mentioned above, the analyzed data of wear for different clad thicknesses is tabulated in Table 11.7 below.

These findings led to the conclusion that the substrate material might be given good wear resistance properties with just a 5 mm Stellite clad thickness.

11.8 CONCLUSIONS AND SCOPE FOR FUTURE WORK

The investigations as discussed in the present work indicated the metallurgical and wear behavior of different Stellite 6 claddings overlaid on the substrate material. The main conclusions that can be drawn based upon this experimental work are given below:

1. Optical microscopy of different clad overlays indicated that double layered clads possessed finer grain microstructure as compared to bottom layer.

2. Wear performance of Stellite 6 cladded pins tested at a high temperature of 300°C indicated that wear loss at this testing condition was less compared to when tested at room temperature, which could be attributed to the formation of an oxidative layer that in turn decreases the frictional resistance between the two mating surfaces. This was further evident from lower coefficient of friction corresponding to 300°C testing temperature. The wear loss for the identical sort of pin tested at room temperature and tested at high temperature was 70 mg and 34 mg, respectively. The wear loss for the Stellite 6 cladded pin during testing at 300°C showed 51.4% less wear.
3. Different sets of wear testing conditions indicated the use of Stellite 6 claddings and it was found that loading condition beyond 4.5 kg was not recommended for Stellite 6 cladding due to the reason that the beneficial effect of oxide formation was completely lost.
4. Wear performance of Stellite 6 pins indicated that mass loss corresponding to single layer pins was higher as compared to double layer pins. Furthermore, remelted passes improved the wear performance of Stellite 6 claddings due to the reason that microhardness associated to the remelted surfaces also increase.
5. According to the findings of wear tests performed on various cladding layer thicknesses, the wear was 29% lower for 4 mm of cladding than for 3 mm of cladding. Additionally, it was discovered that the wear for 5 mm clad thickness was 20% lower than that for 4 mm clad thickness, while the wear loss for clad thicknesses of 5 mm and 6 mm was nearly same.
6. Finally, the results of the current work allow us to draw the conclusion that double layer cladding techniques employing the GTAW process, followed by the remelting of the top layer of class pass, can aid in improving the microhardness and, subsequently, its wear performance at high temperatures.
7. The friction coefficient of the cladding at 300°C is found as 0.299 which is smaller than that of at room temperature and is 0.365; hence, validating less frictional force as well as less wear in between the mating surfaces at 300°C.
8. The wear for 3 mm clad thickness was found as 199.80 mg, while for 4 mm clad thickness the wear reduced to 155 mg. Wear further for 5 mm and 6 mm clad thickness was found almost same i.e., approximately 126 mg.
9. A clad thickness of 5 mm is found as the critical clad thickness during weld overlay of Stellite 6 over mild steel under tested loading and working conditions after which the wear rate of Stellite 6 cladding becomes uniform and the effect of dilution vanishes.

11.8.1 Scope for Future Work

1. Surface morphology study can be extended by TEM examination.
2. Loading and temperature conditions can be further explored.
3. Remelting of the cladded surface using laser beam welding process can be studied.

REFERENCES

[1] Kusmoko, Ȧ. A., Dunne, Ȧ. D., Li, Ȧ. H. (2014) A comparative study for wear resistant of Stellite 6 coatings on nickel alloy substrate produced by laser cladding, HVOF and plasma spraying techniques. *Int J Curr Eng Technol* 4:32–36.

[2] Synthesis V. P. (1940) Chapter 1 Introduction. *Synthesis (Stuttg)*: 1–9

[3] Khanpara B., Rathod P. (2017) A review on hard surface coatings applied using stellite hardfacing techniques. *Ijetmas* 5:132–139.

[4] Gholipour A., Shamanian M., Ashrafizadeh F. (2011) Microstructure and wear behavior of stellite 6 cladding on 17-4 PH stainless steel. *J Alloys Compd* 509:4905–4909. 10.1016/j.jallcom.2010.09.216

[5] Birol Y. (2010) High temperature sliding wear behaviour of Inconel 617 and Stellite 6 alloys. *Wear* 269:664–671. 10.1016/j.wear.2010.07.005

[6] Atamert S., Stekly J. (1993) Microstructure, wear resistance, and stability of cobalt based and alternative iron based hardfacing alloys. *Surf Eng* 9:231–240. 10.1179/sur.1993.9.3.231

[7] Liu D. S., Liu R. P., Wei Y. H., Pan P. (2013) Properties of cobalt based hardfacing deposits with various carbon contents. *Surf Eng* 29:627–632. 10.1179/1743294413Y.0000000162

[8] Motallebzadeh A., Atar E., Cimenoglu H. (2015) Sliding wear characteristics of molybdenum containing Stellite 12 coating at elevated temperatures. *Tribol Int* 91:40–47. 10.1016/j.triboint.2015.06.006

[9] Jing Q. F., Tan Y. F. (2013) Microstructure and tribological properties of cobalt-based stellite 6 alloy coating by electro-spark deposition. *Mater Res* 16:1071–1076. https://doi.org/10.1590/S1516-14392013005000082

[10] Sreedhar B. K., Albert S. K., Pandit A. B. (2015) Improving cavitation erosion resistance of austenitic stainless steel in liquid sodium by hardfacing - comparison of Ni and Co based deposits. *Wear* 342–343:92–99. https://doi.org/10.1016/j.wear.2015.08.009

[11] Pradeep G. R. C., Ramesh A., Durga Prasad B. (2012) Hardfacing of AISI 1020 steel by arc welding in comparison with TIG welding processes. *J Sci Res* 5:119–126. 10.3329/jsr.v5i1.11899

[12] Kapoor S. (2012) A thesis on "High-Temperature Hardness and Wear Resistance of Stellite Alloys" submitted to Carleton University Ottawa: Ontario.

[13] Altan T., Deshpande M. (2011) Selection of die materials and surface treatments for increasing die life in hot and warm forging. 1–23. https://www.forging.org/uploaded/content/media/Altan_paper_Die_materials_and_surface_treatments6.pdf

[14] Ferozhkhan M. M., Duraiselvam M., kumar K. G., Ravibharath R. (2016) Plasma transferred arc welding of stellite 6 alloy on stainless ateel for wear resistance. *Procedia Technol* 25:1305–1311. 10.1016/j.protcy.2016.08.226

[15] Zhu Z. Y., Ouyang C., Chen J. H., Qiao Y. X. (2019) Microstructure and mechanical properties of stellite 6 alloy powders incorporated with Ti/B4C using Plasma-Arc-surfacing processes. *Mater Tehnol* 53:3–8. 10.17222/MIT.2018.044

[16] Kusmoko A., Dunne D., Li H., Nolan D. (2013) Deposition of Stellite 6 on nickel superalloy and mild steel substrates with laser cladding. *Int J Adv Mater Manuf Charact* 3:469–473. 10.11127/ijammc.2013.07.01

[17] Apay S., Gulenc B. (2014) Wear properties of AISI 1015 steel coated with Stellite 6 by microlaser welding. *Mater Des* 55:1–8. 10.1016/j.matdes.2013.09.056

[18] Prakash A., Shahi A. S. (2020) Investigations on high temperature wear and metallurgical characteristics of Stellite 6 GTA (Gas Tungsten Arc) weld claddings. *Mater Res Express* 7, 2: 026509. 10.1088/2053-1591/ab6e2b

[19] Ratia V. L., Zhang D., Carrington M. J., et al (2019) The effect of temperature on sliding wear of self-mated HIPed Stellite 6 in a simulated PWR water environment. *Wear* 420–421:215–225. 10.1016/j.wear.2018.09.012

Chapter 12

Parametric Selection and Quantitative Optimization of Process Parameters in Milling Using HDD

R. Rekha, P. V. Rajesh, N. Baskar, and P. Jothi Palavesam
Saranathan College of Engineering, Tiruchirapalli, India

CONTENTS

12.1 INTRODUCTION

The search for novel and improved materials is continually an imperative topic for current industrial and scientific demands and for producing a product at the lowest possible cost, which is an indispensible customer desire. To fulfill safety and operating standards, new materials are created and material properties are enhanced lined up with present technological breakthroughs. Composites are one such novel material that is extensively employed in industrial appliance because of their elevated characteristics like higher strength, wear resistance, fracture toughness, corrosion resistance and outstanding thermal characteristics [1,2]. Typically, composite materials are made by distributing one or more reinforcement material that are identified as discontinuous phases, in a matrix which is a continuous phase. The reinforcements will be held together by the matrix material and also maintains the integrity of the material, whereas the reinforcement material is responsible for the composites' main structure. The structure

DOI: 10.1201/9781003436072-12

and characteristics of the matrix-reinforcement contact gas has a great effect on the mechanical as well as physical characteristics of the composite material [3]. Based on the various material types utilized as the matrix, the composite materials are categorized as: polymer, ceramic and metal. By adding reinforcement to these matrix materials, metal matrix composites (MMCs), polymer matrix composites (PMCs) and ceramic matrix composites (CMCs) are developed.

MMCs have always been a prominent choice in technologically advanced research and applications, demonstrating significant promise in automotive, military applications, aerospace, electronic instruments and various other fields. The most essential quality of MMCs is that it can be customised to meet unique needs, which distinguishes it from other materials. Aluminium-based metal matrix composites (Al-MMCs) have a lot of potential for manufacturing composites with the characteristics needed for specific applications using a range of reinforcing elements. Based on specification and application requirements, Al-MMCs have been developed to achieve good mechanical and tribological qualities while remaining lightweight. Aluminum is one of the most often used metals. Aluminum is easy to work with, lightweight and corrosion resistant. It also has a high strength and can be manufactured economically using a variety of processes. Aluminum has a high specific strength, but due to its low melting point, it performs poorly at high temperatures [1,3]. Aluminum and its alloys possess excellent thermal and electrical properties, with low hardness and wear resistance. Owing to these properties, it is the most commonly utilized matrix materials in MMC manufacturing. Reinforcing materials are commonly utilized to improve the current attributes of aluminium and its alloys, such as fracture toughness, ultimate tensile strength, wear resistance, liquidus temperature, thermal and electrical stability. Al_2O_3, SiC, TiC, Si_3N_4, TiB_2, B_4C, graphite and metal particles are the most often utilized reinforcing particles. Boron carbide has a Vickers Hardness greater than 30 GPa and it is one among the hardest materials, next to cubic boron nitride and diamond. It is a covalent boron–carbon ceramic that is utilized in tank armor, bulletproof vests, engine sabotage powders and a variety of other industrial purposes. A new composite is an aluminum metal matrix bonded with boron carbide (B_4C). This composite is widely used in the automobile applications because of its exceptional characteristics like strength-to-weight ratio, wear resistance, stiffness and enhanced temperature toughness. In comparison with other reinforcements like Al_2O_3 and SiC, boron carbide has unusual features, including neutron absorption. Boron carbide (B_4C) is a highly hard, abrasive substance that ranks with diamond and cubic boron nitride in terms of hardness [4]. Traditional abrasives, such as Al_2O_3 and SiC, have simply been surpassed due to their superior and cost-effective efficiency. The liquid state technique can be used to make complex shaped goods specifically for manufacturing metal matrix composites. Stir casting is one of the liquid state manufacturing processes that is a easy and cost-efficient technique for producing MMCs with particle reinforcement [5].

Although there are a lot of advantages of MMCs, complete accomplishment of MMCs is cost-prohibitive to some extent due to the material's poor machinability. Hence, researchers have attempted in the study of machinability of MMCs under various machining operations. By using the stir casting manufacturing method, B. Vijaya Ramnath et al. created hybrid composites with aluminum alloy (LM25) as the matrix material, graphite (Gr) and boron carbide (B_4C) as reinforcement materials [6]. This study investigates the impact of controllable factors like pulse on, pulse off and current on performance measures viz. surface roughness, tool wear rate (TWR) and material removal rate (MRR) in the wire electrical discharge machining (WEDM) process of aluminium alloy (LM25), graphite (Gr) and boron carbide (B_4C) hybrid composites. While drilling Al2618, Al2618 with 10% boron carbide (B_4C) and Al2618 with 10% B_4C and 5% graphite (Gr) hybrid composites, B. N. Sharath et al. investigated the influence of controllable factors on responses such as burr height, thrust force and torque. A liquid metallurgical procedure is used to create the composites. Feed rate is found to have a substantial influence on burr height and thrust force. In comparison to the Al2618 + 10% B_4C composite, the Al2618 + 10% B_4C +5% Gr-reinforced hybrid composites have a lower thrust force and burr height. The significant graphite surface lubrication qualities were responsible for the lowered thrust force, torque, and burr height. When machining graphitic composites, the chips produced are discontinuous, which is advantageous [7]. H. Guo et al. examined the effect of controllable factors of wire electrical discharge machining taper cutting on aluminium boron carbide MMCs. On operational measurements such as angular error and surface roughness, the influence of significant operating factors such as current, pulse on off, percentage of boron carbide, wire feed, taper angle, wire tension and part has been investigated [8]. According to B. C. Kandpala and his associates, due to the strong ceramic reinforcement in the metal matrix, ordinary machining of composites is extremely challenging [9]. Even unconventional methods like EDM and laser jet machining cause significant damages to the subsurface as well as a heat-impacted zone on the object. Hence, the composite material was machined using electro-chemical machining (ECM), which is unconventional machining technology employed to create automobile and aerospace parts, as well as dies and moulds. Fine abrasive particles are mixed with electrolytes to boost the rate of material removal and the work's surface quality. The combination of abrasive particles with anodic dissolution can speed up the material removal process.

Optimization is an important function to ponder upon, while machinability analysis of different combination of specimens are taken into account [10]. The best possible and practically feasible specimen need not have the highest test values always. Nowadays, multi-criteria decision making plays a more intriguing role in optimization as more number of contrasting objectives are to be satisfied in optimizing machinability of milling process, as represented in the below mentioned research work. Response surface methodology, one of the robust and unique optimization techniques, is increasingly used in research circle due to its flexibility and accuracy [11]. It is classified into three phases: Design analysis to check the feasibility of model, evaluation to employ appropriate confidence limit to prepare ANOVA and optimization prediction to set the goal and find out optimized results.

The machinability study of MMCs during milling operation and optimization using historical data design (HDD) is shown to have a large amount of research potential, according to the review of the literature. Hence, in this paper, an attempt has been made to fabricate aliminium boron carbide metal matrix composites using a gravity die casting process as well as to study the significance of milling operational factors over the performance measures.

12.2 MATERIALS AND EXPERIMENTAL METHODS

12.2.1 Selection of Materials

Material selection is indeed an inalienable part in the fabrication of composites. It is mandatory to select the major and minor constituents of a composite by evaluating different materials available and selecting the most suitable one before the production. For a MMC with aluminium as a base metal, arguably the reinforceable minor constituent will be a ceramic powder of fine particle size [4,9].

12.2.1.1 Matrix

Normally in an AMC, aluminium alloy will be the major constituent or matrix. In this research work, precipitation treated cast aluminium alloy LM25 is selected as matrix, due to its excellent workability and weldability with moderate wear rate and superior corrosion resistance. It is one of the multi-purpose aluminium alloys, whose practical applications cover, but not limited to, automobile wheels, cylinder blocks and heads and engine body parts. The elemental composition and general properties are given below in Tables 12.1 and 12.2.

12.2.1.2 Reinforcement

Reinforcement is the strengthening agent in AMCs. This incorporating material may be in the form of continuous fibres, flakes or fine particles/powders. In this research work, boron carbide, one of the most popular ceramics, is used. In terms of ceramic hardness, it is ranked third behind diamond and cubic boron nitride (CBN). Finely pulverized B_4C powder of

Table 12.1 Elemental Composition of Aluminium Alloy LM25

Elements	*Percentage (%)*
Magnesium (Mg)	0.970
Silicon (Si)	0.620
Copper (Cu)	0.392
Iron (Fe)	0.125
Manganese (Mn)	0.108
Chromium (Cr)	0.079
Others (Total)	0.040
Aluminium (Al)	97.666

Table 12.2 Common Characteristics of Aluminium Alloy LM25

Properties	*Unit*	*Value(s)*
Modulus of Elasticity	MPa	68.9
Ultimate Tensile Strength	MPa	130
Melting Point	°C	582–652
Density	g/cm^3	2.7
Thermal Conductivity	W/m-K	167
Coefficient of Thermal Expansion	m/°C	23.6×10^{-6}
Brinell Hardness	–	45

particle size 100 μm will be reinforced inside a metallic matrix during the casting procedure. Its low density (2.5 g/cc) ensures uniform miscibility inside the metal matrix and the high melting point provides thermal stability [12].

12.2.2 Fabrication Method

Production of MMCs is frequented by a gravity die casting (GDC) technique. The technique starts with the preheating of a mild steel casting eie (Dimensions: 200 * 150 * 15 mm prepared by electric discharge machining) in an electric oven for 15 minutes. This activity is followed by pre-heating boron carbide ceramic powder for about 10–20 minutes in a muffle furnace [12]. In the first phase, the LM25 alloy of required weight proportion bought in the form of cylindrical rods is charged in a graphite crucible and fired using a pit furnace. When it attains a semi-solid state by going beyond melting point,a degassing tablet (normally hexa chloro-methane) is added to avoid formation of air-bubbles. Then, the molten metal is stirred with a motorized propeller continuously for 5 minutes to create vortex [8,13–15]. The pre-heated ceramic powder is poured into the molten metal alloy at a constant pour rate and thoroughly mixed together, while remaining inside

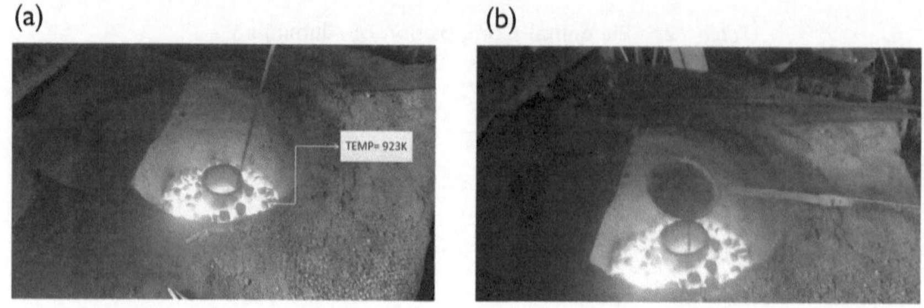

Figure 12.1 Gravity die casting.

the furnace. Finally, the hot composite mixture is put inside a mould cavity to attain the desired shape and size [13]. Figure 12.1 indicates the step-by-step procedure for gravity die casting.

12.3 EXPERIMENTAL SETUP AND PROCEDURE

In this investigation, vertical milling machine, as shown in Figure 12.2, is used to estimate milling performance on aluminium boron carbide composite. The significance of spindle speed, tool feed and depth of cut on machining time, metal removal rate and cutting force is studied in the investigation. The carbide composite tool is use for carrying out the milling process on aluminium boron carbide composite [15]. The cutter has 40 mm diameter and 8 cutting edges. Table 12.3 shows the range and levels of machining parameters used to carry out the experimental work.

A 80 × 100 × 100 × 22 mm dimension of a workpiece was gripped in the machine vise. On the machine's spindle, a 75 mm T-Max cutter with a carbide composite tool insert was fitted. Automatic feed was used during the milling process. Using a stopwatch, the total amount of time required for the machining process is calculated. Evaluating the difference between the volume of the material before and after machining yields the quantity of material removed. The weighing machine is used to measure this [16]. The metal removal rate (MRR) of the workpiece during milling is determined by the volume of metal removed per computed time. The formula for calculating MRR is expressed as Equation (12.1):

$$\text{MRR} = \frac{w_i - w_f}{\rho w T} \tag{12.1}$$

w_i - weight of the work piece before milling
w_f - weight of the work piece after milling
ρ_W - density of work piece material
T - machining time

Figure 12.2 Vertical milling machine.

Table 12.3 Levels and Ranges of Controllable Factors of Face Milling Operation

S.No.	*Parameters/Factors*	*Units*	*Level 1*	*Level 2*	*Level 3*
1	Spindle Speed (A)	rpm	357	520	727
2	Tool Feed (B)	mm/min	0.08	0.12	0.17
3	Depth of Cut (C)	mm	0.5	1	1.5

The cutting force is measured by using a dynamometer. Normally, the milled specimen is placed in the milling tool dynamometer for measuring various cutting forces namely: axial force (X), radial/thrust force (Y) and tangential force (Z). The final resultant force is calculated by the succeeding Equation (12.2):

$$F_R = [X^2 + Y^2 + Z^2]^{1/2} \tag{12.2}$$

Table 12.4 L_9 Historical Data Design of Experiments for Milling

Experiment number	*Input parameters*		
	Spindle speed (A) rpm	*Tool feed (B) mm/rev*	*Depth of cut (C) mm*
1	357	0.08	0.5
2	357	0.12	1
3	357	0.17	1.5
4	520	0.08	1
5	520	0.12	1.5
6	520	0.17	0.5
7	727	0.08	1.5
8	727	0.12	0.5
9	727	0.17	1

12.3.1 Experimental Design Using Historical Data Design (HDD)

It is one of the most vibrant and widely used variants in response surface methodology, performed using Design Expert v.10 software. This optimization technique is specifically more user-friendly, as it allows different number of levels to be used for various factors created at irregular intervals between them not withstanding the number of factors (i.e., each varying factor can have unequal number of levels created therein) [17–19]. An L_9 matrix comprising of three factors and three levels is created as an experimental design with the help of software that is given in Table 12.4.

12.4 RESULTS AND DISCUSSION

12.4.1 Experimental Outcome

Table 12.4 contains the experimental design combinations for face milling of an aluminium boron carbide composite with a carbide composite tool in a vertical milling machine. The experiments were conducted based on L_9 historical data design by the recommendation of DOE. Before conducting the milling process, the material is sized to the required dimensions and face milling is carried out to remove the outer irregular surfaces. The workpiece materials are weighed repeatedly to check the equal weight of the trial piece [16,20]. The selection of input process parameters plays a crucial role for attaining a high metal removal rate and good cutting force [21]. Based on the machine specifications, the important milling process parameters influencing the metal removal rate and cutting force are selected. After completing the milling process, the responses obtained are shown in Tables 12.5 and 12.6. The face-milled aluminium boron carbide composite test specimen is shown in Figure 12.3.

Table 12.5 Experimental Results for Material Removal Rate

E.NO	*Speed (rpm)*	*Feed (mm/rev)*	*DOC (mm)*	*Before weight (kg)*	*After weight (kg)*	*Before volume (m^3)*	*After volume (m^3)*	*M/C time (min)*	*MRR (mm^3/rev)*
1	357	0.08	0.5	0.4536	0.4428	1.68E-04	1.64E-04	5.6	714.285
2	357	0.12	1	0.4428	0.4212	1.64E-04	1.56E-04	3.45	2318.84
3	357	0.17	1.5	0.4212	0.3888	1.56E-04	1.44E-04	2.25	5333.33
4	520	0.08	1	0.3888	0.3672	1.44E-04	1.36E-04	5.67	1410.93
5	520	0.12	1.5	0.3672	0.3348	1.36E-04	1.24E-04	3.5	3428.57
6	520	0.17	0.5	0.3348	0.324	1.24E-04	1.20E-04	2.15	1860.46
7	727	0.08	1.5	0.324	0.2916	1.20E-04	1.08E-04	5.63	2131.44
8	727	0.12	0.5	0.2916	0.2808	1.08E-04	1.04E-04	3.37	1186.94
9	727	0.17	1	0.2808	0.2592	1.04E-04	9.60E-05	2.2	3636.36

Table 12.6 Experimental Results of Cutting Force Measurement

Ex.NO.	*Speed (rpm)*	*Tool feed (mm/rev)*	*DOC (mm)*	*Cutting force (N)*			*Resultant force (N)*
				X	Y	Z	
1	357	0.08	0.5	5	2	0	5.385
2	357	0.12	1	7	3	1	7.681
3	357	0.17	1.5	12	2	3	12.529
4	520	0.08	1	4	6	2	7.483
5	520	0.12	1.5	6	7	3	9.695
6	520	0.17	0.5	14	5	0	14.866
7	727	0.08	1.5	5	9	2	10.488
8	727	0.12	0.5	4	7	1	8.124
9	727	0.17	1	11	8	1	13.638

12.4.2 Optimization Analysis Using Historical Data Design (HDD)

12.4.2.1 Significance of Face Milling Controllable Factors on MRR

From the tabulation of results given in Table 12.7, obtained from the determination of MRR for all the nine specimens, it is apparent that the metal removal rate escalates with a rise in spindle speed when the feed is kept at its lowest level. As the workpiece material is ductile in nature, high MRR that is in the form of continuous chip rolls provides a good surface finish [22–24].

Transformation: none

Type 3: Partial Sum of Squares (Analysis of Variance) ANOVA Model for Material Removal Rate

Figure 12.3 Aluminium boron carbide composite test specimen after milling process.

Table 12.7 Results of Milling Experiments Used in HDD Analysis

M/C parametric combinations	*Factors/process parameters*			*Output test responses*	
	A:spindle speed rpm	*B:tool feed mm/rev*	*C:depth of cut mm*	*MRR mm^3/rev*	*Resultant force N*
1	357	0.08	0.5	714.285	5.385
2	357	0.12	1	2318.84	7.681
3	357	0.17	1.5	5333.33	9.529
4	520	0.08	1	1410.93	7.866
5	520	0.12	1.5	3428.57	9.695
6	520	0.17	0.5	1860.46	10.934
7	727	0.08	1.5	2131.44	10.638
8	727	0.12	0.5	1186.94	13.648
9	727	0.17	1	3636.36	14.488

The MRR is highest with a spindle speed 357 rpm, tool feed 0.17 mm/rev and depth of cut 1.5 mm. While the spindle speed is increased to 520 rpm from 357 rpm along with depth of cut, MRR forms a 'V-trend' that is evident from the experimental results. In the case of a spindle speed kept at its highest i.e. 727 rpm, a buckling trend of inverted 'V-trend' is achieved with increasing depth of cut. The change in MRR with respect to tool feed seems inconclusive, as it doesn't follow a specific pattern.

The ANOVA for Response Surface 2FI Model for material removal rate results are shown in Table 12.8. The model is significant if the model F-value is high. The model has a very small risk of failing. As the Prob > F values of tool feed and depth of cut are less than 0.05, they are influencing terms. Particularly with Prob > F value 0.0007, the depth of cut is the most dominating influencer. The error signal charecterized as a noise signal is weak.

The adjusted R^2 value of 0.9982 is not far from the predicted R^2 value of 0.9709. It suggests that the fit is good. An indication of signal-to-noise (S/N) ratio is adeq precision. Its value of 84.397 > 4 shows that the signal is strong enough.

Regression equation for MRR in correlation with factors

$$\begin{aligned} \text{MRR} = {} & 106.15296 + 0.10203 * \text{spindle speed} + 2316.21698 * \text{feed} \\ & - 1214.45048 * \text{depth of cut} - 5.12005 * \text{spindle speed} \\ & * \text{feed} + 0.75248 * \text{spindle speed} * \text{depth of cut} \\ & + 25541.96975 * \text{feed} * \text{depth of cut} \end{aligned}$$

Figure 12.4(a) shows that there is a steep increase in MRR value with an increase in tool feed, whereas the MRR remains almost flat with an increase in spindle speed. Further, it can be concluded that the MRR is not improved by spindle speeds with greater values. Figure 12.4(b) shows that there is a gradual increase in MRR with increasing limits of depth of cut. The phenomenon shows a linear positive trend.

12.4.2.2 Significance of Face Milling Controllable Factors on Resultant Cutting Force

Similar to MRR, resultant cutting forces increase with increase in spindle speed and tool feed as well [10,11,19,25,26]. When the tool rotating speed is increased and impinges over the surface of the workpiece material, more amount of impact force is felt. In this case, the influence of depth of cut on resultant force cannot be singled out and it depends upon spindle speed, as the combinatorial design suggests. The amount of force used to machine the specimen depends on the interaction between spindle speed and cut depth.

Transformation: none

Type 3: Partial Sum of Squares (Analysis of Variance) ANOVA Model for Resultant Cutting Force

The outcomes of the ANOVA for the Response Surface 2FI Model for resultant force are shown in Table 12.9. The model is significant, as shown by the high model F-value (42.64). There is only a fractional 0.0231 chance for the model to fail. As the Prob > F values of tool feed and spindle speed

Table 12.8 Response 1: Material Removal Rate – ANOVA for Response Surface 2FI Model

Descriptives	*Sum of squares*	*DoF*	*Mean square values*	*F-value*	*p-value*	*Inferences*
Model	16824878.43	6	2804146	737.4748	0.0014	Significant
Spindle Speed (A)	5413.176824	1	5413.177	1.423635	0.3552	
Tool Feed (B)	4398298.344	1	4398298	1156.728	0.0009	
Depth of Cut (C)	5316407.399	1	5316407	1398.185	0.0007	Predominantly influencing factor
Interaction between A & B	3155.049289	1	3155.049	0.82976	0.4585	
Interaction between A & C	7522.532142	1	7522.532	1.978384	0.2948	
Interaction between B & C	509289.9915	1	509290	133.9404	0.0074	
Residual	7604.724621	2	3802.362			
Correlation Total	16832483.16	8				
	Std. Dev	61.6633	R-Squared	0.999548		
	Mean	2446.795	Adj R-Squared	0.998193		
	C.V% %	2.520166	Pred R-Squared	0.970885		
	PRESS	490084.3	Adeq Precision	84.39679		

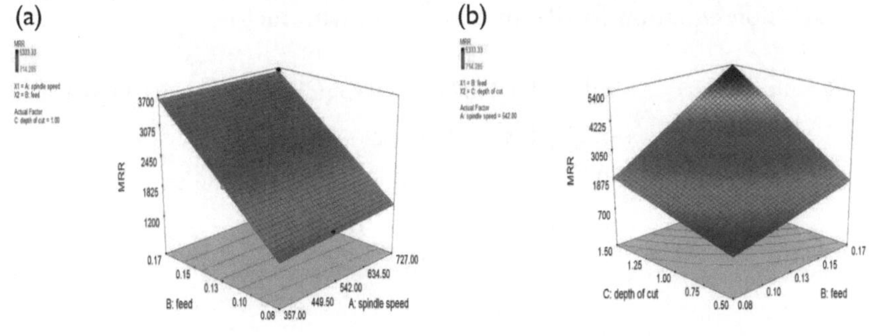

Figure 12.4 3D Interaction graph between factors influencing MRR.

Table 12.9 Response 2: Resultant Force – ANOVA for Response Surface 2FI Model

Descriptives	*Sum of squares*	*DoF*	*Mean square values*	*F-value*	*p-value*	*Inferences*
Model	65.75817	6	10.9597	42.64225	0.0231	significant
Spindle Speed (A)	26.49517	1	26.49517	103.088	0.0096	predominantly influencing factor
Tool Feed (B)	7.828755	1	7.828755	30.46031	0.0313	
Depth of Cut (C)	0.000941	1	0.000941	0.003661	0.9573	
Interaction between A & B	0.004622	1	0.004622	0.017985	0.9056	
Interaction between A & C	0.724344	1	0.724344	2.818295	0.2352	
Interaction between B & C	0.028074	1	0.028074	0.109233	0.7724	
Residual	0.51403	2	0.257015			
Correlation Total	66.2722	8				
	Std. Dev.	0.506966	R-Squared	0.992244		
	Mean	9.984889	Adj R-Squared	0.968975		
	C.V.% %	5.077337	Pred R-Squared	0.449824		
	PRESS	36.46134	Adeq Precision	20.34553		

are less than 0.05, they are influencing terms. Particularly, with Prob > F value 0.0096, spindle speed is the predominant influencer. The error signal charecterized as a noise signal is weak.

The difference between the predicted R^2 value of 0.4498 and the adjusted R^2 value of 0.9690 is greater. There is still a minor fit issue. An indication of signal-to-noise (S/N) ratio is adeq precision. A strong signal is indicated by the estimated value of 20.346 > 4.

Regression equation for RF in correlation with factors

$$
\begin{aligned}
\text{Resultant force} = {} & -4.96694 + 0.021616 * \text{spindle speed} + 24.10870 * \text{feed} \\
& + 3.22069 * \text{depth of cut} + 6.19738\text{E} \\
& - 003 * \text{spindle speed} * \text{feed} - 7.38386\text{E} \\
& - 003 * \text{spindle speed} * \text{depth of cut} \\
& + 5.99691 * \text{feed} * \text{depth of cut}
\end{aligned}
$$

Figure 12.5(a) shows that maximum increase in resultant force happens when the spindle speed is maximized, whereas the resultant force experiences a not-so-great improvement with high feed. The change in resultant force is drastic at a superior tool rotation speed, as the workpiece requires more energy to withstand a fast rotating tool [27,28].

Figure 12.5(b) shows that there is a gradual increase in resultant force with rising values of depth of cut. The phenomenon shows a linear positive trend with less impact when compared to the spindle speed.

Table 12.10 below indicates the target and range of factors and responses.

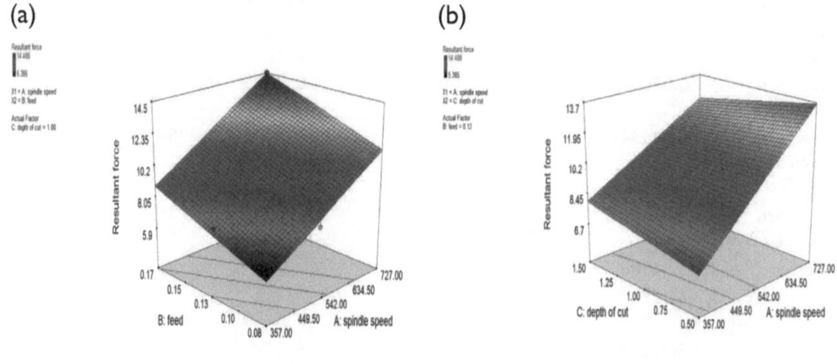

Figure 12.5 3D Interaction graph between factors influencing resultant cutting force.

Table 12.10 Satisfying Conditions Obtained from HDD Analysis

	Satisfying conditions		
Terms	*Target*	*Minima (-α)*	*Maxima (+α)*
Spindle Speed	is in range	357	727
Tool Feed	is in range	0.08	0.17
Depth of Cut	is in range	0.5	1.5
MRR	maximize	714.285	5333.33
Resultant Force	maximize	5.385	14.488

Table 12.11 Optimized Solutions Obtained from HDD Analysis

Priority	*Spindle speed*	*Tool feed*	*Depth of cut*	*MRR*	*Resultant force*	*Desirability index*	*Decision*
1	727	0.13	1.05	2857.24	13.15	0.895847378	Selected

The optimized results obtained from HDD using Design Expert are represented as Table 12.11.

12.5 CONCLUSION

The following conclusions are taken from the aforementioned research work:

- Composites, when compared to their metal alloy peers, have an advantage due to increased mechanical and surface features such high hardness, improved impact strength, and greater wear resistance. Particularly aluminium matrix composites reinforced with a light density ceramic namely boron carbide possess excellant mechanical properties when compared with an individual alloy.
- Gravity die casting is a hassle-free process of fabricating MMC specimens with utmost precision and material surface integrity.
- Milling, carried out in a vertical milling machine in the above-mentioned research work is one of the most naive and commonly used macro-surface material removal process, in which dimensional clarity is obtained from the uniform removal of fine particles.
- Material removal rate positively influences the surface finish of composite components. The more the MRR, the higher the surface finish.
- For composite specimens, in which the metal alloys are strengthened by the presence of different proportions of ceramic powders high energy is required to mill off surface layers. Hence, resultant cutting forces are high, when the parameters like feed and depth of cut are high, as composites are harder to machine compared to individual metal alloys.
- Optimization cannot be misinterpreted as the one that gives maximum values as results but optimum feasible values as results. As both MRR and cutting forces fall under maximizing criteria, this optimization comes under single objective category.
- The best and optimal specimen is selected from a set of composite combinations through an optimization technique called HDD, made real by Design Expert software v.10.

REFERENCES

[1] T. Rajmohan, K. Palanikumar, M. Kathirvel, Optimization of machining parameters in drilling hybrid aluminium metal matrix composites. *Transaction Nonferrous Metals Society of China* 2012, 22, 1286–1297.

[2] Y. C. Liang, S. Wang, W. D. Li, X. Lu, Data-driven anomaly diagnosis for machining processes *Engineering*. 2019, doi:10.1016/j.eng.2019.03.012.

[3] R. S. Keshav Singh, R. A. Pandey, Fabrication and Mechanical properties characterization of aluminium alloy LM24/B_4C composites. *Materials Today: Proceedings* 2017, 4, 701–708.

[4] B. Manjunatha, H. B. Niranjan, K. G. Satyanarayana, Effect of mechanical and thermal loading on boron carbide particles reinforced Al-6061 alloy materials. *Science & Engineering A* 2015, 632, 147–155.

[5] B. Ravia, B. Balu Naikb, J. Udaya Prakash, Characterization of aluminium matrix composites (AA6061/B_4C) fabricated by stir casting technique. *Materials Today: Proceedings* 2015, 2, 2984–2990.

[6] B. Vijaya Ramnath, C. Elanchezhian, M. Jaivignesh, S. Rajesh, C. Parswajinan, Evaluation of mechanical properties of aluminium alloy-alumina-boron carbide metal matrix composites. *Materials and Design* 2014, 58, 332–338.

[7] S. Karabulut, H. Karakoc, R. Cıtak, Influence of B_4C particle reinforcement on mechanical and machining properties of Al6061/B_4C composites. *Composites Part-B Engineering* 2016, 101, 87–98.

[8] H. Guo, Z. Zhang, Processing and strengthening mechanisms of boron-carbide-reinforced aluminum matrix composites. *Metal Powder Report* 2017, 73(2), 80–90.

[9] B. C. Kandpala, K. H. Singhc, Fabrication and characterisation of Al_2O_3/aluminium alloy 6061 composites fabricated by Stir casting. *Material Today: Proceedings* 2017, 4, 2783–2792.

[10] G. Campatelli, L. Lorenzini, A. Scippa, Optimization of process parameters using a response surface method for minimizing power consumption in the milling of carbon steel. *Journal of Cleaner Production* 2014, 66, 309–316. doi:10.1016/j.jclepro.2013.10.025.

[11] H. Öktem, T. Erzurumlu, H. Kurtaran, Application of response surface methodology in the optimization of cutting conditions for surface roughness. *Journal of Materials Processing Technology* 2005, 170(1-2), 11–16. doi:10.1016/j.jmatprotec.2005.04.96

[12] K. Shirvanimoghaddam, H. Khayyam, H. Abdizadeh, M. Karbalaei Akbari, A. H. Pakseresht, Boron carbide reinforced aluminium matrix composite. *Physical, mechanical characterization and mathematical modeling Materials Science &Engineering* 2016, 658, 135–149.

[13] K. Rajkumar, P. Rajan, J. Maria, A. Charles, Microwave heat treatment on aluminium 6061 alloy boron carbide composites. *Procedia Engineering* 2014, 86, 34–41.

[14] I. Sudhakar, G. Madhusudhan Reddy, K. Srinivasa Rao, Ballistic behavior of boron carbide reinforced AA7075 aluminium alloy using friction stir processing - An experimental study and analytical approach. *Defence Technology* 2015, 12(1), 12–31.

[15] A. Nieto, H. Yanga, L. Jiang, J. M. Schoenunga, Reinforcement size effects on the abrasive wear of boron carbide reinforced aluminum composites. *Wear an International Journal on the Science and Technology of Friction Lubrication and Wear* 2017.
[16] J. Udaya Prakash, T. V. Moorthy, S. Perumal, Optimization of machining parameters in drilling of aluminium matrix composites using Taguchi technique. *International Journal of Applied Engineering Research* 2015, 10(68), 526–529.
[17] R. Kumar, R. Rana, S. Lata, R. kumar Sonkar, Optimization of process parameters on over-cut in drilling of AlB_4C MMC. *International Journal Of Modern Engineering Research* 2015, 5, 24–30.
[18] R. Rekha, N. Baskar, M. R. A. Padmanaban, A. Palanisamy, Optimization of cylindrical grinding process parameters using meta-heuristic algorithms. *Indian Journal of Engineering & Materials Sciences* 2020, 27, 389–395.
[19] P. V. Rajesh, A. Saravanan, Desirability approach-based optimization of process parameters in turning of aluminum matrix composites. *Computational Technologies in Materials Science*, (1st ed.). CRC Press, 2021, 71–100. doi: 10.1201/9781003121954. eBook ISBN: 9781003121954.
[20] I. Kerti, F. Toptan, Microstructural variations in cast B_4C-reinforced aluminium matrix composites (AMCs). *Materials Letters* 2008, 62, 1215–1218.
[21] S. E. Shin, Y. J..Ko, D. H. Bae, Mechanical and thermal properties of nanocarbon-reinforced aluminum matrix composites at elevated temperatures. *Composites Part B* 2016, 106, 66–73.
[22] A. Taskesen, K. Kütükde, Experimental investigation and multi-objective analysis on drilling of boron carbide reinforced metal matrix composites using grey relational analysis. *Measurement* 2013, 47, 321–330.
[23] S. Magibalan, P. Senthil kumar, P. Vignesh, Aluminium metal matrix composites – A review. *Transactions On Advancements In Science and Technology* 2017, 38(1), 55–60.
[24] M. K. Sathish Kumar, B. Sravan Kumar, K. Srinivasan, Manufacturing and machining of aluminium metal matrix composites- An overview. *International Journal of Applied Engineering Research* 2015, 10(9), 9312–9315.
[25] A. Palanisamy, R. Rekha, S. Sivasankaran, C. Sathiya Narayanan, Multi-objective optimization of EDM parameters using grey relational analysis for titanium alloy (Ti–6Al–4V). *Applied Mechanics and Materials* 2014, 592, 540–544.
[26] W. Sukthomya, J. D. T. Tannock, Taguchi experimental design for manufacturing process optimization using historical data and a neural network process model. *International Journal of Quality & Reliability Management* 2005, 22(5), 485–502. doi:10.1108/02656710510598393.
[27] F. De Santis, Using historical data for Bayesian sample size determination. *Journal of the Royal Statistical Society: Series A (Statistics in Society)* 2007, 170(1), 95–113. doi:10.1111/j.1467-985x.2006.00438.x.
[28] S. J. Shin, J. Woo, S. Rachuri, Energy efficiency of milling machining: Component modeling and online optimization of cutting parameters. *Journal of Cleaner Production* 2017, 161, 12–29. doi:10.1016/j.jclepro.2017.05.013.

Index